Integrated Farming System Practices

AF640802

NIPA® GENX ELECTRONIC RESOURCES & SOLUTIONS P. LTD.
New Delhi-110 034

www.nipaers.com
Browse, Search, Read & Buy...
Print Books eChapters
eBooks Publishers
Forthcoming Subject Catalogues
New Titles
Online Resources on:
✓ Current Affairs
✓ Reasoning, Logic & Aptitude
✓ Competitive Examinations
✓ English Language Lab
✓ Writing & Pronunciation Tools
✓ Personality Development
✓ Online Programmes for Professionals
✓ Interview Preparation
Pay Using
PhonePe
Google Pay
paytm
PayPal
UPI
CC Avenue

Integrated Farming System Practices

Challenges & Opportunities

Edited by

Prof. Sankarsana Nanda

Dean, Extension Education & Research
Orissa University of Agriculture and Technology
Bhubaneswar (Odisha)

NIPA® GENX ELECTRONIC RESOURCES & SOLUTIONS P. LTD.
New Delhi-110 034

NIPA® GENX ELECTRONIC RESOURCES & SOLUTIONS P. LTD.

101,103, Vikas Surya Plaza, CU Block
L.S.C. Market, Pitam Pura, New Delhi-110 034
Ph : +91 11 27341616, 27341717, 27341718
E-mail: newindiapublishingagency@gmail.com
web: www.nipabooks.com

For customer assistance, please contact
Phone: + 91-11-27 34 17 17
Fax: + 91-11- 27 34 16 16
E-Mail: feedbacks@nipabooks.com

ISBN: 978-81-19215-91-1

Composed and Designed by NIPA®.

Preface

Throughout the world, agriculture is faced with an immense challenge: how to increase yields to feed a growing population from depleted natural resources such as soil, water and biodiversity etc. in the present face of climate change. The capacity of the global food system to support a rising world population while preserving healthy ecosystems is the subject of much debate. But, agriculture plays a very important role in the economy of most of the nations of the developing world. For India, agriculture is the base of nation's economy, largely rained and dominated by smallholders. As per reports from a number of governments, non-government and intergovernmental organizations, the countries and agriculture especially practiced by smallholders are particularly vulnerable to climatic changes and they have least capacity to adapt to these changes. In most of the developing countries of the world, the agriculture is not merely a matter of cultural practice but it is the source of livelihoods of millions of smallholders and their dependents. Changing climate as predicted will not only affect the agricultural productivity but also heavily hit the lives and livelihoods of the smallholder families and ultimately lead to food insecurity to the concerned nation and the world. Business-as-usual scenarios of population growth and food consumption patterns indicate that agricultural production will need to increase by 70 percent by 2050 to meet global demand for food. The impacts of climate change will reduce productivity and lead to greater instability in production in the agricultural sector i.e.crop and livestock production, fisheries and forestry that already have high levels of food insecurity and environmental degradation and limited options for coping with adverse weather conditions. The ongoing rise in global temperatures, increasingly changeable weather patterns and greater competition for land, energy and water will affect the food system as well as the ecosystem services which now underpin agriculture and the natural environment. However, the prosperity of any country depends upon the prosperity of farmers. Presently in India, the marginal and small farmers are literally illiterate, financially handicapped, their holdings are small and scattered, not suited for high-tech agricultural machinery, work in resource poor and quite vulnerable to climate change. To fulfill the basic needs of house hold including food (cereal, pulses, oilseeds, milk, fruit, honey, fish meat etc.) for human, feed and fodder for animals and fuel & fibre for general use warrant an attention about Integrated Farming Systems.

The emergence of Integrated Farming Systems (IFS) has enabled to develop a framework for an alternate development model to improve the feasibility of small sized farming operations in relation to larger ones. Integrated Farming Systems broadly explains more integrated approach to farming as compared to monoculture approaches. This book is thus an attempt to make the readers fully realize the potential of Integrated Farming Systems. This will deliver resources for the integration of suitable farming systems in present day farming to improve the livelihoods of the farmers, without further damaging to the environment and also recruit more degraded lands to the productive system, thereby helps in food security in the eve of climate change. Further it, focuses on sustainable production from specific time and space, conservation of natural resources, improves in carbon sequestration, low emission agriculture, conservation of tillage and addresses many issues of Integrated Farming Systems in the present context of food security and climate change.

The author is extremely thankful to Dr.V.P Singh, Senior Policy Advisor, World Agroforestry Centre (ICRAF), South-Asia, New Delhi; Dr. Dennis Garrity, Former Director General and Dry Land Ambassador, World Agroforestry Centre (ICRAF), Nairobi; Dr. S. Ayyappan, Director General, Indian Council of Agricultural Research, New Delhi; Prof. M.S.Swaminathan, Founding Director, MSSRF, Chennai; Dr. Peter Kenmore, FAO Representative in India, New Delhi; Dr. Peter Holmgren, Director General, Centre for International Forest Research (CIFOR), Indonesia; Dr. Robert.S.Zeigler, Director General, International Rice Research Institute, Philippines; Dr. William. D. Dar, Director General, ICRISAT, Hyderabad, India; Prof. Eric Tollens, Former, Chairman, Board of Trustee, ICRAF and Member, Board of Trustee, International Institute of Tropical Agriculture (IITA), Ibadan; Dr. Anupam Joshi, Senior Environment Specialist, the World Bank in India, New Delhi and Dr. B. Gangwar, Former Project Director, Farming Systems Research, ICAR, Modipuram for their academic inputs and support. I strongly acknowledge Prof. A.K.Sahoo for his constant scientific inputs and critical assessment, which gave a strong base for developing such a publication of its kind. The author is also extremely thankful to Prof. M. Kar, Vice Chancellor, Orissa University of Agriculture and Technology for his all support and moral encouragement. Further, the author also strongly acknowledges all chapter contributors for their intellectual stimulations and valuable inputs. Finally, the author acknowledges and thankful to the publisher for bringing out this edition with success.

1st July, 2015 **S.S.Nanda**

Contents

Integrated Farming System Practices: Challenges & Opportunities, pp 1-34

New India Publishing Agency, New Delhi, India

1

Integrated Farming Systems: Prospects & Practices

S.S. Nanda

Introduction

Agriculture is a dynamic sector affected by rapid changes in environment and climate, technologies, development priorities, impact of changes in other sectors and social changes such as family structure, migration and international policies such as globalization and liberation. The Indian Green Revolution of the 1960s and 1970s, with its package of improved seeds, farm technology, better irrigation and chemical fertilizers, was highly successful in meeting its primary objective of increasing crop yields and augmenting food supply. However, the Green Revolution as a development approach has not necessarily translated into benefits for the lower socioeconomic strata, particularly the rural poor and farm women, in terms of greater food security or greater economic opportunity. It has increased the need for cash incomes in rural households to cover the costs of technological inputs which has forced women to work as agricultural laborers and increased the need for unpaid female labor for farming tasks.

In the era of burgeoning population and shrinking natural resources integrated farming system may be the panacea. Integrated Farming system approach envisages the integration of agroforestry, horticulture, dairy, sheep and goat rearing, fishery, poultry, pigeon, biogas, mushroom, sericulture and by-product

Directorate of Extension Education
Orissa University of Agriculture & Technology, Bhubaneswar – 751 003, Odisha, India

utilization of crops with the main goal of increasing the income and standard of living of small and marginal farmers. Integrated systems are about bringing crops and livestock into an interactive relationship with the expectation that together, as opposed to alone, they will generate positive effects on outcomes of interest, such as profitability overall productivity and conservation of non-renewable resources. It is, however, much more than this. The "system" includes the environment, soil characteristics, landscape positions, genetics and ecology of plant and animals. It involves management practices, goals and lifestyles of humans, social constraints, economic opportunities, marketing strategies and externalities including energy supplies and costs and impacts of farm policies. Systems also reflect available natural resources and the impact on their use, wildlife issues, target and non-target plant and animal species, micro-organisms and indeed all of the definable and indefinable factors that ultimately interact to result in an outcome that is never constant. The aims of integrated farming system is to increase productivity, profitability, sustainability, balanced food, clean environment, recycling of available resources and income round the year. Besides we can also think for adoption of new technology, solving energy crises: fuel and fodder crises, avoiding deforestation, increasing employment generation by taking more than one component, input-output efficiency and enhancing opportunity for agriculture oriented industries as well as uplifting living standard of the farmers (Gill *et al*, 2009).

With industrialization and market based economy, the farmers are forced to become crop-specific farmers depending on presence of agri-based industries like sugar factory, ginning mill, soya processing plant, rice mill, oil mill, dal mill, feed mill etc. Besides; factors like urbanization, industrialization, commercialization and mechanization are responsible for selection of crop, cropping pattern and other enterprises. Dairy farming, poultry keeping, growing vegetable & fruit, bee keeping and mushroom cultivation have become independent enterprises, particularly around cities and industrial establishments to explore the market potential. In the commodity oriented market scenario, the focus is usually on a singular production system. The crop based research and development approach further isolated different farming system from each other.

In the present day context, there is limited possibility of increasing additional area under cultivation. It is of immense importance to produce more for ever-increasing population in the years to come. Efficiently integration of crops with animals (cow, buffalo, pig, goat, sheep, fish), birds (poultry, pigeon, duck), multipurpose trees and agro-forestry systems and other enterprises (bio-gas, apiary, mushroom etc.) clearly showed the best advantages over conventional system of cropping under irrigated and rainfed/dry land conditions as well as in tribal areas (Patel and Dutta, 2004).

Since single crop production enterprises are subject to high degree of risk and uncertainty for the farmers, it is necessary to develop suitable integrated farming systems models. Farming system creates an opportunity for developing diversified models for different types and categories of farmers.

Indian economy is predominantly rural and agriculture oriented where the marginal and small farmers constitute 76.2 % of farming community. Due to failure of monsoon, the farmers are forced to judicious mix up of agricultural enterprises like dairy, poultry, pigeon, fishery, sericulture, apiculture etc., suited to their agro-climatic and socio- economic condition. The farming system, as an concept, takes into account the components of soil, water, crops, livestock, labour, capital, energy and other resources, with the farm family at the center managing agricultural and related activities. The farm family functions within the limitations of its capacity and resources, the socio-cultural setting, and interaction of these components with physical, biological and economic factors.

Over the years farmers of India have taken agriculture as their way of life. Traditionally, Indian farmers adopted integrated farming system approach for their livelihood. To fulfill the food demand of the family, cultivation of rice, wheat, pulses, oilseeds, vegetables, fruits etc. are given due importance. Subsequently, other components like dairy, poultry, goatery, piggery and pisciculture were also added into the system. With industrialization, farmers were forced to become commodity oriented with development of agri-based industries like sugar factory, ginning mill, soya processing plant, rice mill, oil mill, dal mill, etc. In present day situation, farmers no more depend on their own farm resources to meet all the family demands. Several items like vegetables, fruits, animal feeds, etc. are purchased from the market.

Growth in farm income and household food security has lagged behind the expectations of the 1996 World Food Summit, which committed to having the number of the undernourished by 2015. Increases in agricultural productivity required to boost farm income and food security depend ultimately on farm management decisions in relation to choices of enterprises, production technologies and agricultural inputs among others.

Within the village there is generally a range of farm and household sizes that may still produce a similar range of agricultural products, in spite of using different technologies. An understanding of the wide spectrum of farming systems that exist across the developing world can contribute to effective rural development strategies and agricultural development policies. Farm household systems and their immediate external rural environment, including local effects of policies and institutions, markets and information linkages, are interdependent and over time, co-evolve in response to changes in population, markets, technologies, policies, institutions and information flows.

Differences in the farming system account for a major part of the variation in farm management decisions. By adopting a livelihood approach to define farming systems-recognizing multiple sources of livelihoods, including cash crops, aquaculture, self-consumption and off-farm income-it is possible to consider the local institutional environment, including farm- gate price ratios, local markets, credit availability to farmers and arrangements for resource-sharing as an integral part of the local system.

An ideal framework would allow for the tremendous diversity of agricultural settings to be simplified and codified, without eliminating important differences that need to be taken into account by development practitioners and would be hierarchical in order to meet the different needs of decision makers at different levels.

During the last 4-5 decades of agricultural research and development in India, major emphasis has also been given to component and commodity based research projects for developing crop variety, animal breed, farm implement which were mostly conducted in isolation and at the institute level (Behera, 2010). This component, commodity and discipline based research has proved largely inadequate in addressing the multifarious problems of small and marginal farmers (Jha, 2003). These farm families and agricultural labourers have to face unemployment and under employment due to seasonal work in crop production (Swaminathan, 1981) and feel unsecured due to the natural calamities occurring at one or the other seasons of the year, as single crop production enterprises are subjected to high degree of risk and uncertainty. Fluctuation with the market trends for a single commodity and their dependence on external inputs is the most important constraint for the commodity based farming. Due to intensive agriculture several ills have also appeared in Indian agriculture, such as declining factor productivity, degradation of natural resource base, environmental degradation including ground water depletion & contamination and declining farm profitability & productivity (Sharma and Behera, 2004). To tackle such problems, farming systems approaches to research has been widely recognized, where whole farm is viewed as a system and interactions among the various components are taken into consideration.

Definition

The term farming system, in its broadest sense, is any research that views the farm in holistic manner and considers interactions (between the components and of components with the environment) in the system (CGIAR, 1978). This type of research is most appropriately carried out by interdisciplinary teams of scientists who in association with extension specialists, continually interact with the farmers in the identification of problems and in devising ways to solve them. It aims at generating and transferring technologies to increase the resource productivity for an identified group of farmers. Farming System research is an approach to agricultural research and development that view the whole farm as a system and focus on (1) the interdependencies between the components under the control of members of the household and (2) how these components interact with each other in respect of physical, biological and socioeconomic factors not under the household's control (Shaner *et al*, 1982).

Farming system is a complex inter-related matrix of soil, plants, animals, implements, labour, capital and other inputs controlled in parts by farming families and influenced to varying degrees by political, economic, institutional and social forces that operate at many levels (Mahapatra, 1994). It is focused round a few selected inter-dependants, inter-related and often inter-linking production systems based on a few crops, animals and related activities.

The farming system, as a concept, takes into account the components of soil, water, crops, livestock, labour, capital, energy and other resources with the farm family at the center managing agricultural and related activities. The farm family functions within the limitations of its capability, resources, socio-cultural setting and interaction of these components with physical, biological and economic factors (Behera and Sharma, 2007). It creates an opportunity for developing diversified models for different types and categories of farmers. Farming system is a resource management strategy to avail maximum efficiency of a particular system (Gill *et al.*, 2009).

As per Jayanthi *et al.* (2001) crops, animals and trees are the major components of any integrated farming system. Crop may have subsystems like mono crop, mixed / intercrop, multi-tier crops of cereals, pulses, oilseeds, forages, vegetables, flowers, etc. Animal components may include cattle, goat, sheep, poultry, silk worm, honey bees etc. Timber, fuel, fodder and fruit trees may either collectively or individually included in the system.

Philosophy

The entire philosophy of farming system revolves around better utilization of time, money, resources and family labour. The farm family gets scope for gainful employment round the year thereby ensuring good income. The results of investigation on the development of suitable farming systems for semi-arid tropical situations in Haryana indicated that mixed farming systems of crops and animals are more efficient, remunerative and generate more employment, than arable farming systems under small land holdings for irrigated and dry land conditions (Singh *et al.*, 2009).

Objectives

The basic aim of integrated farming system is to derive a set of resource development and utilization practices, which lead to substantial and sustained increase in agricultural production (Kumar and Jain, 2005). It is possible through generating and transferring agricultural technologies appropriate for farmers (Biggs, 1985). The primary objective of FSR is to improve the well being of individual farming families by increasing the productivity of their farming system, given the constraints imposed by resources and the environment (Patil *et al.*, 2014).

i. ***Productivity:*** Farming system provides on opportunity to increase economic yield per unit area per unit time by virtue of intensification of crop and allied enterprises.

ii. ***Profitability:*** The system as a whole provides an opportunity to make use of produce/waste material of one enterprise as an input in another enterprise at low/no cost. Thus, by reducing the cost of production the profitability and benefit cost ratio works out to be high.

iii. ***Potentiality:*** Soil health, a key factor for sustainability is getting deteriorated and polluted due to faulty agricultural management practices viz., excessive use of inorganic fertilizers, pesticides, herbicides, high intensity irrigation etc. In farming system, organic supplementation through effective use of manures and waste recycling is done, thus providing an opportunity to sustain potentiality of production base for much longer time.

iv. ***Balance food:*** In farming system, diverse enterprises are involved and they produce different sources of nutrition namely proteins, carbohydrates, fats and minerals etc from the same unit of land, which helps in solving the malnutrition; problem prevalent among the marginal and sub-marginal farming households.

v. ***Environment safety:*** The very nature of farming system is to make use or conserve the byproduct/waste product of one component as input in another component and use of bio control measures for pest and disease control. These eco-friendly practices bring down the application of huge quantities of fertilizers, pesticides and herbicides, which pollute the soil water and environment to an alarming level whereas, IFS will greatly reduces environmental pollution.

However, the broad objectives of farming system models are :

- To develop farm-household systems of rural communities on a sustainable basis

- To improve efficiency in farm production
- To raise farm and family income
- To increase welfare of farm families and satisfy basic needs
- To provide meaningful employment to the family members
- Appropriate utilization of available resources like land, water, labour, animal power, climatic condition and finance.
- *In situ* recycling of organic residues
- Decrease in cost of cultivation through enhanced input use efficiency
- Integration of primary or secondary produce/waste of one component for the benefit of other component(s)
- Upgrading soil and water productivity
- Nutritional security
- Environmental security
- Continuous flow of income & employment throughout the year.

The farming system research advocates about (i) development of relevant and viable technology for small farmers should have the full knowledge of the existing farming system and (ii) that technology should be evaluated not solely in terms of its technical performance but in terms of its conformity to the goals, needs and socio-economic circumstances of the targeted small farm systems with special reference to profitability and employment generation.

Research and extension efforts with a farming systems perspective have varying objectives ranging from increasing the body of knowledge about farming systems to solving problems in different farming situations. Expectations are high in its problem solving role where the aim is to increase productivity of the farming system by generating and transferring agricultural technologies appropriate for farmers (Biggs, 1985). Primary objective of FSR is to improve the well being of individual farming families by increasing the productivity of their farming system, given the constraints imposed by resources and the environment.

Principles

Farming system research is based on the following basic principles.

- Make the farm household self-sufficient and make the farm free of vulnerable from external forces.

- Enterprise diversification to increase income, employment, risk minimization, improvement in natural resources, environment and diet of farm families.
- The interaction between the components and the components with the environment was taken in consideration and the synergy among the interacting components needs to be explored.

Advantages

- Security against complete failure of a system.
- Minimization of dependence for external inputs
- Optimum utilization of farm resources like labour, animal power, machinery, finance etc.
- Efficient use of natural resources like land, water, sunlight, etc.
- Employment generation for the family members round the year.
- More income from limited resources

Characteristics of farming systems

The Farming System Research (FSR) activities are to be farmer oriented, system oriented, problem solving approach, inter-disciplinary, compliments mainstream disciplinary research, test the technology in on-farm trials and provide feedback from the farmers. The strategy of FSR should emphasize that the research agenda should be determined by explicitly define farmers needs through an understanding of the extension farming systems rather than its perception by the researcher. The farming systems research and extension should be dealt in holistic manner on farmers participatory mode with problem solving approach, keeping genders activity, inter disciplinary and interactive approach. It should emphasize extensive on-farm activities and complement the experimental on-station research and acknowledges the location specificity of technical solutions and document the inter dependencies among multiple clients. Greater importance is placed on feedback to modify the content of subsequent on farm trials, if necessary, by changing research priorities focusing policy shifts based upon micro level analysis (Patil *et al.*, 2014).

Thus, farming system represents an appropriate combination of farm enterprises, viz., cropping systems, horticulture, livestock, fishery, forestry, poultry and the means available to the farmer to raise them for profitability. It interacts adequately with environment without disclosing the ecological and socio-economic balance on one hand and attempts to meet to national goal on others. Hence there is a need to do research on farming systems for sustainable income along with the maintenance of natural resources. It is designed to understand farmers; priorities, strategies and resource allocation decisions. Characteristics of farming system are as under:

- Farmer oriented holistic approach
- Effective farmers participation
- Unique problem solving system
- Dynamic system
- Gender sensitive
- Responsible to society
- Environmental sustainability
- Location specificity of technology
- Diversified farming enterprises to avoid risks due to environmental constraints
- Provides feedback from farmers

Farming system focuses on

- Holistic approach & system-oriented
- Eco-friendly
- Farmer participatory
- Problem-solving approach
- Interactive
- Interdisciplinary in nature
- Recognizes inter dependencies among multiple clients
- Complements mainstream disciplinary research
- Acknowledge location specificity
- Tests technology in on-farm trials and through adaptive research.
- Provides feedback from the farming community

It is of immense importance to produce more for ever-increasing population in the years to come with dwindling resource base and declining per capita availability of land. Efficient integration of crops with animals (cow, buffalo, pig, goat, sheep, fish), birds (poultry, pigeon, duck), multipurpose trees and agroforestry systems and other enterprises (bio-gas, apiary, mushroom etc.) clearly showed the best advantages over conventional system of cropping under irrigated and rainfed conditions as well as in tribal areas (Patel and Dutta, 2004). To make farmer self sufficient and meet its family demand, it needs to adopt integrated farming system with several components.

In India the majority of farmers are under small and marginal category with small areas of land. With such small holdings, it would not be possible to sustain a farm family of five members with a single rainfed crop. To fulfill the nutritional demand of a family, it is necessary to take up various crops along with allied enterprises to obtain food grain, oil seed, vegetables, fruits, milk, fish, egg, meat etc.

Components of farming system

- Field crops
- Vegetable crops
- Fruit crops
- Floriculture
- Livestock (Cow, buffalo, sheep, goat, pig)
- Pisciculture
- Poultry
- Duckery
- Agro-forestry
- Bee keeping
- Mushroom cultivation
- Bio-gas
- Vermicompost

Animal components in farming system

Mixed farming systems (where crops and animals are integrated on the same farm) form the backbone of small-scale Asian agriculture. Livestock in these systems not only provide 90% of the milk, 77% of the ruminant meat, 47% of pork and poultry meat, and 31% of eggs, but also fill an important economic and ecological niche throughout Asia. Poor farmers own animals for a variety of advantageous reasons:

- Diversification in the use of production resources and reduction of risk
- Promotion of links between system components (land, crops and water)
- More efficient use of household labour
- Production of high quality protein food for household food security and for the market
- As an investment that generates return through reproduction, as well as from their other goods and services
- Economic security (assets that can be converted into cash or exchanged for goods and services when necessary)
- Supply of draught power for cultivation, transportation and haulage operation
- Contribution to soil fertility through nutrient cycling (manure)

Factors governing choice of enterprises and resource allocation

- Climatic conditions
- Soil types
- Farmers preferences
- Size of the farm
- Knowledge, skill and technology
- Storage, transport and marketing
- Resource mobilizing power
- Credit facilities available
- Socio-economic conditions
- Customs, sentiments and believes

Possible outcome by adoption of IFS

- Increased productivity, profitability and sustainability
- Balanced food
- Efficient resource recycling
- Money round the year

- Employment generation
- Improves skill of the farmers
- Improves standard of living of small and marginal farmers.
- Sustain soil health.
- Provides opportunity for agri-oriented industries

Livelihood security through integrated farming system

Integrated Farming System is one of the five main strategies to improve livelihood namely,

- Intensification of existing production patterns;
- Diversification of production and processing to make it market-oriented with value-addition;
- Increased operated farm, herd or enterprise-size and expansion of the agricultural frontier
- Increased off-farm income, both agricultural and non-agricultural to supplement or replace farming activities
- Integration of several components in a farming system to increase income.

Sustainability of farming system

Sustainability is the objective of the farming system where production process is optimized through efficient utilization of inputs without infringing on the quality of environment with which it interacts on one hand and attempt to meet the national goals on the other. The concept has an undefined time dimension. Farm as a single unit is to be considered and planned for effective integration of the enterprises to be combined with crop production activity, such that the end-products and wastes of one enterprise are utilized effectively as inputs in other enterprise. For example, the wastes of dairying viz., dung, urine, refuse etc. are used in preparation of FYM or compost which serves as an input in cropping system. Likewise, the straw obtained from crops (maize, rice sorghum etc.) is used as a fodder for dairy cattle. Further, in sericulture the leaves of mulberry crop as feeding material for silk worms, grain from maize crop are used as a feed in poultry etc. (Patil *et al.*, 2014)

Cropping system vis-a-vis farming system approach

A Cropping System signifies the sequence of crops grown over a specific piece of cultivated land over a period of time and seeks to increase the benefits from the available physical resources. Therefore, the basic approach in an efficient cropping system leading to higher production and better economic returns suggests efficient utilization of inputs. The latter includes selection of suitable crop varieties having their growth duration fitting in the frame of the rotation and their adaptability to the growing conditions. Further, tillage practices, planting methods, plant population, crop interaction, effect of crop combinations and cropping sequences on pest incidence, efficient use of fertilizers and irrigation water, and utilization of farm power are the other management inputs affecting success of cropping system. Socio-economic factors like size of holding and its consolidation, timely and adequate input availability, farmers' demand for a part of the produce for self consumption, nearness to the centres of disposal of the farm produce and above all economic returns of marketable surplus influence the decision for including specific crops in a cropping system. The increased use of chemical fertilizers coupled with the introduction of short-duration high-yielding varieties and improved irrigation practices have resulted in spectacular increases in the yield of crops and thereby added a new dimension to the productivity of the land. Cropping systems research over the last few decades has led to identification of system-based production technologies, which have made a visible impact in enhancing crop productivity. System-based approach emphasizes on integration of component technologies to provide stable farm income, low-cost farm technology, high productivity, low risk, better utilization of available resources, complementarities to existing cropping programme and additional employment generation.

Due to such approaches, several ills have appeared in Indian farming, such as decreasing factor productivity, resource use efficiency, and declining farm profitability and productivity (Sharma and Behera, 2004). Environmental degradation including ground water contamination and entry of toxic substances into the food chain has become a significant problem. To tackle such problems, farming systems approaches to research has been widely recognized. Farming system refers to particular arrangement of enterprises that are managed in response to physical, biological and socio-economic environment and in accordance with farmer's goals, preferences and resources. For research in farming system approach, the whole farm rather than single farm enterprise is taken in to consideration, while decisions are taken for technology adoption and production activities.

Farming system category

Based on available natural resource base and pattern of farm activities, farming systems have been categorized as follows (John and Aidan, 2001):

- Irrigated farming systems, embracing a broad range of food crops;

- Wetland rice based farming systems, dependent upon monsoon rains supplemented by irrigation;
- Rainfed farming systems in humid areas of high resource potential, characterised by a crop activity (notably root crops, cereals, industrial tree crops-both small scale and plantation-and commercial horticulture) or mixed crop-livestock systems;
- Rainfed farming systems in steep and highland areas, which are often mixed crop-livestock systems;
- Rainfed farming systems in dry or cold low potential areas, with mixed crop-livestock and pastoral systems merging into sparse and often dispersed systems with very low productivity;
- Dualistic (mixed large commercial and small holder) farming systems, across a variety of ecologies and with diverse production system;
- Coastal fishing, often mixed farming systems; and
- Urban based farming systems, typically focused on horticultural and livestock production.
- Landless farmers having only back yard area.

Farming systems modules

Several farming system modules have been developed at different places to work out the possibility of profitable agricultural activities. Under NAIP by OUAT, Bhubaneswar; the 0.8 ha model gave rice equivalent yield of 29.15 t, net return of Rs.1,37,907 and created employment opportunities of 555 human days as against rice equivalent yield of 4.05 t, net return of Rs.12,739 and employment opportunities of 204 human-days in case of conventional system of cropping alone. The 1.6 ha model gave rice equivalent yield of 41.45 t, net return of Rs.1,98,168 and created employment opportunities of 899 human-days as against rice equivalent yield of 7.21 t, net return of Rs.17,052 and employment opportunities of 400 person-days in case of conventional system of cropping alone.

Farming System modules for small farmers in dryland studies of the farming systems modules on micro-watershed basis conducted on Alfisols during 2005-08 indicated that a farming system module for 1.1 ha area with arable crops (0.4725 ha), agro-forestry (0.3496 ha), vegetables (0.1150 ha), grasses (0.1256 ha) and bushes (0.0890 ha) gave the highest gross income of Rs 16,080, and net income of Rs 9,793 and a benefit: cost ratio of 2.38. The individual enterprises of arable cropping, agro-forestry, vegetables, grasses and bushes contributed 38.2, 10.3, 27.2, 7.1 and 17.2%, respectively, to the total net income.

In Chhatishgarh, a model having 2 bullocks + 1 cow + 1 buffaloes + 10 goats + 10 poultry + 10 ducks along with crop cultivation was the best with a net income of Rs 33076 per year against arable farming (crop farming) alone (Rs 7843 per year) with a cost returns of 1: 2.238 and employment generation of 316 days (Ramrao *et al.* 2006).

In Chhatishgarh, a mixed farming (crop-livestock) module of 1.5 acre small scale holders was developed with the employment generation of 571 mandays, net income of Rs. 58456 per year against crop farming alone with employment generation of 385 mandays and net returns of Rs. 18300 per year only (Ramrao *et al.* 2005).

Madhava Swamy (1985) observed that the net returns were higher by Rs 620, 5198 and 1598 in diversified farms of farming, poultry and sheep rearing, respectively over the crop enterprise farm in Karnool district of Andhra Pradesh.

Integrated Farming System comprising the components like cropping, vegetables, fishery, poultry and goat rearing was undertaken at Agricultural Research Station, Siruguppa, Karnataka during the wet and dry seasons of 2003- 04 and 2005-06 to study the productivity, profitability, energy flow, employment generation and water requirement of IFS over conventional rice-rice system in Tungabhadra project area of Karnataka. Integrated farming system approach recorded 26.3 and 32.3 per cent higher productivity and profitability, respectively over conventional rice-rice system. Among the components evaluated, the highest net returns was obtained from crop (63.8 %), followed by goat (30.9 %), fish (4.0 %) and poultry (1.3 %), respectively. Employment generation and water requirement was 275 man days/ha/year and 1247 mm, respectively under the integrated farming system (Channabasavanna et al.2009).

Farming System models help to generate employment for family members including farm women. As observed by Sanjev Kumar (2013), there may be employment generation up to 690 man days in a year from one acre of land. For Urban and peri- urban areas, where the land is available and facility for gardening is available, vegetable based farming system will be the best option for better economic returns.

Scope of integrated farming system in Odisha

Farming System approach or Integrated Farming System (IFS) seems to be the only alternative to ensure food, nutritional and economic-security, particularly to the large chunk of small and marginal farmers. The per capita land availability in the state, which was 0.39 ha in 1950 has been reduced to 0.12 ha in 2000 and is projected to further reduce to 0.05 ha by 2020. The farmers are prone to several challenges like flood, cyclone, draught etc. In order to mitigate the situation it has become necessary to go either for multiple cropping systems or Integrated Farming System with available natural resources in a particular locality so that there will be a holistic contribution from various sectors like field crops, horticultural crops, plantation crops, dairy, fishery, poultry, duckery, mushroom cultivation, apiculture, agro-forestry etc. The small and marginal farmers comprising 84 per cent of the farming community own 52 per cent of the farm land and rest being owned by the medium and large farmers. The farmers of the state operate under complex-diverse-risk prone environment with threat from biotic and abiotic sources. Farming is mainly rice-based where more than 52 per cent of gross cropped area is under rice of which more than 61 per cent is rainfed. The per capita food grain production is 156 kg and the share of food grain to India is 2.04 per cent (Nanda *et al.*, 2007a).

Odisha is endowed with vast natural resource including water. It is a finite and vulnerable natural resource, but is depleted and degraded very fast. Water must be used properly, efficiently, economically, environmentally, optimally and judiciously to derive maximum benefit from it. Out of total cultivated area of 61.80 lakh ha, about 47.2 per cent is high land, 28.4 per cent medium land and 24.4 per cent low land.

In Odisha, small and marginal farmers comprising 84 per cent of the farming community own 52 per cent of the farm land and rest being owned by the medium and large farmers. Farming is mainly rice-based where more than 52 per cent of gross cropped area is under rice of which more than 61 per cent is rainfed (Nanda *et al.*, 2007a). There is strong need to develop suitable integrated farming system models to mitigate the risk due to single crop production and uncertainty due to seasonal and irregular income and employment. Systematic efforts have been initiated to start with integration of farming system components in order to generate nutritional security and dependable employment. The state is divided into 10 agro-climatic zones based on physiography and cropping situations.

Table 1: Agro-climatic zones of Odisha

S. N.	Agro Climatic Zone	Major soil type	Major crops
1.	North Western Plateau	Lateritic mixed red and yellow	Rice, Black gram, Arhar, Groundnut, Rapeseed and mustard, sesame, vegetables
2.	North Central Plateau	Red & yellow	Rice, Niger, Maize, Arhar, Back gram, vegetables II
3.	North Eastern coastal Plain	Alluvial Red Lateritic Coastal saline	Rice, Black gram, Groundnut, green gram, Vegetables
4.	East & South Eastern Coastal Plain	Alluvial Coastal Saline Lateritic	Rice, Black gram, Groundnut, Jute, Vegetable, Green gram
5.	North Eastern Ghat	Brown Forest Red and yellow Lateritic	Rice, Maize, Niger, Rapeseed and mustard, vegetables
6.	Eastern Ghat	Sandy clay loam Mixed red and black Mixed red & Yellow	Ragi, Niger, Maize, Arhar, Vegetables
7.	South Eastern Ghat High Land	Red Loam, Lateritic Black loam	Rice, Ragi, Niger, Maize
8.	Western Undulating	Red, Black, Red and Yellow	Rice, Black gram, Groundnut, Sesame, Vegetables
9.	West Central Table Land	Red, lateritic, Black, Mixed red and yellow	Rice, Green gram, Groundnut, Sesame, Vegetables
10.	Mid Central Table Land	Red, lateritic, alluvial, black	Rice, Green gram, Groundnut, Sesame, Vegetables

No single farm enterprise is likely to support the small and marginal farmers for generation of adequate income and gainful employment year round (Mahapatra, 1994). Therefore, there is need to develop suitable integrated farming systems for such farmers since single crop production enterprises are subject to a high degree of risk and uncertainty because of seasonal, irregular and uncertain income and employment to the farmers. Under such situation integrated farming system models will be of much helpful to the farmers of the state.

- Small and marginal farmers account for 86 % of total number of holdings and average size of holding is very less to meet the family need.
- Thirty two point six per cent of people are below poverty line during 2011-12.
- Varied agro-climatic conditions.
- About 47% of land in the state is upland which provide a wide scope for Integrated Farming Systems.
- The state is subjected to recurrent natural disasters and Integrated Farming Systems act as insurance against these.

- Integrated Farming Systems will provide employment for the rural youth and will ensure nutritional security to the small and marginal farming community.
- The state has a well equipped extension support system through Krishi Vigyan Kendras and various Government Departments to provide technical guidance to the practicing farmers.
- Good scope for marketing of diversified products in the state.
- There is great scope for pisciculture and ancillary enterprises like mushroom, apiary and suitable value addition for enhancing profitability of Farming Systems in the state.

Table 2: Integrated Farming System identified for different Agro-climatic Zones of Odisha

Sl.No	Agro-climatic Zones	Farming System Modules
1	North Western Plateau (Sundergarh, Deogarh)	Crop (rice-mustard/green gram)-horticulture-dairy-goatery-poultry-agroforestry
2	North Central Plateau (Keonjhar, Mayurbhanj)	Crop (rice/maize/sabai grass-pulse mustard)-horticulture-dairy-goatery-poultry-apiculture-agroforestry
3	North Eastern Coastal Plain (Balasore,Bhadrak,Jajpur)	Crop (rice-pulse/oilseed/vegetable)-dairy pisciculture-mushroom
4	East & South Eastern Coastal Plain (Kendrapada, Jagatsinghpur, Khurda, Puri, Nayagarh, Cuttack)	Crop (rice-pulse/oilseed/vegetable)-dairy-poultry- pisciculture-mushroom
5	North Eastern Ghat (Kandhamal, Rayagada Gajapati, Ganjam)	Crop (rice/maize-pulse/oilseeds/ vegetable/ fruits)-goatery-sheep-poultry-agroforestry –apiculture
6	Eastern Ghat Highland (Nabarangpur, part of Koraput)	Crop (rice/maize-niger/pulse)-horticulture goatery-sheep-poultry- agroforestry-apiculture
7	South Eastern Ghat (Malkangiri, part of Koraput)	Crop (rice/maize/ragi/til-vegetable)-poultry goatery-sheep- agroforestry-apiculture
8	Western Undulating Zone (Kalahandi, Nuapada)	Crop (rice/cotton-pulse/oilseeds) horticulture-dairy-poultry- goatery-mushroom
9	Western Central Table Land (Bargarh, Bolangir, Boudh, Sonepur, Jharsuguda, Sambalpur)	Crop (rice/pulse/oilseed/vegetable)-dairy poultry-piggery-goatery –mushroom
10	Mid Central Table Land (Angul, Dhenkanal)	Crop (rice/pulse/oilseed)-poultry-dairy apiculture-goatery-mushroom

Integrated farming systems for different farming situations

Farming system models have been developed for all the agro-climatic zones of the state, mostly executed by the farmers under direct supervision of the scientists working under Krishi Vigyan Kendras and All India Coordinated Research Project on Integrated Farming Systems. Some of the results are depicted in the following tables.

East and south eastern coastal plain zone

Pisciculture is said to be the most profit making enterprise in the integrated farming system models as evidenced from a pond based farming system. With intensive management and appropriate care the profit from pisciculture can be enhanced.

In east and south eastern coastal plain zone, a marginal farmer in the village Rampada, Bhapur, Nayagarh earned net return of Rs 1,44,850/year with B:C ratio of 2.56 from a pisciculture-horticulture-poultry system in an area of 0.94 ha (Table 3). Maximum net return of Rs 99, 955 (69%) was obtained from pisciculture with B:C ratio of 3.83.

Table 3: Performance of Pisciculture-Horticulture-Poultry system, Nayagarh

Components		Area (ha)	Expenditure (Rs)	Gross return (Rs)	NMR (Rs)	B-C ratio
Pisci culture	Mixed culture	0.56	20,500	71,000	50500	3.46
	Stunted fish fingerling	0.18	6200	15,975	9775	2.58
	Fish earling		3700	32,880	29,180	8.89
	Prawn culture	0.06	4900	15,400	10,500	3.14
	Sub total	**0.80**	**35,300**	**1,35,255**	**99,955**	**3.83**
Horti culture	Banana	0.08	5800	28,000	22,200	4.83
	Papaya	0.04	800	1935	1135	2.42
	Vegetable	0.02	1200	4200	3000	3.50
	Sub total	**0.14**	**7800**	**34,135**	**26,335**	**4.38**
Poultry	Layer	40 No.	3600	12,000	8400	3.33
	Broiler	500 No.	46,000	56,160	10,160	1.22
	Sub total	**540 No.**	**49,600**	**68,160**	**18,560**	**1.37**
	Grand total	**0.94167**	**92,700**	**2,37,550**	**1,44,850**	**2.56**

Nanda *et al.* (2007 b)

Integrated Farming System conducted in village Dorada of Athagarh in Cuttack district under East & South Eastern Coastal Plain Zone indicated that rice-based cropping system coupled with horticultural crops and rearing of dairy animals enhanced the productivity and profitability substantially as compared to conventional system and was 9.84 times higher in terms of net monetary return (Table 4). The B:C ratio was higher in horticultural crops(2.91) compared to field crops(1.84) and dairy animals(1.79). The result indicated that as compared to field crops, the net profit from horticulture crops are more (Rs 72,650) from the same land area (1.2 ha). The residue of animal component is used for increasing production of both horticultural and field crops.

Table 4: Crop based farming system at Dorada, Athagarh, Cuttack

Components		Area (ha)	Expenditure (Rs)	Return (Rs)	NMR (Rs)	B-C ratio
Crop component	Rice-wheat	0.40	15,222	29,582	14,360	1.94
	Rice-pumpkin	0.80	49,104	88,704	39,600	1.81
	Sub total	**1.20**	**64,326**	**1,18,286**	**53,960**	**1.84**
Horticulture	Diascorea	0.80	20,000	48,400	28,400	2.42
	Chilli	0.40	18,000	62,250	44,250	3.46
	Sub total	1.20	38,000	1,10,650	72,650	2.91
Animal component	Desi cow	1 no.	1500	2304	804	1.54
	C B Cow	2 nos.	13,300	24,240	10,940	1.82
	Sub total		**14,800**	**26,544**	**11,744**	**1.79**
	Grand total	2.40	1,17,126	2,55,480	1,38,354	2.18
Conventional Cropping system						
	Rice (local)	2.40	22,000	33,800	11,800	1.54
	Desi cow	1 no.	1200	2160	960	1.80
	Total	**2.40**	**23,200**	**35,960**	**12,760**	**1.55**

Integrated Farming System, in broader sense, can include several medicinal plants to enhance the system profitability from a specified area. Among various medicinal plants Aloe vera and Bacha can be taken to broaden the components in the model. A farming system model in Athagarh in Cuttack district, with a holding size of 5 ha, with components like rice, sugar cane, vegetable, ginger, Aloevera and cow earned a net profit of Rs 101700 from five ha land against the profit of Rs 10,500 from Conventional Cropping System (Table 5).

Table 5: Crop-medicinal plant-animal system at Dhaipur in Athagarh

Sl. No	Enterprises under IFS	Area in ha/No	Labour in MD	Expenditure (Rs)	Return (Rs.)	Net Return (Rs)	B-C Ratio
1. Crop Component							
	HYV Ric	2.40	130	15,200	29,300	14,100	1.92
	Sugarcane	0.85	110	24,800	43,500	18,700	1.75
	Yam	0.25	40	3,700	8,500	4,800	2.30
	Total	3.50	280	43,700	81,300	37,600	1.86
2. Medicinal plants							
	Ginger	0.25	50	4,500	10,900	6,400	2.42
	Aloe vera	1.15	150	65,000	1,20,000	55,000	1.84
	Bacha	0.10	25	1,500	2,800	1,300	1.80
	Total	1.50	225	71,000	1,33,700	62,700	1.88
3. Animal Component							
	Desi cows	2nos	50	3,100	4,500	1400	1.45
	Total	2	50	3,100	4,500	1,400	1.45
	G.T.	5.0/2no	555	1,17,800	2,19,500	1,01,700	1.86
Conventional Cropping System							
	Rice(Local)	5.0	130	19,600	29,500	9,900	1.51
	Desi cows	2nos	35	1,800	2,400	600	1.33
	Total	5	165	21,400	31,900	10,500	1.49

Integrated Farming System model with pisciculture, goatery, mushroom, poultry, vegetables and paddy cultivation in 1.15 ha area at Nimapara in Puri district generated net return of Rs 4, 53, 595/year as compared to Rs 38, 480 in the conventional method involving only rice cultivation and pisciculture (Table 6). The B:C ratio was the maximum (4.5) from banana cultivation followed by pisciculture (4.1). The residue obtained from rice was utilized in mushroom cultivation and the residues of live stock and horticultural crops were utilized in vermin composting. The poultry litters were partially used in pisciculture.

Many times, farming system activities become source of employment generation for the family members and hire labourers providing employment throughout the year. In the present system, from multiple components, 804 man days are generated of which 493 for male and 311 for female from an area of 1.15ha. Well designed farming system models will able to provide sustainable employment in the rural sector of the state.

Table 6: Performance of vegetable-pisciculture-poultry-dairy system at Nimapara

Component	Area (ha/No*.)	Expenditure (Rs)	Net returns		B.C. ratio	Employment generation	
			(Rs)	% contribution		Male	Female
Crop component							
Paddy	0.4	22,000	23,080	5.1	2.0	90	60
Horticulture component							
Vegetables	0.2	27,000	60,500	13.3	3.2	105	45
Banana	0.1	5,090	17,810	3.9	4.5	25	18
Papaya	0.05	775	425	0.1	1.5	3	4
Animal component							
Dairy	5*	64,800	1,51,200	33.3	3.3	151	30
Goatery	5*	7,000	15,500	3.4	3.2	25	10
Poultry	100*	15,520	23,980	5.3	2.5	20	18
Other enterprises							
Fishery	0.4	25,000	77,000	17.0	4.1	30	61
Vermicompost	4 units	6400	18,100	4.0	3.8	2	4
Mushroom	1,800*	54,000	66,000	14.6	2.2	40	60
Apiary	3 units	5600	10,600 (in 3 years)	2.3	2.9	2	1
Total	**1.15**	**2,33,185**	**4,53,595**	**100**	**2.9**	**493**	**311**
Conventional method							
Paddy	0.6	12,700	8,480	22.0	1.6	45	25
Fishery	0.4	18,000	30,000	78.0	2.6	52	12
Total	**1.0**	**30,700**	**38,480**	**100**	**2.2**	**97**	**37**

Sometimes, several cropping systems are included within farming system activities. As evident from the present case study, the farming system model is having rice-groundnut and rice-vegetable cropping systems in an area of 0.2 ha, which helped to obtain net profit of Rs 6000 and Rs 8000 respectively. A farmer in Khurda district with 1.2 ha area obtained net profit of Rs 133850 from crop-vegetable-pisciculture-mushroom-duckery system. Among the components maximum net profit of Rs 30000 was received from vegetable cultivation followed by Rs Rs 27000 from pisciculture (Table 7).

Table 7: Crop-vegetable-pisciculture-mushroom-duckery system in Etipur village - Khurda

Enterprises	Area in ha/no.	Family labour in MD			Cost of cultivation	Gross Return (Rs.)	Net Return (Rs)
		M	F	T			
Rice-Groundnut	0.2	15	8	23	6000	12000	6000
Rice- Vegetable	0.2	16	8	24	5000	13000	8000
Sugarcane	0.2	24	8	32	6000	20000	14000
Sub Total	0.6	55	24	79	17000	45000	28000
Plantation Crop							
Coconut	60 No.	16	0	16	7000	30000	23000
Mango	22 trees	8	0	8	5000	20000	15000
Sub Total	82 trees	24	0	24	12000	50000	38000
Pisciculture	0.3 ha	12	0	12	8000	35000	27000
Nursery							
Vegetable	70000	30	10	40	12000	42000	30000
Fruits & Flowers	3000	10	5	15	3000	13000	10000
Sub Total		40	15	55	15000	55000	40000
Mushroom	30 beds	0	3	3	750	1500	750
Apiary	160	0	0	0	200	800	600
Duckery	10 ducks	0	0	0	500	-	-500
	Total	**131**	**42**	**173**	**53450**	**187300**	**133850**

In Odisha, 63.6% of kharif cultivated area utilised for rice cultivation. So, there is need to develop integrated farming system models with rice as major component to enhance the profit of large number of rice farmers in the coastal districts. Understanding this, several rice based farming sytem models are developed in various coastal districts of the state. With adoption of rice based farming system during 2008-09, a medium farmer of village Kaudiabari in Jagatsinghpur district earned a net profit of Rs 1,35,200 with benefit-cost ratio of 2.02. His net monetary return was maximum (50%) from crop components (Table 8).

Table 8: Crop based farming system at Kaudiabari, Jagatsinghpur

Components		Area (ha)	Expenditure (Rs)	Gross return (Rs)	NMR (Rs)	B-C ratio
Crop component	Rice-fallow	2.92	54,000	97,500	43,500	1.81
	Rice-blackgram	1.00	22,500	42,500	20,000	1.89
	Rice-veg.	0.08	4300	9000	4700	2.09
	Sub Total	4.00	80,800	1,49,000	68,200	1.84
Pisciculture	Mixed culture	0.12	6000	16,000	10,000	2.67
	Magur	0.04	4000	10,000	6000	2.50
	Sub Total	0.16	10,000	26,000	16,000	2.60
Dairy	CB cows	2 Nos.	12,000	26,000	14,000	2.17
Poultry	Vanaraja	400 birds	26,000	54,000	28,000	2.08
Duckery	K. Campbel	30 birds	4000	13,000	9000	3.25
Grand Total		**4.16**	**1,32,800**	**2,68,000**	**1,35,200**	**2.02**

North eastern voastal plain zone

Due to consumer preference, vegetable cultivation is picking up in the state especially in the coastal districts throughout the year. Inclusion of vegetable components in the farming system will certainly enhance the ultimate profit of the farmers from limited land area. Several attempts have been made to develop vegetable based farming system model to enhance the income for the small and marginal farmers of the state. A small farmer of village Jamojodi, Tihidi, Bhadrak under north eastern coastal plain zone included vegetable, pisciculture, poultry and dairy in his 1.0 ha farming system model and got Rs 2,13,000 as net monetary return (NMR) with B:C ratio of 2.72 (Table 9). Among the components vegetable cultivation resulted maximum net profit (61%) and B:C ratio of 5.25.

Table 9: Vegetable based farming system at Jamjodi, Tihidi, Bhadrak

Components		Area (ha)	Expenditure (Rs)	Gross return (Rs)	NMR (Rs)	B-C ratio
Vegetable	Kharif	0.36	13,000	70,000	57,000	5.38
	Rabi & Summer		18,000	92,000	74,000	5.11
Pisciculture	Mixed culture	0.64	80,000	1,43,000	63,000	1.79
Poultry	150 birds		7000	20,000	13,000	2.86
Dairy	6 animals		6000	12,000	6000	2.00
	Grand total	1.00	1,24,000	3,37,000	2,13,000	2.72

In the recent days vermin-compost production has become an attractive enterprise, which needs to be included in the farming system models for enhancing net profit. Attempts have been made to include vermin-compost along with other components. A crop based farming system from Balasore district in north eastern coastal plain zone, a marginal farmer with 0.7 ha area obtained net profit of Rs 1,07,900 against Rs 13,550 from conventional practices. There was employment of family members to the tune of 79 male and 48 female totaling to 127 numbers. On the other hand employment generation in the conventional method was only 42 (Table 10). The residues of other components are recycled in vermin compost pits.

Table 10: Crop based farming system from Balasore

Components	Area/ No.	Expenditure (Rs.)	Net Profit (Rs.)	B.C. ratio	Man days Male	Man days Female
Crops	0.4 ha	7,300	8,400	2.15	22	18
Horticulture	0.2 ha	6,800	20,600	4.02	21	5
Pisciculture	0.1 ha	3,000	15,000	6	12	-
Animal	50 nos poultry	3,600	22,500	7.25	-	6
Husbandry	50 nos ducks	3,000	21,400	8.13	-	6
Vermicompost	3 pits	18,000	20,000	2.11	24	13
Total	0.7 ha	41,700	1,07900	3.58	79	48
Conventional methods	0.6 ha	9,680	13,550	2.39	22	20

North central plateau zone

In some areas there is scope for adopting multiple components like crop, horticulture, pisciculture, poultry, dairy, so that all the available resources can be appropriately utilized. This will ultimately add to income of the farm family by utilizing the family labour round the year. Paddy shared higher NMR followed by poultry fruit crops (Table 11)

Table 11. Rice based Farming System at Mayurbhanj

Components	Area/No.	Gross income	Expenditure	Net Profit	B:C ratio
Paddy	5.2 ha	225000	91000	134000	2.47
Wheat	0.4 ha	11000	5000	6000	2.2
Pulses	4.8 ha	84000	40000	44000	2.1
Mango	150 nos.	75000	25000	50000	3.0
Coconut	270 nos.	38000	10000	28000	3.8
Other fruits	112 nos.	26000	6000	20000	4.33
Dairy	05 nos.	36500	15000	21500	2.43
Poultry	1,000 Per batch	230000	110000	120000	2.09
Duckery	800 per batch	144000	64000	80000	2.25

Rearing dairy family has become a profitable vocation in some coastal districts due to the heavy demand of milk and milk products in the markets. Steps are initiated in some cases which practiced farming system models with dairy as a principal component in Mayurbhanj district (Table 12)

Table 12: Dairy based Farming System at Baunsvilla, Mayurbhanj

Components	Area (ha)	Expenditure (Rs)	Gross return (Rs)	NMR (Rs)	B:C ratio
Rice	1.60	14,000	28,800	14,800	2.06
Banana	130 Nos.	2000	13,000	11,000	6.50
Pisciculture	1.00	27,000	70,000	43,000	2.59
Poultry	80 Nos.	3000	7680	4680	2.56
Dairy	12 C.B. cows	1,80,000	2,92,000	1,12,000	1.62
Total	2.60	2,26,000	4,11,480	1,85,480	1.82

Combination of several enterprises not only in crease the net profit of the farmers but also provide employment to the family members round the year. Rural employment can be ensured with large scale adoption of farming system models even in the undeveloped districts of interior Odisha. As a result this will discourage migration of labour to other districts and states which is rampant in tribal dominated districts of the state (Table 13).

Table 13: Pond based Farming System in Keonjhar

Sl.No.	Enterprises	Area/No.	Expenditure (Rs)	Gross return(Rs)	NMR(Rs.)	B:C.Ratio	Employment generation		
							Male	Female	Total
1. Crop components (Paddy)									
i	Paddy	0.6 ha	15,100	27,180	12,080	1.8	38	82	120
	Paddy	0.6 ha	15,100	27,180	12,080	1.8	38	82	120
2. Horticultural crop									
i	Banana	50 plants	2,500	5,200	2,700	2.08	10	8	18
ii	Seasonal vegetables (Brinjal, Cole crops, tomato etc.)	0.32 ha	21,200	61,820	39,620	2.91	128	55	183
	Total	0.36 ha	23,700	67,020	42,320	2.82	138	63	201
3. Livestock									
i	Diary (Cross breed)	2 no.	32000	49000	17000	1.7	61	12	73
ii	Goatry	12 no.	10500	36400	25900	3.46	42	10	52
	Total	14	42500	85,400	42900	2.00	103	22	125
4. Poultry									
i	Duckery	50	4,000	10,800	6,800	2.7	3	7	10
	Total	50	4,000	10,800	6,800	2.7	3	7	10
5. Pisciculture									
i.	Carp polyculture (Carps & Prawn)	0.24 ha	22,100	53900	31800	2.43	36	17	53
ii.	Nursery pond) (Fry to fingerlings	0.08 ha	5,800	18,100	12,300	3.61	10	7	17
	Total	0.32 ha	27,900	72,000	44,400	2.58	46	24	70
	Grand total	1.28 ha	113200	2,62,400	1,48,500	2.31	328	198	526
Conventional method									
i	Paddy	0.6 ha	14,000	20,200	6,800	1.44	38	82	120
ii	Carp polyculture	0.24 ha	11,000	19500	8500	1.77	30	12	42
	Total	0.84 ha	25,000	39,700	15,300	1.58	68	94	162

West central table land zone

In some areas of the state rice is the dominant crop effecting the farm practice and decision making process in a larger way. In such areas farming system models with rice as the major component can be made popular among the farming community (Table 14).

Table 14: Rice based Integrated Farming System in Baragarh

Component	Area/ No.	Expenditure(Rs)	Net Profit (Rs)	B:C Ratio
Paddy	0.6 ha	16800	16680	1.99
Fish	0.2 ha	12000	8000	1.67
Banana	80 nos.	3200	4800	2.5
Papaya	80 nos.	4800	8200	2.7
Vegetables	0.8 ha	20000	30000	2.5
Total	1.6 ha	56800	67680	
Conventional method	1.6 ha	37600	39200	

Mid central table land zone

Nutritional garden, though practised in small area can be included in farming sytem models to get quality vegetable produciton contineously round the year. Addition of vermicompot as a component will certainly enable the farmer to use the products for enhancing profit of nutritional garden.

Table 15: Crop based Farming System at Shyamasunderpur, Angul

Components		Area (ha)	Expenditure (Rs)	Gross return (Rs)	NMR (Rs)	B:C ratio
Crop component	Rice - pulse	1.20	31,000	62,000	31,000	2.00
	Rice-oilseed	0.40	6000	12,000	6000	2.00
Horticulture	Nutritional garden	0.02	4000	10,000	6000	2.50
Pisciculture	0.80	34,000	70,000	36,000	2.06	
Vermi-compost	3 units	7000	27,000	20,000	3.86	
	Grand total	**2.42**	**82,000**	**1,81,000**	**99000**	**2.21**

Crop component with Rice-Pulse system showed better performance in rainfed/ partially irrigated sistuation. However with availability ponds, pisciculture was the best performing one (Table 15). Inclusion of Horticultural Crops with animal component was most beneficial in rainfed areas of Dhenkanal district with higher net return and benefit cost ratio (Table – 16).

Table 16: Crop based Farming System in Dhenkanal

Components	Area (ha)	Expenditure (Rs)	NMR (Rs)	B-C ratio
Horticulture	1.20	38,000	72,650	2.91
Animal component	0	14,800	11,744	1.79
Total	3.60	1,16,300	1,57,894	2.36
Conventional system	2.40	23,200	12,760	1.55

Western undulating zone

Profit from Integrated Farming System is multiplied when several components are included in the system. Inter cropping, mixed cropping in mingled with poultry, dairy and pisciculture increase the net monetary return and benefit cost ratio (Table 17). The NMR in IFS in farmers' field was 82 times higher than conventional practices.

Table 17: Crop based Farming System in Kanakpur, Kalahandi

Components		Area (ha)	Expenditure (Rs)	Gross return (Rs)	NMR (Rs)	B-C ratio
Crop component	Rice-rice	1.0	20,000	56,500	36,500	2.83
	S.cane+veg	1.2	45,000	2,05,000	1,60,000	4.56
	Maize-wheat	0.4	8000	21,400	13,400	2.68
	Banana+veg	2.4	1,55,000	4,00,000	2,45,000	2.58
	Sub total	5.00	2,28,000	6,82,900	4,54,900	3.00
Animal component	C B cow	2 Nos	7200	17,280	10,080	2.40
	Desi cow	3 Nos	9000	21,600	12,600	2.40
	Sub total	0	16,200	38,880	22,680	2.40
Poultry	Chicks	40 Nos	1000	5600	4600	5.60
	Ducks	40 Nos	1000	6000	5000	6.00
	Sub total	0	2000	11,600	9600	5.80
Pisciculture	Mixed culture	0.2	8000	20,000	12,000	2.50
	Sub total	0.2	8000	20,000	12,000	2.5
	Grand total	5.2	2,54,200	7,53,380	4,99,180	2.96
Conventional cropping system						
	Rice-mung	1.2	12,000	15,000	3000	1.25
	Desi cow	1 No	600	720	120	1.20
	Pisciculture	0.2	500	3500	3000	7.00
	Total	1.4	13,100	19,220	6120	1.47

The pond based farming system in Kalahandi with the components of field crops, vegetable, floriculture, fruits, poultry etc. have resulted a net profit of Rs 75567 from 0.73 ha land with B:C ratio of 2.73 (Table 18).

System of Rice Intensification (SRI) when included in the system, the profit is enhanced. Pisciculture is the single most component which results maximum profit (Rs. 32200) as evidenced from the present case study (Table 18).

Table 18. Pond based Farming System in KVK, Kalahandi

Enterprise	Area (ha)	Expenditure (Rs)	Gross return	NMR (Rs)	B:C ratio
Pisciculture	0.23	19800	52,000	32200	2.63
Vegetables	0.02	3940	14,832	10892	3.76
Maize & sunhemp	0.02	400	1850	1450	4.63
Floriculture	0.02	250	1200	950	4.80
Fruits	90 no.	3400	9840	6440	2.89
Poultry	80 no.	7300	13650	6350	1.87
SRI paddy	0.46	8600	25885	17285	3.01
Total	0.73	43690	119257	75567	2.73

North eastern ghat zone

Integrated Farming System with components of crop-horticulture-dairy in Gajapati district in North eastern ghat zone had recorded a net profit of Rs 2,90,000 with B:C ratio of 4.58 (Table 19). Vegetables shared maximum net profit (Rs 1,90,000) in the system. On the contrary, the net return from conventional system was Rs 146000 from crop, vegetable and dairy enterprises.

Growing vegetables both during kharif and rabi season and to the profit significantly. The residue from dairy is used as mannure for cultivation of vegetables and field crops.

Table 19. Crop-horticulture- dairy based Integrated Farming System from Gajapati

Component	Area(ha)	Expenditure (Rs)	Return (Rs)	NMR (Rs)	B:C
Crop component	2.5	20,000	80,000	60,000	4.0
Paddy (SRI)+ Sunflower	1.0	8,000	45,000	37,000	5.62
Vegetables (Kharif &rabi)	1.5	50,000	2,40,000	1,90,000	4.8
Dairy (Desi cows)	3.0 cows	3000	6000	3000	2.0
Total	5.0	81,000	3,71,000	2,90,000	4.58
Conventional	5.0	74,000	2,20,000	1,46,000	2.97

With adoption of Integrated Farming System, a small farmer of village Kutulumba in Kandhamal district under North eastern Ghat zone with 1.4 ha area earned a net profit of Rs 57,400 against Rs 21,620 from the Conventional method. The net monetary return was maximum (Rs 20000) from raikia bean followed by garden pea (Rs 18000) (Table 20).

Certain components like Rikia bean which are confined to a specific area can be included in the system to enhance the profit.

Table 20: Crop based Farming System in Kandhamal

Component	Area/No.	Expenditure (Rs.)	Net Profit (Rs)	B.C. Ratio	Employment Generation	
					Male	Female
Turmeric	0.2 ha	9,000	13,400	2.4	12	25
Raikia Bean	0.8 ha	16,000	20,000	2.2	40	35
Garden Pea	0.4 ha	12,000	18,000	2.5	25	10
Mango	15 Nos	2,100	1,900	1.9	10	02
Apiary	4 Boxes	3,200	1,600	1.5	10	02
Backyard Poultry	25 Nos	1,250	2,500	3.0	-	06
Total	1.4 ha		57,400		97	80
Conventional method	1.4 ha		21,620		60	55

Crop based Farming System in Rayagada district in North Eastern Ghat Zone could create employment avenue for 396 man days (136 male & 260 female) in an area of 2.84 ha. The net profit from the system was Rs 95,250 against Rs 38825 in the conventional systems (Table 21). Among major crops rice, maize, wheat, sugar cane, vegetable, banana etc. contributed towards the net profit.

Inclusion of several components in the integrated farming system model will certainly benefit for employment generation as evidenced from the present example.

Table 21. Crop based Farming System in Rayagada

Component	Area/ No.	Expenditure (Rs.)	Net Profit (Rs.)	B.C. Ratio	Employment Generation	
					Male	Female
Homestead	0.04 ha	150	650	5.33	-	-
Paddy	0.8 ha	13,000	21,000	1.61	24	60
Oilseed	0.8 ha	18,000	21,600	2.16	25	90
Arhar	0.4 ha	5,000	16,000	4.2	15	30
Vegetable for market	0.4 ha	24,000	41,000	2.7	55	68
Mango	0.2 ha	2,000	2,900	2.45	5	5
Pisciculture	0.2 ha	2,000	5,100	3.55	12	4
Total	2.84 ha	64,150	95,250		136	260
Conventional method	3.0 ha	44,075	38825		55	52

Eastern ghat highland zone

In the forest dominated area inclusion of tree components in the farming system module can be much helpful. Several tree crops can be taken with vegetables (Table 22) to earn short term and long term profit from small piece of area.

This will be beneficial to the forest depending tribal people to earn their livelihood from vegetables and tree cultivation.

Table 22: Silvi- Horticultural Agroforestry Model in Koraput

Sl. No.	Components	Area (ha)	Expenditure (Rs)	Gross Return (Rs)	Net Profit (Rs)	B-C ratio
1.	Teak+Chilli	1.0	31,888	60,000	28,112	1.88
2.	Teak+Brinjal	1.0	36,892	57,600	20,708	1.56
3.	Teak+Tomato	1.0	31,144	46,800	15,656	1.50

Conclusion

Sustainable agriculture in farming with efficient use of natural resources for increased productivity and production would result in improved farm income, maintenance of ecological balance, easy accessibility to food and social benefits and improved quality of life for farming communities (Varughese and Thomas, 2009). Appropriate utilization of resources like human labour, water, land, use of byproduct of one component for production of other components with suitable integrated farming system model will help in increasing profit of small and marginal farmers of the state. Scientific management of available resources and recycling of agricultural waste by way of integrating different enterprises will make farming more profitable. Per hectare profitability was comparatively low when rice occupied major area in the system in the integrated farming system modules. Higher B-C ratio was obtained from vegetable cultivation and pisciculture. Hence for higher profit, emphasis should be given on inclusion of allied enterprises in the rice based farming system modules.

Future thrust

- Need to design Integrated Farming System models so that the rural producers can reach to the urban consumers.
- Need to create the database in relation to type of farming system, infrastructure, economics, sustainability, etc. under different farming situations.
- Quantification of biomass production and its overall efficiency in attaining sustainability.
- Need to develop research modules under different holding size with varying economically viable and socially acceptable systems.
- Development of indigenous technical know-how existing in the farming community and its scientific validation and popularization.

- The assessment and refinement of technologies at cultivator's field.
- Need to prepare a contingent planning to counteract the climate threats under different farming situations.
- Need to prepare a policy draft for the consideration of planners for its promotion at large scale with nominal financial assistance either through short/medium/long term loans and other promotional advantage.
- Use of efficient microorganism for recycling of crop residues.

Action plans need

- There is need to enhance relationships with farmers, strengthening of farmer organizations at grass root levels, support democratization of policy formulation and good governance and ensure greater equity in gender relationships.
- Interactions between crop farming, fishing, forests and global climate change is needed for holistic approaches .
- The world's growing population seeks to meet its economic and social needs without preventing future generations from doing the same.
- Integrated farming systems need to enhance productivity and profitability for the farmer, yet still preserve the quality of the environmental resources – soil, water and air.

References

Behera, B., Neducchezhiyan, M., Mohanty, S. And Panda N. 2014. Final report, sustainable rural livelihood and food security to rainfed farmers of Orissa, National Agricultural Innovation Project (Component 3). Orissa University of Agriculture and Technology, Bhubaneswar, India, 120 p.

Behera, U.K. 2010. Manual on Farming System. Division of Agronomy, IARI, New Delhi.

Behera, U.K. and Mahapatra, I.C. 1999. Income and employment of small and marginal farmers through intrgrated farming systems. Indian Journal of Agronomy. 44(3):431-439.

Behera, U.K. and Sharma, A. R. 2007. Modern concepts of agriculture. Indian Agricultural Research Institute, New Delhi. Pp. 24.

Biggs, S.D.1985. A farming system approach. Some unanswered questiones. Agricultural administration and extension.18: 1-18.

CGIAR. 1978. Farming systems research at International Agricultural Research Centres. TAC Secretariate, Rome, Italy.

Channabasavanna, A. S., Biradar, D. P., Prabhudev K. N. and Mahabhaleswar Hegde. 2009. Development of profitable integrated farming system model for small and medium farmers of Tungabhadra project area of Karnataka. Karnataka J. Agric. Sci., 22(1) : pp 25-27.

Gill, M.S., Singh, J.P. and Gangwar, K.S. 2009. Integrated farming system and agriculture sustainability. Indian Journal of Agronomy. 54(2): 128-134.

Jayanthi, C., Balusamy, M. and Mythili, S. 2001. Integrated farming system in rainfed lands. A chapter in the book Land use planning and watershed management in rainfed agriculture. Published by Department of Agronomy, TNAU, Coimbatore. Pp. 202.

Jha, D. 2003. An over view of farming system research in India. Annals of Agricultural Research. 24(4): 695-706.

John Dixon and Aidan Gulliver. 2001. Farming Systems and Poverty Improving Farmers' Livelihoods in a Changing World. FAO and World Bank Publication. Pp. 9.

Kumar, S. And Jain D.K. 2005. Are linkages between crops and livestock important for the sustainability of the farming system ? Asian economic Review.47: 90-101.

Mahapatra, I.C. 1994. Farming system research- A key to sustainable Agriculture. *Fertilizers News* 39(1):13-25.

Madhava Swamy G. 1985 Effect of diversified farming on income and employment. Indian Journal of Agricultural Economics. 40(3): 333.

Nanda, S.S., Mishra, S.N. and Mohanty, M.2007a. On farm farming system for income and employment generation in Kalahns district of Orissa. J. Farming Systems Research and Development 13(1):56-59.

Nanda, S.S. Mohanty, M, Pradhan K.C. and Mohanty, A.K. 2007b. Integrated nutrient management for sustainable production, economics and soil health in rice-rice system under acid lateritic soil of costal Orissa. J. Farming Systems Research and Development 13(2): 186-190.

Patel, R.H. and Dutta, S. 2004. Integrated farming system approach for sustainable yield and economic efficiency – A review. Agricultural Reviews.25(3). http://www.indianjournals.com /ijor. aspx?target = ijor:ar&volume =25 & issue=3&article=006 viewed on 31.1.10

Patil S. K., Urkurkar J.S., Rathore A.L. and Nandeha K.L. 2014. Integrated farming system models for marginal and small farmers. Direstorate of Extension Services, IGKVV, Raipur.

Ramrao, W. Y., Tiwari, S. P. and Singh P. 2006 Crop-livestock integrated farming system for the Marginal farmers in rainfed regions of Chhattisgarh in Central India. Livestock Research for Rural Development 18 (7). http://www.lrrd.org/lrrd18/7/ cont1 807.htm.

Ramrao W Y, Tiwari S P and Singh P 2005 Crop-livestock integrated farming system for augmenting socio-economic status of smallholder tribal of Chhattisgarh in central India. Livestock Research for Rural development. *Volume 17, Article# 90* Retrieved May 17, 2006, from http://www.cipav.org.co/lrrd/ lrrd17/8/ramr17090.htm.

Shaner, W.W., Phillip, P.E. and Schmehl W.R. 1982. Farming system research and development: guidelines for developing countries. West view Press, Boulder, and Colarado, USA.

Sharma, A.R. and Behera U.K.2004. Fertilizer use and option for diversification in rice-wheat cropping systems in India. Fertilizer News. 49 (12): 115-131.

Singh, K.P., Singh, S.N. and Kadian, V.S. 2009. Employment generation potential of integrated farming systems on small farm situations in India. www. conference.ifas.ufl.edu/ifsa/ posters/Singh.doc viewed on 31.1.10

Swaminathan, M. S. 1981. Indian agriculture - Challenges for the Eighties, Agriculture Situation of India. 36(6): 349-59.

Varughese, K. and Thomas, M. 2009. Integrated farming system for sustainability in coastal ecosystem. Indian Journal of Agronomy. 54 (2):120-127.

Integrated Farming System Practices: Challenges & Opportunities, pp. 35-43

New India Publishing Agency, New Delhi, India

2

Integrated Farming Systems in Single Objective Framework using Linear Programming

U.K. Behera

Introduction

Decision making is the most important aspect of any business and industry. Farming is a business and agriculture is also an industry. Hence, decision making plays an important role with regard to the problems concerning production of commodities. The main questions before the producer or the production manager are: (i) What to produce, (ii) How to produce, and (iii) How much to produce.

Linear programming (LP) is a modeling tool that can assist in the solution of many problems in agriculture. It is useful in selecting the best alternative from a number of available course of actions.

LP models are designed to "optimize" a specific objective criterion subject to a set of constraints. The quality of the resulting solution depends on the completeness of the model in representing the real system.

Systems analysis techniques such as optimization proved to be useful for efficient resource allocation under various constraints (Loucks *et al.*, 1981; Vedula and Mujumdar, 2005; Taha, 2005). Optimization models optimize the use of farm resources, costs/profits and can analyse farm response to policy changes in an effective way (Loucks *et al.*, 1981). Linear Programming is one of the most

Division of Agronomy, Indian Agricultural Research Institute, New Delhi – 110012 applied solution methodology in agricultural planning to determine the optimal/ compromise policy (Loucks *et al.*, 1981; Raju, 1995; Vedula and Mujumdar, 2005) in single and multiple objective framework. However, their application to IFS is limited (Mahapatra and Behera, 2004).

Components of linear programming

A Linear Programming problem consists of three major components: (i) Decision variable, (ii) Objective function, and (iii) Constraints

Decision variable

The decision variables relate to the decision that must be made and are expressed with algebraic symbols. The decision variables (x_j, j = 1, 2, 3---n) can be defind as:

x_j = number of hectares of land that should be planted in crop j if profits are to be maximized.

Objective function

In the crop planning the objective function will be established by first defining the profits for each crop j, on a per hectare basis, as Cj. The objective function can be expressed as

$$\text{Maximize } Z = C_1 X_1 + C_2 X_2 + C_3 X_3$$

Or

$$\text{Maximize } Z = \sum_{j=1}^{n} C_j X_j$$

The value of the objective function will be expressed in rupees because the unit of Cj are rupees per ha for crop j and the units on the decision variables (Xj) are hectares of crop j. Thus, the combination of X's that yields the largest Z is the optimum cropping plan.

Constraints

The objective function could be increased without limit, if there were no constraints on the problem.

Suppose, the farmer has only 200 ha of land and 300 hours of labour time to devote to the crops in the equations. The constraints must be expressed in algebraic equation. In case of the constraints, the amount of each resource needed to produce one unit of output must be known.

The term aij is used to describe the amount of resource 'i' needed to produce one unit of product j. The term bi, is used to represent the total amount of resource i available.

Thus, the constraints for a linear programming problem, with j = 1, 2, 3 and i = 1, 2 (as in above example), can be expressed as:

$$a_{11}\ x_1 + a_{12}\ x_2 + a_{13}\ x_3 \leq b_1$$

$$a_{21}\ x_1 + a_{22}\ x_2 + a_{23}\ x_3 \leq b_2$$

The less-than-or-equal-to signs in these equations stem from the fact that the land and labour limits are not to be expressed. In other words the farmer may use less than the maximum of one resource, if the optimum solution seems it best to do so.

Formulation of a linear programming problem

The linear programming techniques identify the combination of the products which will maximize the profit without violating the resource constraints. When a problem of management is expressed in terms of the decision variables with appropriate objective function and constraints we say that the problem has been formulated. Identifying and formulating linear programming problems much as an art as it is a science. The more one formulate linear programming problems, the better one's ability to support decision making process.

Problem formulation

A farmer is planning for the upcoming growing season. She has three crops, she is considering planting on her 200 ha of land. Considering all factors, the farmer estimates that 300 hrs of labour time can be devoted to these three crops. Assuming that machines and capitals are not limiting, the farmer wants to know how many hectares she should plant in each crop in order to maximize her profits. She receives Rs. 50,000 profit per ha for groundnut (Crop 1), Rs. 35000 for the pigeonpea (Crop 2) and Rs. 40,000 for maize (Crop 3). It is estimated that crop 1 requires 5 labour hours, each hectare of crop 2 requires 3 labour hours, and each hectare of crop 3 requires 4 labour hours. The total linear programming formulation of the crop planning example can be stated as follows:

Maximize $Z = 50{,}000\, x_1 + 35{,}000\, x_2 + 40{,}000\, x_3$

Subject to:

$x_1 + x_2 + x_3 \leq 200$ (Land constraint)

$5x_1 + 3x_2 + 4x_3 \leq 300$ (Labour constraint)

and

$x_1 \geq 0,$

$x_2 \geq 0,$

$x_3 \geq 0$

Steps/stages followed in linear programming

Linear programming is a multistage process that starts with problem definition and determination of study objectives.

Problem definition and determination of study objectives

The first stage of any study is to define the decision problem and determine the study objectives. Until this is done it is impossible to consider the formulation of the model. Nevertheless, there is sometimes a tendency for studies to be carried out in reverse order. One should remember that the technique and the model must always be subservient to the problem. This does not mean, however, that the study objective may not change as the study progresses. As the study develops, new insights may be gained as to the real nature of the decision problem and the objectives may need to be adapted accordingly.

Formulation of the model

The first question to answer here is whether or not linear programming provides an appropriate methodology for investigating a decision problem of this nature. Once the investigator is satisfied that linear programming is the most suitable technique in relation to the objectives and available study resources then care is needed to ensure that the model is adequately defined. It is easy to over constrain the model and in this way pre-empt the optimal plan. The analyst needs a combination of modeling experience, farming knowledge, judgment and intuition to avoid an inadequately formulated model.

It is good to list the activities or areas of the model that seem important. Having made the list, it is now easier to make decisions as to the required structure and level of detail that is appropriate. These decisions will depend, of course, on the study objectives and the time that is available to understand the study. In making these decisions the principle to always follow is "start simple". Only add detail where necessary to capture the essential elements of the problem. Examples of the decisions required are:

1. Should labour constraints be built in to the model?
2. Should land requirements be included in the model?
3. Should capital requirements be included in the model?

Working capital constraints should be included into the model, if it appears that availability of finance might be a limiting factor. A student undertaking a linear programming project, a farm advisor, and a private farm management consultant are all likely to make different decisions from one another on account of their different study objectives and time availability.

Collection of data

The data should be suitable and relevant for incorporation into modeling studies. The data may be checked with the experts to see if they "look reasonable". These experts may be researchers, consultants and also farmers themselves. Farmers often have direct experience of what it is possible to achieve on their land and it is both foolish and arrogant to ignore these experiences. In some cases it may be possible to supplement available information by undertaking a survey or trial work.

Model testing

Model testing can be viewed as a multistage process comprising verification, validation and sensitivity analysis.

Verification

Once the initial matrix has been constructed, the logic of the model should be rechecked. Resource requirements and supplies should also be checked for recording errors. The data are then ready to be entered into the computer.

The model should then be run to see if a mathematically optimal linear programming solution can be generated. If error have been made in formulating the model or in entering the data to the computer, it may be possible to generate a feasible solution. Most computer packages print out messages which assist in locating the part of the matrix where the error has occurred. Some of the more common errors that create infeasibilities are listed below:

(i) Co-efficient placed in the wrong row or column of the matrix. This often results when the table is not written out in a tidy manner.

(ii) Co-efficient given the wrong sign. This is particularly common in transfer rows.

(iii) Co-efficient not defined consistently. Co-efficient in all cells of a row must be calculated on the basis of a common unit.

For example, it is possible for a feed supply and demand relationship to be started in either a kilograms of drymatter or megajoules of energy, but not both. If kilograms of dry matter is the chosen unit then all co-efficient in that row must relate in a column must relate to one unit of the activity.

(iv) The model may be inconsistently constrained. This can be caused by contradictory statements, that is, constraint equations. For example, the following situation is infeasible since it is impossible to simultaneously satisfy all constraints.

Statement 1 : The number of ewes must exceed 1500.

Statement 2 : The number of hoggets must exceed one third the number of ewes.

Statement 3 : The total number of sheep (ewes plus hoggets) can not exceed 2000.

Infeasibilities also commonly occur where pre conceived ideas as to that is optimal are built in as maximum or minimum constraints, resulting in insufficient flexibility in the model.

(v) Activity levels may be unbounded. For example, attempt to run the model will identify only those errors that cause a solution to be mathematically impossible to achieve.

Validation

The purpose of the test so far discussed is to verify that the model is operating according to design specifications. Validation is much more subjective than verification and is an ongoing procedure. We should be constantly asking ourselves whether the model structure is sufficiently realistic to be providing useful answers to the questions posed, and whether the results appear reasonable in relation to expectations whereas verification can be measured in absolute terms, validation can only be assessed relative to specified study objectives.

One useful validation test is to see whether it is possible to come up with a feasible solution by adding additional equality constraints to force the existing activities undertaken on the farm in to the computed plan at their present levels. This should give a similar financial return and require similar levels of resource use as the existing farm plan. If the farm under investigation is similar to other neighboring properties, then it should also be possible by the same procedure to recreate the current farming systems on these other properties.

Sensitivity analysis

One way of gaining confidence in a linear programming model is to vary, one at a time, those costs, prices, resource supplies and input-output coefficients which seem on a priorty grounds to be important but for which either there is uncertainly as to their true value or about which there is likely to be variability in the real world. This type of procedure is called sensitivity analysis.

In case of resource supplies and activity returns some guidance as to whether or not a particular parameter value is critical can be obtained from the linear programming solution analysis. This analysis will show how stable the plan is to changes in activity costs and returns and whether changing the supply of a resource will affect the optimal plan.

The general rule is to vary only one parameter at a time. Otherwise, there is a likelihood of obtaining a "confounded" result with no indication as to the relative importance of each parameter change.

The exception to this general rule is whether two or more costs, activity returns, or input-output coefficients are correlated such that a mis-specification of one inevitably means a mis-specification of the other.

Implementation of the model

After the model is tested and processes, viz. verification, validation and sensitivity analysis are performed satisfactorily, the model is ready for application in the field. The case studies or other farming problems may be attempted with confidence.

Interpreting the results

The computer provides some data or information towards facts. The findings or output of the analysis needs to be interpreted. The final step in the study is to write-up the results in the understandable format or language. Farmers for which the result is meant may not be able to understand the output of the analysis. The essential "message" derived from a linear programming study can often be summarized in a few pages or even paragraphs.

Use of output in decision- support

The final stage in a study is to interpret the management implications of the model output and to express the conclusions and recommendations in terms that are meaningful to practicing farmers.

Assumptions

Linearity

The term linear refers to the fact the straight line relationships are employed in linear programming. For example if z = 2x, then the variable "z" increases in direct proportion to the magnitude of the variable 'x' for "x" values of 1, 2, 3 & 4, the values of "z" are 2, 4, 6, 8 respectively.

Additivity

It is assumed that the activities being modeled are additive in the sense that when two or more are used, their total product is the sum of their individual products. No interaction among activities are allowed.

Divisibility

It is assumed that factor can be used and commodity can be purchased in fractional units resources and activities are divisible and can be expressed in fractional units.

Finiteness

It is assumed that there are only a finite number of activities and constraints so that a feasible solution may be determined. The solutions are finite. There is limit to the number of alternations and resources.

Homogeneity

It is assumed that all units of the same resource or activity are identical.

Positivity

All activities are non-negative.

Determinism

It is assumed that all co-efficient are known with certainty.

Proportionality

The returns and resource requirements are independent of the level of activity. The two assumptions, i.e. additivity and proportionality define linear relationship among the activities and the inputs, and they also lead to an aggregate production function that has constant returns to scale. It means that if all fixed resources are increased by a factor of proportionality K, then the value of production will also increase by the same factor.

Application of linear programming

The application area of linear programming are:

(i) Transportation problem

(ii) Military applications

(iii) Operation of system of dams

(iv) Personnel assignment problem

(v) Farming system management

(vi) Other applications, viz. manufacturing plants, distribution centres, production management and manpower management.

Limitations of linear programming

(i) Linear programming is applicable only to problems, where the constraints and objective functions are linear i.e., where they can be expressed as equations, which represent straight lines. In real life situations, when constraints or objective functions are not linear, this technique cannot be used.

(ii) Factor such as uncertainty, weather conditions etc. are not taken in to consideration.

References

Loucks, D. P., Stedinger, J. R., and Haith, D. A., 1981. *Water Resources Systems Planning and Analysis*, Prentice-Hall, Englewood Cliffs, New Jersey.

Mahapatra, I.C. and Behera, U.K., 2004. Methodologies of farming systems research. In: Panda, D., Sasmal, S., Nayak, S.K., Singh, D.P. and Saha, S. (Eds), *Recent Advances in Rice-based Farming Systems*, 17-19 November 2004, Cuttack, Orissa, Central Rice Research Institute, pp79-113.

Raju, K. S.,1995 Studies on Multicriterion Decision Making Techniques and Management of Irrigation Systems. PhD thesis, Indian Institute of Technology, Kharagpur.

Taha, H.A., 2005. *Operations Research: An Introduction*, Prentice- Hall of India Private Limited, New Delhi.

Vedula, S. and Mujumdar, P.P., 2005. *Water Resources Systems, Modeling Techniques and Analysis*, Tata Mcgraw-Hill Publishing Company Limited, New Delhi.

Integrated Farming System Practices: Challenges & Opportunities, pp. 45-66

New India Publishing Agency, New Delhi, India

3

Crop Diversification in Integrated Farming Systems

L.M. Garnayak and S.K. Swain

Indian agriculture is characterized by small farm holdings. The average per capita land holding is speculated to squeeze to 0.10 ha by 2050. The proportion of marginal holdings is to swell more due to the growing population, inheritance laws of division of land and diversion of agricultural land to non-agricultural purposes. But we need around 457 Mt food grains to feed our nation by 2050 as against the present level production of about 257 Mt. Significant changes are also taking place in domestic and international demand for crop products due to improvement in income and standard of living, fast urbanization, and changing life styles and food preference patterns. The food demand is projected to grow annually at 2.0 % for cereals as against 3-4 % for edible oils and pulses, 4-5% for milk and milk products, meat, fish, eggs, fruits, vegetables, sugar and gur. In contrast, association of emerging problems like declining per capita land availability, increasing nutritional insecurity, decrease in factor productivity, increasing risk with individual farm component, unsustainability of cropping systems and unwillingness of young farmers towards farming lead to an apprehension in achieving the target. Farming system which aims at long term productivity, profitability, recycling of resources and employment generation, comes out as a major strategy to tackle all such problems. In farming system, various components of farming are integrated based on the cardinal principles

Directorate of Research, Orissa University of Agriculture & Technology, Bhubaneswar-751003, Odisha, India of minimizing the competition and maximizing the complementarity between the enterprises so as to improve productivity and profitability as well as resource conservation along with maintenance of the environment. A survey conducted by the AICRP on Integrated Farming Systems indicates existence of 19 predominant farming systems with majority (85%) as crop + livestock (Gangwar and Ravisankar, 2014). Crop dominant farming systems are existing in most of the states such as Andhra Pradesh, Bihar, Chattisgarh, Goa, Haryana, Jammu and Kashmir, Jharkhand, Kerala, Karnataka, Madhya Pradesh, North-East, Maharashtra, Odisha, Punjab, Tamil Nadu, Uttar Pradesh and Uttaranchal while livestock dominant systems, are present in Rajasthan and Parts of Gujarat. West Bengal, parts of Odisha and Assam states have the fisheries as a major source of income to the existing farming systems. The scope for promotion of horticulture (fruit) based systems exists in Jammu and Kashmir, Himachal Pradesh, Maharashtra, parts of Uttar Pradesh and in Sikkim while plantation dominant systems, are available in Andaman and Nicobar Islands and Kerala. Further, the contribution of crop and livestock to gross income of existing farming systems in 732 marginal households in 30 districts of 20 states in various NARP zones of India indicates that in majority of the place crop component contributes more than 50 % while at few districts such as Samba (Jammu), Aurangabad (Maharashtra), Mehsana and Panchmahal (Gujarat), livestock component contributes either equally or more. Thus cropping systems play a pivotal role in most of the farming systems of India.

Concept of crop diversification

Traditionally, increased food production has come from putting more land under cultivation. However, in large areas of the world, especially in Asia, all the land that can be economically cultivated is already in use. In future, most of the extra food needs must come from higher production from land already being farmed. As estimated, 80% of total food grain requirement in India would come from increase in productivity and only 20% through expansion of arable land. A major share of this increase is likely to come from increasing the number of crops produced per year on a given land using improved crop cultivars and cropping systems with sound management of farm resources. The concept of cropping systems is as old as agriculture. Farmers preferred mixed cropping, especially under dry land conditions, to minimise the risk of total crop failure. Even in Vedas, there is a mention of first and second crops, indicating the existence of sequential cropping. Cropping pattern refers to the yearly sequence and spatial arrangement of crops or of crops and fallow on a given area while cropping system involves the cropping patterns used on a farm and their interaction with farm resources, other farm enterprises, and available technology which determine their makeup. Intensive cropping entails growing a number of crops on the same piece of land during the given period of time and the turn-around period between one crop and another is minimised through modified land preparation. It is possible when the resources are available in plenty. Crop intensification technique includes intercropping, relay cropping, sequential cropping, ratoon cropping, etc. All such systems come under the general term multiple cropping.

Crop diversification is a concept, which is opposite to crop specialization. This diversification may be a shift of a crop to another crop or a shift from less profitable and sustainable cropping system to more profitable and sustainable cropping system and use of resources in best possible way by changing and modifying the degree, trend and time options of crop/cropping activities. It may be also shift from single crop farming to multiple crop farming, from subsistence farming to commercial farming or from low value food crops to high value food or non-food crops.This may be achieved by varietal substitution, crop diversification or crop intensification.

The ability of a country to diversify in order to attain various goals, will depend upon the opportunities for diversification and responsiveness of farmers to these opportunities. New opportunities that would benefit crop diversification are technological breakthroughs, changes in demand pattern, changes in government policy, development of irrigation and other infrastructure, development of new trade arrangements, and others. Similarly, challenges and threats necessitating crop diversification result from market and price risks, risk associated with existing crop management practices, adverse changes like degradation of natural resources and the environment, and socio-economic needs like employment generation, attaining self-sufficiency in some crops and earning foreign exchange from others. In general, it is presumed that higher the level of agricultural technology, lesser the degrees of diversification. As agriculture in less developed region is more dependent upon nature, the risk of crop loss is very high. In the areas where the variability of rainfall is high and adequate sources of irrigation are not available, farmers grow several crops in a season to get something from their fields in case of extreme weather. It has been also observed that whenever, a cropping system becomes unsustainable due to any reason, it is replaced by a new emerging system. A farmer is interested in cropping system that has a balance mix of higher biological productivity and economic returns with lesser risk and which possesses more stability over a period of time under fluctuating environmental situations.

Diverse benefits of crop diversification

Diversification is considered essential to reap economic benefits arising out of complementary and supplementary enterprises. Crop diversification has been recognized as an effective strategy for achieving the objectives of food security, nutrition security, income growth, reduction of risks from aberrant weather, poverty alleviation, employment generation, judicious use of land and water resources, sustainable agricultural development and environmental improvement. Crop diversification fulfils basic needs for cereals, pulses, oilseeds and vegetables and regulates farm income of the farmer. Diversification of enterprises generates more employment and spread out labour more uniformly as the farmers are engaged in different activities like sowing, weeding, harvesting and marketing of different crops round the year. Crop rotations or growing crops with different families and growth behaviours can break the incidence of insects and disease cycles, reduce weeds, curb erosion, supplement soil nutrients, improve soil structure and conserve soil moisture. Diversification can also soften impacts on environmental resources, exploit market potentials, create new industries based on agriculture, strengthen rural communities and aid the domestic economy by enabling producers to grow export potential crops.

Options for diversification of crops and cropping systems

India has a wide divergence with respect to rainfall, climate and soil types. Accordingly the country has been delineated into 15 agro-climatic regions by the Planning Commission, 127 agro-climatic zones by NARP and 20 agro-ecological regions by the NBSS&LUP. A large number of agricultural items are produced due to its diverse agro-climatic conditions. Various cropping systems exist in different agro-climatic zones (Table 1). The major cropping systems (Fig.1) are rice-wheat (10.5 Mha), rice-rice (5.89 Mha), rice-pulses

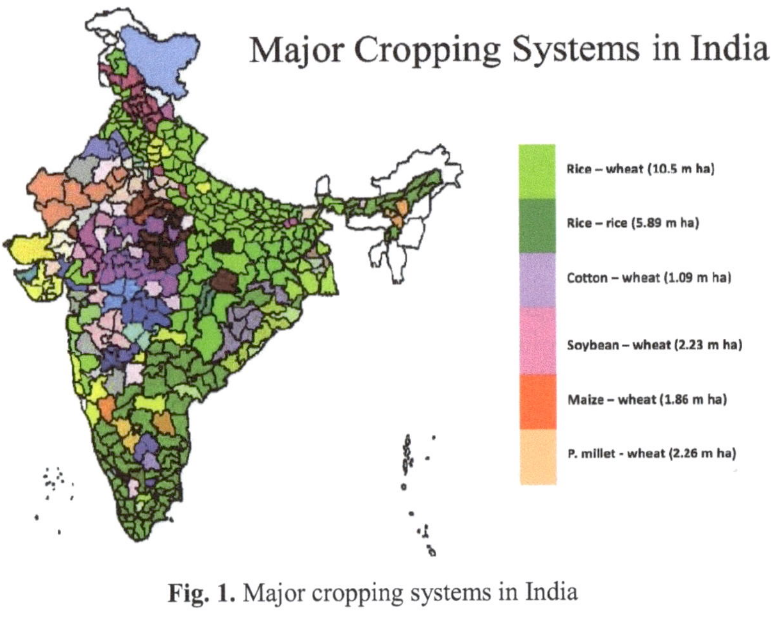

Fig. 1. Major cropping systems in India

Table 1: Predominant cropping systems in different agro-climatic regions of India

Sl. No.	Agro-climatic region	Rainfall, mm	Climate	Soils	Predominant cropping systems
1.	Western Himalayan	1650-2000	Cold arid to humid and sub-humid	Hill soils, sub-mountain meadow skeletal	Rice-wheat, maize-wheat, rice-potato-potato
2.	Eastern Himalayan	1840-3528	Per humid to humid	Brown hills, acidic soils, alluvial, tarai soils, red sandy, laterite soils	Rice-fallow, rice-rice, rice-pulses/ oilseeds
3.	Lower Gangetic Plains	1302-1607	Moist sub-humid to dry sub-humid	Recent alluvial, red,yellow loamy soils	Rice-rice, rice-wheat, rice-potato-jute rice-potato-vegetables
4.	Middle Gangetic Plains	1211-1470	Moist sub-humid to dry sub-humid	Alluvial, calcareous, tarai soils	Rice-wheat, rice-maize, rice-potato-sunflower
5.	Upper Gangetic Plains	721-979	Dry sub-humid tosemi-arid	Alluvial, tarai soils	Rice-wheat, sugarcane-ratoon-wheat, pearlmillet-mustard
6.	Trans Gangetic Plains	360-890	Semi-arid to dry sub-humid and arid	Alluvial, calcareous soils	Rice-wheat, cotton-wheat, pearl millet-wheat
7.	Eastern Plateau and Hills	1296-1436	Dry sub-humid tomoist sub-humid	Red, yellow, sandy loam to laterite	Rice-black gram/niger/linseed, rice-groundnut, rice- vegetables
8.	Central Plateau and Hills	490-1570	Semi-arid and dry sub-humid	Mixed red and black,medium black , grey,brown and alluvial soils	Maize/sorghum+ soybean-wheat, sorghum+pigeonpea-Bengal gram/ linseed, maize+soybean-wheat, sorghum-wheat
9.	Western Plateau and Hills	602-1040	Semi-arid to dry sub-humid	Medium to deep black, shallow red loamy soils	Cotton-wheat, sorghum+ soybean-Bengal gram/ durum wheat, pearl millet+ blackgram/ green gram-safflower
10.	Southern Plateau and Hills	677-1001	Arid, semi-arid to dry sub-humid	Medium to deep black, red sandy to loamy coastal and deltaic alluvium	Sorghum-cotton-groundnut, rice-Bengal gram/green gram, rice-rice, rice-*rabi* maize (Fodder)

Contd.

Sl. No.	Agro-climatic region	Rainfall, mm	Climate	Soils	Predominant cropping systems
11.	East Coast Plains and Hills	780-1287	Moist sub-humid to semi-arid	Deltaic coastal alluvial, laterite red and medium black soils	Rice-groundnut-green gram, rice-green gram/black gram, rice-rice
12.	West Coast Plains and Hills	2226-3640	Per humid to dry humid	Lateritic and coastal alluvium, red loamy soils	Soybean-wheat, cowpea/ groundnut/rice-greengram, cotton+ foxtail millet-green gram, rice-rice
13.	Gujarat Plains and Hills	340-1793	Semi-arid to drysub-humid	Deep black, coastal alluvium, medium black and brown soils	Castor-chillies-fallow, pearl millet -mustard/isabgol/cumin, cotton-wheat, pearl millet+moth/guar-barley-fallow
14.	Western Dry	55-395	Arid to extremely arid	Desert, grey brown soils	Pearl millet+ black gram- mustard, maize+soybean- durum wheat, groundnut- durum wheat- summer bajra, rice-*rabi* maize/groundnut-summer moong, cotton-durum wheat
15.	The Islands	1600-3000	Hot per- humid	Medium to very deep, red loamy and sandy soils	Rice-fallow

Source: Gill *et al*. (2008), Gill *et al*. (2014)

(3.50 Mha), pearlmllet-wheat (2.26 Mha), soybean-wheat (2.23 Mha), maize-wheat (1.86 Mha), rice-vegetables (1.40 Mha), cotton-wheat (1.09 Mha) and rice-maize (0.53 Mha). Each system has its own merits and demerits. The specific issues relating to some important cropping systems as enumerated by Gill *et al.* (2014) are as follows.

Rice-wheat

- Over mining of nutrients from soil
- Disturbed soil aggregates due to puddling in rice
- Decreasing response to nutrients
- Declining ground water table
- Build up of diseases/pests
- Build up of *Phalaris minor*
- Low input use efficiency in north western plains
- Low use of fertilizers in eastern and central India
- Lack of appropriate varietal combination
- Shortage of labour during optimum period for transplanting paddy

Rice-rice

- Deterioration in soil physical conditions
- Micronutrient deficiency
- Poor efficiency of nitrogen use
- Imbalance in use of nutrients
- Non-availability of appropriate transplanter to mitigate labour shortage during critical period of transplanting
- Build up of obnoxious weeds such as *Echinochloa crusgalli* and non-availability of suitable control measures

Cotton-wheat

- Delayed planting of succeeding wheat after harvest of cotton
- Stubbles of cotton create problem of tillage operations and poor tilth for wheat
- Susceptibility of high yielding varieties of cotton to boll worm and white fly and consequently high cost on their control leading to unsustainability

- Poor nitrogen use efficiency in cotton results in low productivity of the system
- Lack of appropriate technology for intercropping in widely spaced cotton

Sugarcane-wheat

- Late planting of sugarcane as well as wheat
- Imbalance and inadequate use of nutrients
- Emerging deficiencies of P, K, S and micro-nutrients
- Poor nitrogen use efficiency in sugarcane
- Low productivity of ratoon due to poor sprouting of winter harvested sugarcane
- Build up of *Trianthema partulacastrum* and *Cyperus rotundus* in sugarcane
- Stubble of sugarcane pose tillage problem for succeeding crops

Maize-wheat

- Delayed sowing
- Poor plant population
- Poor weed management
- Poor use of organic and inorganic fertilizers
- Large area is rainfed
- Declining yield trends in maize-wheat system due to multiple nutrients deficiencies
- Continuously over-mining of nutrients from soil and imbalance on use of fertilizers

Soybean-wheat

- Limited genetic diversity of soybean
- Short growing period hindered agronomy
- Non availability of inputs in time at farm level
- Rain fed nature of crop and water scarcity at critical stage of plant growth
- Insect pests and diseases

- Quality improvement problems
- Inadequate mechanization
- Partial adoption of technology by farmers

Legume-based cropping systems

- No technological breakthrough in respect of yield barriers, particularly in legumes
- Susceptibility of the pulses to aberrant weather conditions especially waterlogging and adverse soils making them highly unstable in performance
- High susceptibility to diseases and pests
- Low harvest index, flower drop, indeterminate growth habit
- Very poor response to fertilizers and water in most of the grain legumes
- Nutrient needs of the system have to be worked out considering N-fixation capacity of legume crops

Paira cropping in rice fallows (Garnayak *et al.*, 2005)

- Improper land levelling during cultivation of *kharif* paddy
- Delayed sowing
- Poor crop stand due to use of poor quality seeds, poor land levelling, poor germination of seeds, improper water management just before the harvest of *kharif* rice and sub-optimum seed rate
- Lack of proper varieties to withstand waterlogged condition in initial stage having resistance to YMV and powdery mildew
- Non-adoption of gap filling i.e., resowing of soaked seeds after harvest of rice
- Water stress at flowering stage
- Lack of weed control
- Poor nutrition
- High pest and disease incidence
- Stray cattle grazing

Rice, the principal crop of the country is grown in about 42.41 Mha area occupying 35% of total area under food grains. Fifteen different rice-based cropping systems are practised in the country, which accounts for the largest area of 27.92 Mha (Mahapatra *et al.*, 2012). The important rice-based cropping systems are rice-rice, rice-wheat, rice-pulse, rice-groundnut, jute-rice, rice-wheat-legume, green manure-rice-wheat, rice-rice-legume, rice-potato/maize/tomato/mustard-rice, jute-rice-wheat/mustard etc. (Sharma and Behera, 2005). Rice being the first crop in most of the cases, the duration of rice variety influences the productivity and profitability of the system. Continuous cultivation of rice for longer periods with low system productivity, and often with poor crop management practices, results in loss of soil fertility with multiple nutrient deficiencies, deterioration of soil physical properties and decline in factor productivity. Diversification and intensification of rice-based systems to increase productivity per unit resource is very pertinent. Hence, selection of alternate and component crops needs to be suitably planned to harvest the synergism among them towards efficient utilization of resource base and to increase overall productivity. Growing of non-rice crops such as pulses, oilseeds, vegetables and tuber crops in place of rainy or post rainy season rice is an alternative approach for realizing higher productivity, profitability and sustainability in rainfed uplands. Non-rice crops exhibited minimum fluctuation in yield over the years (Table 2 & 3) and are also more profitable than rice (Table 4).

Table 2: Performance of rice substituting crops in rainfed highlands of Odisha

Crop	1973	1974	1975	Mean	Fluctuation (%)
Maize	1804	2653	1943	2133	19.9
Jowar	2136	2494	1792	2141	16.4
Ragi	1625	2147	1685	1819	14.3
Mungbean	256	488	299	348	33.0
Rice	1623	327	1210	1054	61.3

Table 3: Performance of crops in normal and drought years in Mid Central Table Land zone of Odisha

Crop	Rice equivalent yield, REY (kg/ha)				
	Drought year (2000)	Normal year (2001)	Drought year (2002)	Mean	SEm
Maize (Grain)	5450	4400	4300	4717	369
Maize (Green cob)	8125	7321	6500	7315	469
Pigeonpea	5550	5081	5268	5300	136
Groundnut	5640	6240	5480	5787	231
Blackgram	4200	4900	3787	4296	325
Cowpea	2800	3600	2400	2933	353
Rice	1010	2850	1215	1692	582
SEm	864	573	696		

Source: Kar *et al.* (2003), Rainfall (mm) - Normal=1442, 2000=1149, 2001=1617, 2002=1002

Table 4: Productivity and profitability of paddy and non-paddy crops in North Eastern Ghat zone of Odisha

Crop	Yield (q/ha)	Net returns (Rs/ha)	Crop	Yield (q/ha)	Net returns (Rs/ha)
Maize	45 (Grain)	5,909	Sweet potato	236 (Tuber)	59,256
Greengram	8 (Grain)	9,773	Elephant foot yam	334 (Tuber)	63,223
Blackgram	9 (Grain)	12,473	Cassava	246 (Tuber)	41,39
Pigeonpea	15 (Grain)	18,284	Arrowroot	152 (Tuber)	36,346
Groundnut	15 (Pod)	9,368	Turmeric	40 (Dry rhizome)	44755
Yam	230 (Tuber)	43,100	Ginger	160 (Fresh rhizome)	95,755
Yambean	167 (Tuber)	60,316	Rice (upland)	25 (Grain)	234

(*Source:* Annual Reports of AICRP on Dry Land Agriculture, Phulbani)

Table 6: Efficient cropping systems identified in different agro-climatic regions of India

Agro-climatic region	Cropping system	System yield (t/ha/yr)	Net returns (Rs/ha)	B : C
Western Himalayan	Rice-marigold-Frenchbean	30.10[1]	1,68,000	2.90
	Rice-radish-potato	26.10[1]	90,815	1.73
	Rice-potato-Frenchbean	20.01[1]	1,19,793	2.78
Eastern Himalayan	Winter rice-onion-cowpea (Fodder)	12.17[1]	53,298	3.06
Trans Gangetic Plains	Maize-potato-onion	27.91[1]	1,25,023	2.50
	Pearlmillet-potato-mungbean	17.91[2]	60,586	2.50
Upper Gangetic Plains	Maize-potato-sunflower	24.16[1]	68,164	1.98
Middle Gangetic Plains	Rice-potato-greengram	18.10[1]	43,200	1.77
	Rice (HYB)-potato-greengram	26.64[1]	1,26,359	1.72
	Rice-lentil-greengram	10.68[1]	24,045	1.02
	Rice-potato-blackgram	12.73[1]	20,670	1.56
	Rice-potato-onion	28.99[1]	83,650	-
Lower Gangetic Plains	Rice-potato-jute	25.38[1]	-	-
	Rice-rice-cowpea (Fodder)	17.62[1]	-	-
Eastern Plateau and Hills	Rice-wheat + potato (1:1)- greengram	27.62[1]	1,24,683	1.67
	Rice-brinjal-green manure	18.12[1]	83,482	3.09
	Rice-potato-cowpea	22.21[1]	81,224	2.15
Central Plateau and Hills	Clusterbean-onion- fallow	15.19[3]	55,693	1.61
	Rice-potato-wheat	11.17[2]	57,352	2.79
	Soybean-potato-lady's finger	08.31[4]	93,626	2.70
Western Plateau and Hills	Soybean-onion	06.58[3]	1,80,878	6.65
	Soybean-mustard-groundnut	06.72[4]	1,26,152	3.48
	Soybean-potato-late wheat	09.16[4]	54,325	2.94

Contd.

Agro-climatic region	Cropping system	System yield (t/ha/yr)	Net returns (Rs/ha)	B : C
Sourthen Plateau and Hills	Maize-onion	12.32[1]	59,564	1.65
	Rice-rice	10.50[1]	27,134	1.66
	Maize-groundnut (Summer)	12.18[5]	44,140	2.33
	Maize-Bengalgram	09.95[4]	33,314	-
East Coast Plains and Hills	Rice-maize-cowpea	17.43[1]	69,020	2.30
	Rice-maize-greengram	14.83[1]	50,800	1.96
	Rice-groundnut-sesame	13.70[1]	41,390	1.76
	Rice-groundnut-greengram	13.60[1]	42,000	1.81
West Coast Plains and Ghat	Rice-rice-lady's finger	14.35[1]	32,434	-
	Rice-rice- amaranthus	42.93[1]	-	-
	Rice-brinjal	22.38[1]	80,071	1.92
	Rice-groundnut + sweet corn (4:1)	19.31[1]	44,642	1.62
Gujarat Plains and Hills	Groundnut-potato-pearlmillet	33.48[3]	1,70,976	3.13
	Rice-onion-cowpea	25.72[3]	72,583	1.92
	Groundnut-wheat-sesame	06.98[6]	1,25,087	2.97
Western Dry	Clusterbean-wheat	03.43[7]	38,712	3.00
The Islands	Coconut + turmeric (Hilly)		39,724	2.28
	Coconut + banana (Hilly)		70,470	2.18
	Coconut based mixed farming system (Hilly)		86,000	1.50
	Rice-rice-maize (Low land)	11.52[1]	12,905	1.91
	BBF system (Okra-amaranth-okra in bed; rice+fish in furrow; brinjal, Moringa and banana in boarder)		1,17,532	2.95

Source: Gangwar and Singh (2011); NB: [1]= Rice equivalent yield, [2]= Wheat equivalent yield, [3]= Pearlmillet equivalent yield, [4]= Soybean equivalent yield, [5]= Maize equivalent yield, [6]=Groundnut equivalent yield, [7]= Clusterbean equivalent yield

Very often it is observed, multiple cropping systems developed through research, rarely get acceptance by the farmers, unless duly supported with inputs and infrastructure. Therefore, in recent years researchers have been trying to design better alternative pro-farmer resource use efficient cropping systems to cater to diverse needs of farmers under different farming situations (Table 6).

Diversification through varietal substitution

Improved high yielding varieties of crops play a key role in raising productivity and production of any cropping system. This is true both for irrigated and rainfed ecosystems.Varietal characteristics that are to be considered for substitution includes, faster rate of growth in initial period, short duration to escape late season drought, strong, deep and penetrating root system to tap water and nutrients in adverse climatic conditions, ability to tolerate mid or late season drought and any aberrant climatic condition pertaining to the specific area where the crops are grown (Table 7).

Table 7: Varietal performance of different crops under same situation in Odisha

Crop	Variety	Duration (days)	Yield (q/ha)
Rice	Vandana	90	25.6
	Heera	88	25.7
	Vanaprabha	93	24.8
	Sneha	88	23.7
	Rudra	92	20.6
	Sankar	91	20.6
	Kalinga-III	95	21.5
	Heera yielded 20-25% more than Rudra, Sankar and Kalinga III		
Blackgram	Pant U-30	72	10.5
	T-9	69	09.5
	Sarala	70	09.6
	LBG-645	81	10.5
	Malabiri local	97	06.1
	Yield improvement due to improved variety : Blackgram 56-62%		
Maize	Navjot	95	40.0
	Kujimaka local	90	22.0
	Yield gain due to improved variety : Maize-80%,		
Groundnut	Smruti	103	16.9
	Lahachana local	120	09.3
	Yield gain due to improved variety : Groundnut – 82%		

Diversification through crop intensification

Crop intensification may be achieved by intercropping systems including several variants such as row, strip, relay, alley, multi-storeyed or mixed (inter) cropping as well as sequential cropping like double, *paira* or relay cropping. Intensification in sequential multiple cropping through introduction of non-conventional crops/ short duration crop cultivars and intensive input management, is a common way of increasing land use efficiency especially in irrigated ecosystems. Through innumerable studies undertaken in different regions and agro-ecological/farming situations of the country in recent years, several alternative cropping systems with high land use efficiency and monetary returns have been identified. Similar to crop sequences, extensive studies carried out under All India Coordinated Research Project on Integrated Farming Systems also have resulted in identification of several land use efficient and profitable intercropping systems in different regions of the country (Yadav *et al.,* 1998). Some of these are intercropping of castor + clusterbean at S.K. Nagar (Gujarat) under arid eco-system; maize + blackgram at Banswara, sorghum + pigeonpea at Indore, pigeonpea + greengram at Bichpuri (Agra) and cotton + groundnut at Junagarh under semi-arid ecosystem; wheat + mustard at Navasari, and mustard + blackgram and maize + cowpea at Karjat under coastal ecosystem. In on-farm trials, intercropping of groundnut + sunflower during *kharif* and groundnut + pigeonpea during *rabi* in Medak, Prakasam and Cuddapha districts of Andhra Pradesh, pigeonpea + groundnut in Ranchi district of Bihar, wheat + mustard and chickpea + mustard in western zone of Haryana, pigeonpea + sunflower in Bellary, fingermillet + pigeonpea in Bangalore district of Karnataka, and wheat + chickpea and chickpea + mustard in Mandla and Morena and chickpea + mustard in Ujjain and Ratlam districts of Madhya Pradesh were more remunerative than sole cropping. Similarly, intercropping of chickpea + mustard in Udaipur and pearl millet + greengram in Jodhpur districts of Rajasthan, pigeonpea + groundnut in north central plateau of Odisha, wheat + mustard in Nasik and Dhule districts of Maharashtra, blackgram + groundnut in Chengalpattu district of Tamil Nadu, and wheat + mustard during *rabi* followed by maize + blackgram and pigeonpea + sorghum during *kharif* in Ghazipur district of Uttar Pradesh were found to be more remunerative compared to their pure cropping under rainfed situation. Yield and economics of some intercropping systems in North Eastern Ghat zone of Odisha are given in Table 8.

Table 8: Performance of intercropping systems in farmers' field in Odisha

Intercropping system	Mean yield(q/ha)	Net returns (Rs/ha)	Returns per Rupee invested (Rs/Re)
Groundnut + pigeonpea (6:2)	9.04 (pods) + 5.72 (seeds)	8,836	1.43
Maize + pigeonpea (2:2)	43.04 (seeds) + 5.33 (seeds)	12,237	1.49
Maize + Cowpea (2:2)	43.73 (seeds) + 16.49 (Green pods)	14,386	1.49
Pigeonpea + okra (2:1)	8.55 (seeds) + 13.99 (fruits)	13,206	1.74
Pigeonpea + radish (2:2)	8.36 (seeds) + 98.23 (roots with tops)	21,726	1.89
Yam + maize (1:2)	94.9 (tubers) + 24600 no. of green cobs/ha	34,952	1.53

Crop diversification in farming systems for specific objectives

Cropping systems for weed management in farming system

Crop substitution/diversification in sequential cropping systems has been found to reduce some obnoxious weeds to a considerable extent, thereby reducing herbicide need to a great extent in areas where such weeds have assumed alarming proportions due to continuous adoption of a certain cropping system. Johnson grass (*Sorghum halepense*) becomes predominant weed in continuous maize cultivation but can be controlled by rotating with cotton. Similarly, a change from rice-wheat and rice-potato system to any other system, not involving rice in rainy reason, tends to reduce population of *Phalaris minor* in wheat considerably. Population of *Phalaris minor* in wheat could be reduced through inclusion of vegetable pea as catch crop before wheat and interruptive cropping (once in 3 years) of mustard instead of wheat (Gangwar and Ram, 2005)

Cropping systems for disease/pest management in farming system

Besides minimizing herbicide use for weed control, insecticide/fungicide use can also be minimized to a considerable extent through cropping system approach. Sorghum ear head fly damage is extremely rare where red gram is planted in alternate rows. Incidence of root rot of cotton caused by *Rhizoctonia solani* fungus is appreciably reduced by intercropping of dew gram (*Phaseolus aconitifolius*). The dew gram intercropping was found to cause moderating effect on soil temperature increase which is unfavourable for the parasitic activity of fungus. Verma *et al.* (1987) while working in intercropping of spices in autumn planted sugarcane observed that coriander, intercropped with sugarcane prevented top borer attack in sugarcane. Garlic and fennel intercropping was also found to reduce incidence of top borer in sugarcane.

Nutrient use efficient cropping systems for farming system

Inclusion of legumes in cropping systems for green manuring, fodder or grain purpose is an assured agro-technique to improve nutrient use efficiency (Yadav *et al.,* 1998). It is now established that 25-50% fertilizer NPK dose of *kharif* crops can be curtailed with use of FYM, green manuring or crop residues. Introduction of green manuring or leguminous crops in the existing rice-wheat system not only increased grain yields but also improved the physico-chemical properties, organic matter contents and nutrients availability in the soil over initial soil fertility (Hegde and Babu, 2002). Incorporation of summer greengram residues after picking pods is as good as green manuring in rice-wheat system. Inclusion of crop like potato in the system shows considerable residual effect on succeeding crops in wide range of cropping systems.

Water-use efficient cropping systems for farming system

Irrigation water is a costly and scarce resource and its availability for agriculture is expected to further go down due to increased demand for domestic and industrial uses. Though water use efficiency can be increased by genetic and environmental manipulation of the crops, it can also be increased by decreasing the evapotranspiration and other losses of water, such as conveyance, application, distribution and deep percolation. In cropping systems perspective, water use efficiency can be increased by identification of appropriate crop combinations in various systems. More remunerative and less water consuming crop rotations have been standardised at different locations in the country (Yadav *et al*., 1998). Sharma and Rajput (1990) found rice-mustard-sesame, rice-mustard-greengram and rice-potato-greengram rotations more water efficient systems in West Bengal. At Kharagpur, rice-wheat, rice-mustard and rice-potato were more viable sequences under lesser water input. At Chiplima (Odisha) net returns per unit of irrigation water (Rs/cm of water) were 28.05 for rice-mustard, 38.07 for rice-wheat-greengram and 52.81 for rice-potato-sesame rotation. In Tarai conditions of Uttar Pradesh, under higher level irrigation, rice-lentil and rice-wheat cropping systems were better while under lowest levels of *kharif* and *rabi* irrigations soybean-wheat proved the best sequence for net returns and benefit. Raised and Sunken Bed (RSB) in rice-based farming systems especially in swampy lands of coastal area aids to both intensification and diversification of cropping, improves water productivity and reduces the salinity problem in degraded land and water (Ravisankar *et al*., 2010).

Crop diversification in farming system in context of climate change

Climate change is real and its effects are experienced worldwide. Indian agriculture is highly prone to the risks due to climate change; especially to drought, because 60% of the agricultural land in India is rainfed, and even the irrigated system is dependent on monsoon. Flood is also a major problem in many parts of the country, especially in eastern part, where frequent flood events take place. In addition, frost in north-west, heat waves in central and northern parts and cyclone in eastern coast also cause havoc. Increase in temperature can reduce crop duration, increase crop respiration rates, alter photosynthesis process, affect the survival and distributions of pest populations and thus developing new equilibrium between crops and pests, hasten nutrient mineralization in soils, decrease fertilizer use efficiencies, and increase evapotranspiration. Climate change also have considerable indirect effects on agricultural land use in India due to availability of irrigation water, frequency and intensity of inter and intra-seasonal droughts and floods, soil organic matter transformations, soil erosion, changes in pest profiles, decline in arable areas due to submergence of coastal lands, and availability of energy.

There is need to identify climate resilient crops and cultivars for different regions to fit in to the cropping systems. Under the climate change scenarios, many of the conventional cultivation practices and strategies may no longer be relevant and farming system approach to diversify the activities will reduce the risk and ensure round the year livelihood. Potential adaptation strategies to deal with the impacts of climate change are developing cultivars tolerant to heat and salinity stress and resistant to flood and drought, modifying crop management practices, improving water management, adopting new farm techniques such as resource conservation technologies (RCTs), crop diversification, improving pest management, better weather forecasts and crop insurance and harnessing the indigenous technical knowledge of farmers. These coupled with improved agro-advisories and weather based crop insurance are likely to help farmers to cope with climate variability and minimize risks.

Development of new crop varieties with higher yield potential and resistant to multiple stresses (drought, flood, salinity) will be the key to maintain yield stability. On-farm water conservation techniques, micro-irrigation systems for better water use efficiency and selection of appropriate crop has to be promoted. Adjustment of planting dates to minimize the effect of high temperature induced spikelet sterility can be used to reduce yield instability so that the flowering period does not coincide with the hottest period. Adaptation measures to reduce the negative effects of increased climatic variability as normally experienced in arid and semi-arid tropics may include changing the cropping calendar to take advantage of the wet period and to avoid extreme weather events (e.g., typhoons and storms) during the growing season. Cropping systems may have to change to include growing suitable cultivars, increasing cropping intensities or crop diversification. For example, there is an urgent need for diversification of the conventional puddled transplanted rice and intensively tilled wheat to other cropping systems such as maize-wheat, pulse-wheat, maize-pulse, oilseed-wheat and direct seeded rice-wheat. The latter systems have less demand for water and nutrient (with legume) and use resources more efficiently thereby increasing farmers' income and exhorting less pressure to the natural resource base.

Cropping systems for resource recycling in farming system

India produces about 500-550 Mt of crop residues annually, which when managed scientifically can sustain soil health and fertility and reduce cost of production of crops. Monoculture systems rely mainly on external inputs while in the integrated system, recycling of nutrients takes place that helps in reducing the cost of production for economic yield. As increasing species diversity enhances the nutrient recycling capacity, adoption of integrated farming system can recycle the crop residues more efficiently than sole crop alone. Thus, selection of the enterprises to ensure recycling of the farm waste so as to depend more and more on on-farm inputs and less on off-farm resources is required for a farming system to be productive, profitable and sustainable. About 95% of nutritional requirement of a system is self sustained through resource recycling (Gill *et al.*, 2009). Rice-sunnhemp system supplemented with mushroom and poultry further enhanced the recycling potential of by 14.9% over rice-sunhemp system alone because the rice straw is used as substrate for mushroom cultivation. The recyclable potential by crop residues available under different systems is about 62% (Korikanthimath and Manjunath, 2005). The supplemented nutritive value of paddy-straw mushroom spent can be utilized for cattle feed to enhance milk yield in rice-mushroom system. In rice+ Azolla-cum –fish culture, the fish in rice field utilized the untapped aquatic productivity of rice ecosystem as the rice bottom is highly fertilised on account of the production of zoo and phytoplankton (Balusamy *et al.*, 2003). The leaves of *Leucaena leucocephala* in agroforestry-fish system, of mulberry trees in seri-fish system and of bamboo shoots in bamboo-fish culture is used as feed for fish and mud from fish ponds to fertilise the crops grown around the ponds. Hill farming is self sustained by flow of organic material and plant nutrients from uncultivated land and adjoining forest. Alternative land use systems such as agri-pasture, agroforestry, silvi-pastural and agri-horticulture are advocated for integrated management of natural resources in arid ecosystem.

Issues and challenges in diversification

Crop diversification helps ameliorating the adverse effects of seasonality on family incomes and peak labour demands, reduce risk due to fluctuating monsoonal patterns, help asset improvement on farms, conserve rainwater and save irrigation water, facilitate easier weed and nitrogen management, reduce water logging, and often result in better yield. Crop diversification and its associated change in tillage, crop establishment, nutrient management, and harvest practices will affect yield and soil fertility status and side by side seeks preparedness to cope with the change. Like any new venture, diversifying a farm posses some new challenges. The major problems and constraints in crops diversification are primarily due to the following reasons (Gill *et al.,* 2014):

- About 60% of the cropped area in the country is rainfed
- Sub-optimal and over-use of resources like land and water, causing a negative impact on the environment and sustainability of agriculture
- Inadequate supply of seeds and plants of improved cultivars
- Fragmentation of land holding less favouring modernization and mechanization of agriculture
- Poor basic infrastructure like rural roads, power, transport, communications, etc.
- Inadequate post-harvest technologies and inadequate infrastructure for post-harvest handling of perishable horticultural produce
- Very weak agro-based industry.
- Weak research - extension - farmer linkages
- Inadequately trained human resources together with persistent and large scale illiteracy amongst farmers
- Host of diseases and pests affecting most crop plants
- Decreased investments in the agricultural sector over the years

However, proper planning is required to surmount most of the obstacles. Some of the most common include:- market development, information on varietal performance, management practices, harvest and post-harvest handling and storage with possible additional costs, availability of seeds and specific pesticides for diversified crops, need to modify or replace farm equipments, modified labour requirement, locating local businesses and infrastructure for handling, transporting, processing, storing and marketing, etc.

Future thrust

Strategy to improve farm productivity as well as family livlihood may include increased operated farm size, intensification of existing production patterns, diversification of production and processing to make it market-oriented with value-addition, increased off-farm income and integration of multiple components in a farming system approach. But in order to design appropriate farming systems suited to diverse farming situations, farmers' participation beginning from planning till execution of research programme will be inevitable. Thus farm family shall be the focal point of all investigations and shall, to the extent possible, have better perception of farm problems and enable them suggest eco-friendly, resource use efficient and economically profitable farming system strategies. Appropriate policies and strategies need to be developed for multi-disciplinary approach of research through inter-institutional/inter-organisational linkages and collaborative programmes to achieve real success in farming systems research.Various researchable issues like documentation of the existing practices of mixed/ intercropping and identification of production constraints, nutrient and water management of inter/ sequence cropping systems, varietal suitability/ plant geometry, inclusion of high value spices, medicinal, aromatic and horticultural crops, identification of resource use (water and nutrient) efficient systems, profitable mixed/ intercropping systems for *rabi* season, standardization production technology of *paira* and bund crop, weed management in mixed/ intercropping system, ITK validation and identification of short duration drought resistant cultivar, etc. should be taken care of for reaping the benefits of diversification.

References

Balusamy, M., Shanmugham, P.M. and Baskaran, R. 2003. Mixed farming an ideal farming. *Intensive Agriculture* 41(11-12): 20-25.

Gangwar, B. and Ram, B. 2005. Effect of crop diversification on productivity and profitability of rice (*Oryza sativa*)-wheat (*Triticum aestivum*) system. *Indian Journal of Agricultural Sciences* 75 (7): 435-438.

Gangwar, B. and Ravisankar, N. 2014. Integrated farming systems for enhancing farm productivity and livelihood security under changing climate. *Souvenir*. National Symposium on Management options for enhancing farm productivity and livelihood security under changing climate held during 29-31 October, 2014 at OUAT, Bhubaneswar.

Gangwar, B. and Singh, A. K. 2011. *Efficient Alternative Cropping Systems*. Project Directorate for Farming System Research, Modipuram, Meerut, India. pp. 339.

Garnayak, L.M., Satpathy, P.C. and Mahapatra, P.K. 2005. *Pulses Production Technology in Orissa*. Directorate of Extension, OUAT, Bhubaneswar.

Gill, M.S., Gangwar, B., Walia, S.S. and Dhawan, A.K. 2014. Efficient alternate cropping systems of India. *Indian Journal of Ecology* 41(2): 219-227

Gill, M.S., Shukla, A. K. and Pandey, P.S. 2008 Yield, nutrient response and economic analysis of important cropping systems in India. *Indian Journal of Fertilizers* 4(4): 11-36.

Gill, M.S., Singh, J.P. and Gangwar, K.S. 2009. Integrated farming system and agricultural sustainability. *Indian Journal of Agronomy* 54 (2): 128-139.

Hedge, D.M. and Babu, S.N. 2002. Strategic issues and technological options for doubling oilseed production. In: *Oilseed-based Cropping Systems-Issues and Technologies*, pp 1-13. Gangwar, B., Sharma, S.K. and Yadav, R.L. (Eds). Project Directorate for Cropping System Research, Modipuram, Meerut, Uttar Pradesh, India.

Korikanthimath, V.S. and Manjunath, B.L. 2005. Resource use efficiency in integrated farming systems. Pages: 109-118. In *Proceedings of a Symposium on Alternate Farming Systems: Enhanced income and employment generation options for small and marginal farmers,* PDFSR, Modipuram, Meerut, Uttar Pradesh, India.

Mahapatra, I.C., Rao, K.S., Panda, B.B. and Shivay, Y.S. 2012. Agronomic research on rice (*Oryza sativa*) in India. *Indian Journal of Agronomy* 57 (3rd ISC Special Issue):9-31.

Ravisankar, N, T. Subrmani, S.K. Ambast and R.C. Srivastava. 2010. Enhancing farm income in Island ecosystem. *Indian Farming* 60 (6):16-19.

Sharma, A.R. and Behera, U.K.2005. Rice-based cropping systems. Pages: 607-647. In. *Rice in Indian Perspective.* S.D. Sharma and B.C.Nayak (Eds). Today and Tomorrow Printers and Publishers (India). P. 666.

Sharma, B.R. and Rajput, A.L. 1990. Sustained rice productivity in eastern region of India through efficient water management techniques. *Indian Journal of Agronomy* 35: 85-90.

Verma, R.S., Motiwale, M.P., Chauhan, R.S. and Tewari, R.K. 1987. Studies on intercropping of spices and tobacco with autumn sugarcane. *Indian Sugar* 31: 451-456.

Yadav, R.L.,Kamta Prasad and Dwibedi, B.S. 1998. Cropping system research. Pages: 193-220. In. *Fifty Years of Agronomic Research in India,* Indian Society of Agronomy, IARI, New Delhi. p. 270.

Integrated Farming System Practices: Challenges & Opportunities, pp 67-105

New India Publishing Agency, New Delhi, India

4

Horticulture based Farming System for Nutritional and Livelihood Security

R.K.Tarai and *D.K.Dora

Introduction

Since times immemorial, world agriculture is following the traditional agri-horticultural methods for livelihood support. With unusual growth of population and faster development of industrial sector, a steep competition surfaced among the domestic, industrial and agricultural sectors for land, water and other natural resources. Though agriculture uses a major chunk of land and water for crop production, it becomes insufficient to provide food, nutritional and aesthetic security to the growing millions at global level in general and the third world nations in particular. Hence, there lies a need to take the avenue of Integrated Farming practices by following the logics of Precision Farming. All over the world, farmers work hard but do not make money, especially small farmers because there is very little left after they pay for all inputs (seeds, livestock breeds, fertilizers, pesticides, energy, feed, labour, etc.). The emergence of Integrated Farming Systems (IFS) based on horticulture farming could help to develop a framework for an alternative development model to improve the feasibility of small sized farming operations in relation to larger ones. Integrated

Krishi Vigyan Kendra, Kalahandi, Odisha

*Department of Fruit Science and Horticultural Technology,College of Agriculture, Orissa University of Agriculture & Technology, Bhubaneswar-751003, Odisha, India

farming system (or integrated agriculture) is a commonly and broadly used word to explain a more integrated approach to farming as compared to monoculture approaches.

Indian agriculture supports 17 % of human population and 11 % of livestock of the world from 2.3 % of global land and 4.2 % of water. It contributes 18.1 % to GDP and livelihood to 650 million people (Singh, 2008). The concerns of Indian Agriculture are - declining farm income and profitability, degradation of soil and biodiversity, safeguard of the dependents for livelihood, regional disparity and uneven growth, declining land, water and capital investment, threat of climate change (temperature, unpredictable wealth, drought and flood) etc. The challenges ahead are to produce more for increasing population , to be competitive in open economy, balancing natural resources, equity and profitability, to create more employment and livelihood and for reducing malnutrition and under nourishment. Considering that nearly 70 per cent of people of India still live in villages, agricultural growth will continue to be the engine of broad-based economic growth and development as well as of natural resources conservation, leave alone food security and poverty alleviation. Accelerated investment is needed to facilitate agricultural development. This would lead agriculture sector on a better path and resurrecting its importance across the sectors will go a long way in making farming a respectable profession.

Horticulture based cropping system have proved ecologically acceptable and commercially viable under rain fed system where cereals & millets are grown as sole crop which often fails due to the vagaries of nature.

Cropping systems are dynamic interactive practices aimed at better use of the production components such as soil, water, air space, solar radiation and all other inputs on a sustainable basis. The productivity of rain fed upland is comparatively low and taking a monocrop in such situation is risky. Future source of growth in Indian agriculture lie in the rain fed areas, which constitute a large chunk of about 90 m.ha. (which is 65 % of the net cultivated area i.e. 147 m.ha.). It is about 30 % of the net rain fed area of Asia and pacific region. However, the most critical and high potential rain fed region occupies nearly 67 m.ha. which is having the mean annual precipitation ranging between 500-1500 mm. This part is characterized by low productivity, slow and poor dissemination of new technologies, large concentration of poor people, high degradation of natural resources including bio-diversity, and poor infrastructural facility. Although the rain fed region is lagging far behind the irrigated and other favorable region, this region has considerable potential and opportunities as it possesses fairly good soil, high precipitation, enough human resources etc.

Integrated approach has several distinct advantages such as security against complete failure of a system, minimization of dependence for external inputs, optimum utilization of farm resources, efficient use of natural resources etc. In order to minimize the risk of the farmers, farming system approach should be encouraged. A proper combination of different farm production systems namely, agriculture, horticulture, livestock, poultry, agro forestry, sericulture and pisciculture need to be promoted.

Government of India immediately after independence gained enough emphasis on the food production to ensure food security to the growing millions which continued up to the 7th Five Year Plan. With the allotment of Rs.1000 crores to the Horticulture sector during the 8th Five Year Plan which happens to be a 40 times hike over the 7th Five Year Plan brought in a visible change in the field of Horticulture. Since then horticulture has never looked back. However, to make an effective use of land and water, the most critical inputs and to make the production system more efficient, horticulture has to become an inevitable segment in the Integrated Farming System. Taking Agro ecological conditions and need and resource base of the people into account, the farming situations considered to be the micro units of the Agro ecological zones need to be addressed.

In general, farmers allocate a relatively small proportion of their land to vegetables and fruits. Agricultural diversification towards high-value crops can potentially increase farm incomes, especially in a country like India where demand for high-value food products has been increasing more quickly than that for staple crops. The comparatively high labour endowments of the small farmers, as reflected in their greater family sizes, induce them to diversify towards vegetables. Although fruit cultivation is also labour intensive (as compared to cultivation of staples), fruits are relatively capital intensive, making them a less advantageous choice for smallholders who tend to have low capital endowments. Furthermore, both the probability of participation in fruit and vegetable cultivation as well as land allocation to horticulture decreases with the size of landholdings in India. Small or medium holders do not appear to allocate a greater share of land to fruits or vegetables. However, the share allocated to vegetables is significantly higher if the family size is bigger, while the reverse is true in the case of fruits.

Need of diversification for small holders

- For reducing the risks due to biotic and abiotic stresses, market price fluctuations and high input costs.

- For meeting the food, feed, fibre, fuel and fertilizer, market demand of diversified products , nutritional requirement of family & increased demand of soil nutrients
- For increasing/improving the income, employment, standard of living & sustainability.

According to Gangwar, (2013), around 85 % of total holdings are under small and marginal farmers having the operational area of 44 %. Marginal farmers have no marketable surplus due to their holding and family size.

Based on 2010-11Agriculture Census out of 137.6 Million Farmers, only 4.9% owns > 4 ha. Smallholders make up about 85%.

i. LARGE > 10 ha (1.0 M)
ii. Medium 4-10 ha (5.8 M)
iii. Semi medium 2-4 ha (13.8 M)
iv. Small 1 -2 ha (24.7 M)
v. Marginal < 1 ha (92.3 M)

Challenges in agriculture and allied sectors

1. Small and Marginal holderes account for 85 % of cultivated land.
2. Average land holding reduced to 1.3 ha
3. 60 % Area would continue to be rain fed
4. Land, water, soil health and biodiversity resources shrinking
5. 65-70 % of animal wealth including small ruminants is with the small holders and land less labourers

Points to be considered for better livelihood security

1. Knowledge based horticulture-Capacity building through training and demonstrations of advanced technologies are proven methods for faster spread of knowledge and skill to the farmers.
2. Adopting suitable varieties tolerant to biotic and abiotic stresses for year round production.
3. Climate change adaptation / mitigation measures
4. Water management
5. Crop protection

6. Precision farming in horticultural crops
7. Eco-friendly farming with agro bio-inputs
8. Off season vegetable cultivation
9. Growing of uncommon exotic vegetables
10. Production of vegetables under protected condition.
11. Post Harvest Management and value addition
12. Pro-Poor value chain development.
13. Linking horticultural farmers with Market

Activities requiring urgent attention

1. Area increase under hybrid/improved varieties (15% to 35-40%)
2. Increased seed replacement
3. Front line demonstration
4. Horticulture based farming system and Crop diversification
5. Precision farming in horticulture
6. IPNM,IPDM
7. Post harvest management

Intensive integrated farming system (IIFS)

It involves agricultural intensification, diversification and value-addition. It helps improve physical and economic access to food, thereby fostering sustainable food security at the level of each individual in a household. The main purposes for such systems are:

1. Recognizing that "poverty" and not "food", is the major constraints to equitable development
2. Matching the production system with the available resources
3. Benefit to the community at large through supply of essential commodities throughout the year, providing year round employment to the farming family and reducing cost of production
4. Insurance against the failure of individual crop/ enterprise etc.
5. Complementarity of crop-animal-fish-birds-multipurpose trees-horticulture
6. Synergistic role of livestock in farming system rather than primary producers.

7. Maximum return per unit area per unit time in terms of produce and money
8. Biodiversity enhancement as the "alternative" feeding system, using mainly local plant resources, which provide comparative advantages to indigenous animal ecotypes and opportunities for greater use of indigenous knowledge
9. Selecting and promoting crops and farming systems which optimize use of natural resources without depleting them.

Possible integrations for Horticulture based farming system are as follows

1. Cereals-Vegetable-Livestock
2. Livestock-Agroforestry-Horticulture
3. AF-Livestock-MFPs-Mushroom
4. Rice-Fish-Azolla-Hedgerow
5. Agri-Horticrops-MPTs-Livestock
6. Agri-Horticrops-Fish-Pig
7. Agri-Horticrops-Fish-Goat
8. Agri-Horticrops-Fish-Duck
9. Agri-Horticrops-Fish-Poultry

Horticulture diversification

Desirable change in the existing system towards more balanced cropping farming system to meet ever increasing demand of food, feed, fibre, fuel and fertilizer on the one hand and maintenance of agro-ecosystem on the other. Diversification is considered to be a good alternative to improve system yield with enhanced profitability. Diversification in Agriculture has given added advantage to achieve the annual growth (5 – 8 %) through horticulture which otherwise, ranges from 2 to 2.5 %. To sustain Agarian growth over 4 to 4.5 %, it is essential to give emphasis on horticulture in a holistic way, so that every sector could be successful (Chadha *et al.*, 2014). Horticulture Diversification in general parlance refers to shift in allocation of land resources in a geographic location from one set of crops to another set of crops. It involves diverse goals. These could be related to out put growth, income enhancement, risk reduction, natural resource sustainability, labour scarcity or any other consideration.

According to Chand *et al*., (2014), Diversification towards horticulture was clearly evident from area share, growth in production and production shares (Table-1).

Table 1: Area wise sharing of crops in India

	1995-96	2005-06	2011-12	1995-96	2005-06	2011-12
Pulses	23.92	22.39	24.46	12.76	11.62	12.53
Oil seeds	25.96	27.86	26.31	13.86	14.46	13.48
Sugarcane	4.15	4.20	5.04	2.21	2.18	2.58
Cotton	9.04	8.68	12.18	4.82	4.50	6.24
Horticulture	10.23	12.81	13.04	5.46	6.65	6.68
i. Fruit & vegetable	2.67	2.90	3.63	1.42	1.50	1.86
ii. Spices & Condiments	7.56	9.92	9.41	4.03	5.15	4.82
Others	17.09	17.58	13.93	9.11	9.12	7.13
All Crops	187.47	192.74	195.25	100.00	100.00	100.00

Source : Chand *et al.*, (2014). *In: Land Use Statistics, Ministry of Agriculture, GOI*

The share of Cereals and Pulses is declining over years whereas, the share of horticultural crops has increased in a big way between 1995-96 and 2011-12. The share was around 22 % during pre-liberalisation period and it increased to around 28 % during recent years. Diversification in Horticulture at state level and relative importance of different states in All India basis showed that Kerala tops and Odisha comes second in terms of diversification towards horticulture (Chand *et al*., 2014). Export of fruits and vegetables put together increased five times in the last 10 years. Simultaneously, import of fruits in to the country also witnessed steep increase. Despite 5 % growth in domestic production, import of fruit has risen faster than export which is a clear pointer to the vast scope of demand for fruits in India. Similarly, Overseas demand has also been rising by close to 20 % per year. Thus both domestic as well as overseas demand strongly favours diversification towards horticulture in the country. (Chand *et al*., 2008).

Agricultural diversification towards high-value crops can potentially increase farm incomes, especially in a country like India where demand for high-value food products has been increasing more quickly than that for staple crops. The comparatively high labour endowments of the small farmers, as reflected in their greater family sizes, induce them to diversify towards vegetables. Although fruit cultivation is also labour intensive (as compared to cultivation of staples), fruits are relatively capital intensive, making them a less advantageous choice for smallholders who tend to have low capital endowments. Furthermore, both the probability of participation in fruit and vegetable cultivation as well as land allocation to horticulture decreases with the size of landholdings in India. Small or medium holders do not appear to allocate a greater share of land to fruits or vegetables. However, the share allocated to vegetables is significantly higher if the family size is bigger, while the reverse is true in the case of fruits.

Diversification of agriculture can be accomplished through cash oriented farming. Beside rice, cultivation of other agricultural products should be encouraged. Such agriculture products may include Cereals, Coarse grains, Sugarcane, Cotton, Pulses, Oilseeds for bio-fuel production, Soybean, Sunflower, Maize, Vegetables, Spices, Garden spices, Ginger, Garlic, Mustard etc. Horticulture may also provide alternative to agriculture. Fruit varieties having good commercial values could be encouraged for national and local markets. The suitable fruits are mango, guava, banana, lemon, sweet lime, pomegranate, etc.

Horticulture based farming system

Horticulture based land use system has proved commercially viable and ecologically acceptable under rain fed system where millets and paddy are grown as sole crops which often fails due to the vagaries of nature. It has been divided into a number of sub-systems. Each system has been given a specific name so that name itself would indicate what it means. This system is profitable where the soil depth is more than one meter and the slope of the land are within 10%. Horticultural based farming system has a huge potential for providing nutrition and livelihood security to farmers and farm women having small land holdings.

Indian Horticulture has emerged as a major producer of horticultural crops (268.82 million tonnes) surpassing the food grain production of 264.77 million tonnes) and is placed second after China in both fruit and vegetable production. This indicates the realization of the Indian Farming community on the significance of horticultural crops in the Integrated Farming System, thereby making the Agri-Horticultural ventures more profitable and sustainable to improve the livelihood of the farming community.

In a country like India, where there is a large vegetarian population, horticulture is likely to provide livelihood security to large rural population. Economic growth and changing food habit and others will keep horticultural crops and their value added products in high demand. Horticulture is a powerful catalyst for economic growth as shift to high value horticultural crops harnesses associated gains. (Ghosh, 2014). Small holder farmer population in high value horticultural crop production is reported to be high (70% of total vegetable production, 55 % of fruits and 50 % of spices) are from farm holding of less than 2 ha. High value fruit crops in conjunction with vegetable crops that provide cash flow until fruit tree begin to produce have much to offer to small holder farmers. Short gestation fruits like banana, papaya, pine apple , strawberry and vegetables including potato and tropical tuber crops and annual crops like ginger, turmeric fit well in small and marginal farm holdings. Many of the vegetable crops fitted well in Rice-Wheat cropping system having irrigation facility while ginger, turmeric, elephant food yam and others having particularly shade tolerance are intercropped in fruit orchards. In Coconut-Arecanut based cropping system, including homestead farming, black pepper, betlevine and yam are trailed with perennial tall species. Multi-tier cropping in coconut with banana, pine apple, black pepper, lemon and fodder grass has been found successful. In Arid zone, multiple cropping to maximise yield gains out of less fertile land and scanty rainfall is a common practice. The intercropping increased the yield of base crop and which in turn gave extra income to the farmers. The biomass produced by the intercrops as well as the production obtained from the intercrop augmented the net return from unit area. It was due to better utilization of land, light and water. The crop combination also helped in checking the soil and water erosion from the sloppy upland while generated additional labour employment. Horticulture production activities (e.g. fruit and vegetable cultivation, vegetable seed production, commercial floriculture, mushroom production etc.) with forward and backward linkages offer an array of agribusiness activities for sustainable livelihood and rural prosperity. Knowledge and skill development are important attributes of successful horti-entrepreneurship for small farm holders. As long as production of horticultural crops remaining relatively more profitable as compared to alternatives, it is feasible to promote crop diversification towards horticultural crops. Pro-poor participatory technology, training and trade support are crucial for livelihood security of small holder farmers. Agribusiness through aggregation of farmer is possible solution to problems of marketing of perishable horticultural crops with back up credit and input supplies.

Various horticulture based land use sub-system suitable for rain fed upland are given below:

Agri-horti system

Through the agri-horticulture system, the rehabilitation of degraded land is feasible. The pressure on the land is increasing and the productivity of the arable land is decreasing. Agri-horticultural system is one of the best forms of farming which has a perennial component of fruit trees. In this system agricultural crops are grown in association with fruit species, e.g., guava + mustard + sunflower, or aonla + mustard + sunflower. The total productivity of mustard and sunflower grown with guava /aolna is higher than that of the sole guava /aonla plantation (Singh 1996).

Silvi-horti system

In this system, forest species are planted in association with fruit species, e.g., guava + eucalyptus + subabul or Bael + eucalyptus + phalsa or aonla + eucalyptus + subabul. These models are considered very useful for production of fruits, fuel and fodder.

Silvi-oleri system

In this system, vegetable crops are grown in association with forest species. The crop combinations are teak + brinjal + bottle gourd and teak + okra + bitter gourd.

Horti-pastoral system

In order to keep the options open with regard to the farmers liking in the region, a viable alternative land-use system may be worked out keeping horticultural as well as pastoral components in the system. There is also an urgent need to give equal thrust to agri-horti pastoral system for overall satiability of the farming community so as to make use of the marginal and degraded lands in the rainfed agro-ecosystem. Intercrops are selected in the orchards according to the soil depth. Millets perform much better than sorghum on light and shallow soils. Where as sorghum is favoured on deep heavier soils. Fodder sorghum is preferred in areas with 600-800 mm rainfall, where as Pearl millet grows better in areas with high rainfall areas (>800 mm).

In marginal and sub-marginal lands, several perennial grasses and legumes can be grown as intercrops in the orchards. Depending upon rainfall, several range grasses and legumes can be grown which will augment the much needed forages in these areas, some legumes such as *Desmodium spp., Macroptilium artopurpurem, Clitoria ternatea* and *stylosanthes spp* are suitable as perennial forages. Amongst different *stylosanthes* spp, *stylosanthes hamata* is found to be most suitable in regions with 600-1000 mm rainfall. Stylosanthes guinensis performs well in areas receiving high rainfall (850-1400mm). In low rainfall regions (400-750 mm.), *stylosanthes scabra* is most suitable. Several of the fodder trees such as *Albizzia lebbek, A.procera, A. amara, A. acacia* are useful as fodder trees for silivi-pasture.

Fruit based intercropping system

In this system, vegetable crops are grown in association with fruit species, e.g., mango +cowpea/French bean, aonla / guava + tomato and mango + bottle gourd. Thus the interspaces between plant rows of an orchard can be utilized by growing vegetables, which can provide additional income in the initial years till the orchard becomes productive. Intercropping is a practice of growing two or more crops at the same time in the same field and intensified in terms of both time and space. It is often practiced in long duration crops i.e. fruit plants in order to make use of the early available light and space. It can generate more income per unit area per unit time. The intercrops serve as insurance against crop failure.

Fruit crop based intercropping system is one of the important cropping systems recommended to mitigate the aberrant climatic conditions. It ensures supply of balanced food, feed and cash needs of marginal farmers without extra expenses. Experimental evidences have also proved that yield stability is greater with intercropping than sole cropping. Due to uncertainty of monsoon and frequent dry spells, sole cropping in rainfed / dry land areas is risky.

Advantage of fruit based farming system

1. Fruit trees are efficient enough in providing higher economic return even under stressed growing conditions prevailing under the upland situations than the other annual crops.
2. The approach aims at improving productivity by effective utilization of air space which is not utilized in single tier system. The multitier system aims at sustainable management of natural resources like soil, water, space and environment.
3. Sensitivity to short term fluctuation is reduced by decreasing the risk of pest and diseases and by spreading those risks through species diversity. If one plant component fail to produce, the production of other plant components may compensate for it.
4. The summation of productivity of the component crops can increase the total productivity of the land where poor soil fertility coupled with low water holding capacity of the soil contributes towards low productivity of any crop in this region.
5. Higher labour requirement per unit area of multitier system contribute towards creation of job opportunities at site.
6. High return per unit area under upland conditions is the ultimate result of fruit based multitier cropping system.

The concept of fruit based multitier cropping system

The fruit based multitier cropping system is a self-sustainable system where solar energy can be harvested at different heights, soil resources can be efficiently used and cropping intensity is increased. The system consists of three main components viz. main crop, filler crop and inters crops which occupy three different tiers in space of the production system.

The main crops

The main crops are the fruit species having a larger canopy size and prolonged juvenile as well as productive phase. They utilize the upper most layer of the multitier system from which the economic productivity is obtained. Generally the crops utilize the entire land after 20-25 years whereas only 25-30 % of land is effectively used up by the main crop up to 10 years. Under the multitier system, these plants are planted at wider spacing.

The filler crops

The filler crops are the fruit species which are precocious in nature, prolific bearers having short stature. They utilize the middle layer of the multitier system from which economic productivity is obtained. These plants are planted with the purpose to generate additional income from the land during the initial 10 years of the orchard (during the juvenile and initial bearing stage of the main crops) by utilizing the unused land and space. The plants are generally hardy in nature and have shorter economic life than the main crops. The filler crops are planted within the main crop at a closer spacing. The filler plants can be removed after the main crops attain effective canopy size for yielding economically.

Inter crops

The intercrops occupy the lower most layer of the multitier system and are grown in the remaining unused land of the multitier system. Generally the intercrops are the location specific annual crops, selected as per the climatic and socio-economic suitability. The inter crops also include the dependant crops like creepers which are grown with the support of main or filler crops. During initial years of the multitier system any crops can be taken whereas during the later years shade tolerant crops can be grown as inter crops.

Role of HYVs and Hybrids for increasing the productivity of horticultural crops

It is apparent that the development of high yielding varieties of crops that possesses a broad spectrum of resistance to pests, disease and to diverse soil stress, coupled with good management, had helped to raise the crop productivity to high levels. The large-scale cultivation of improved crop varieties together with efforts to maintain good soil fertility and water management helps to increase production through higher yield per hectare. Land is a shrinking resource for agriculture. A rational land use plan is needed to increase agricultural production by achieving higher yields per ha through intercropping, multiple cropping and increasing cropping intensity.

With the increasing pressure of population on agriculture land, it is now increasingly being felt at the international level, that for meeting the food requirements of the increasing population, more land will have to be brought under cultivation. But India is fortunate to have good potential for increasing productivity, as the productivity in our farming system is far behind its actual potential. The country will have to accord high priority to reducing the gap between reliable and actual yields in farmers' fields by identifying and removing the constraints responsible for the yield gaps. After independence, our agricultural policies were influenced greatly by the needs of big landlords. As a result, the needs of the large majority of small peasants were neglected. That is why even after 67 years of independence, three-fourth of our agricultural land remains uneconomical. Because of this lacuna in our agricultural policies, small farmers remain below the poverty line, and our country has not prospered agriculturally. More than 80 % of Indian farmers own two and half acre or less land. Their share of cultivated land is about a third of the total available agricultural land in the country. Over time due to high population growth that caused a division of land holdings and a very slow growth rate of the rural economy, the pressure on land has been steadily increasing and the number of small and marginal farmers has been growing. These farmers can play a leading role in the development of the country by contributing to the nation's capital formation, if their uneconomic holdings are converted into economic ones. However, with the traditional cropping system, small and marginal farmers are finding it difficult to produce adequate food to feed their families. The only way to convert these holdings into profit-making ones is through the intensive use of land through diversification of crops. The interventions of Precision Farming not only elevate the input use efficiency, but also considerably reduces the gap between the genetic potential and the realised yield of a crop variety.

Table 2: Gap in Productivity (t/ha) of Vegetable crops: A comparison of India Vs advanced countries

Vegetables	World	India	Top 2 countries
Brinjal	17.60	16.40	45.30 (Spain), 34.00(Japan)
Cabbage	21.60	21.20	63.00 (Korea), 42.40 (Poland)
Cauliflower	18.90	17.10	22.6 (China), 19.80 (USA)
Peas	8.10	7.10	18.20(Belgium),16.30 (Netherlands)
Tomato	18.40	12.70	54.20(USA), 52.80(Spain)
Okra	6.40	9.80	14.20(Egypt) 10.00(Burkina Faso)

Source: Peter *et al.*, (2008)

More and Sharma (2008), recommended some suitable varieties in various fruits and vegetable crops for inclusion in the cropping system as follows:

Table 3: Some new released varieties recommended in fruit crops

Fruit crops	Existing Commercial Varieties	New Released varieties
Ber	Gola, Seb, Umran, Banarasi Karaka, Kiathali, Mundia	Goma Kirti, Thar Bhubharaj, Thar Sevika
Aonla	Banarasi, Chakaiya, Francis, Kanchan, Krishna, Anand-1, Anand-2,	NA-6, NA-7, NA-10, Lakshmi-52, Goma Aishwarya
Bael	Kagzi, Mirzapur Seedling, Etawah, Gonda, Ayodhya, Pant Aparna, Pant Urvashi, Pant Shivani, Pant Sujata,	NB-5, NB-9,CISH- Bael-1 & 2
Pomegranate	Ganesh, Dholka, G-137, Jalore Seedless, Jyoti, Bassein Seedless.	Mridula, Phule Arakta, Bhagawa, Ruby Amlidana,
Fig	Poona Fig, Dinkar	Dianna, Conadrio, Excel
Tamarind	Aurangabad Selection, Periakulum Selection	PKM-1 & 2, Pritisthan
Custard apple	Balanagar, Mammoth, Island Gem	Arka Sahan,
Guava	Allahabad Safeda, Sardar Guava, Lucknow-49, Chatted, Apple Colour, Seedless, Red Fleshed, Kohir Safed*, Kohir Jam, Arka Amulaya, Allahabad Surkha, Allahabad Surkh, Hybrid-16·Hisar Safed	Lalit, Pant Prabhat, Sweta

Table 4: Some new varieties developed in horticultural crops for increased productivity

Crops	Existing commercial varieties	New varieties
Tomato	Punjab Chuhhara, Pusa Ruby, HS-102, Sweet-72, S-12, Mangla, Pusa Early Dwarf	Arka Ahuti; Arka Saurabh, Arka Vishal, Arka Vardan, Arka Abha, Arka Alok, Arka Shrestha, Arka Abhijit,
Brinjal	Pusa Purple Long, Pusa Purple Round, Pusa Purple Cluster, Pusa Kranti, Pusa Anmol, Arka Kusumakar, Arka Sheel, Arka Shirish, Arka Navneet	Punjab Sadabahar, Arka Nidhi, Arka Keshar, Pant Rituraj, Pusa, Uttam
Chilli	Pusa Jawala, Pant C-1, Mathania, Sindhur	Arka Gaurav, Arka Mohini, Arka Basant, Bharat, Indira,
Round melon	-	Arka Tinda
Pumpkin	-	Arka Chandan, Arka Suryamukhi, Pusa Vishwas
Bottle gourd	Pusa Summer Prolific Long, Pusa Summer Prolific Round	Thar Samridhi, Pusa Meghdoot, Pusa Manjari, Thar Samridhi
Bitter gourd	Pusa Do Mausami, Coimbatore Long	Arka Harit, Pride of Gujarat
Pea	Early Badger, Arkel, Asauiji, Bonneville, Jawahar Matar-4	Organ Sugar Podded
Onion	Nasik Red, Patna Red, Arka Pragati, Pusa Red, Pusa Ratnar, Pusa White Round, Pusa White Flat, Early Grano, Punjab Selection, Agrifound Dark Red	Arka Kalyan, Arka Niketan, Arka Lalima, Arka Kirtiman , Bhima Shakti, Bhima Super, Bhima Kiran, Bhima Red
Radish	Japanese white, Pusa Himani, Pusa Chetki, Pusa Reshmi, Punjab Safed, Chinese Pink, Kalayani White	Arka Nishant
Cowpea	Pusa Dofasli, Pusa Phalguni, Pusa Ritu Raj , Pusa Barsati	Arka Garima, Pusa Komal
Cauliflower	Pusa Deepali, Punjab Kunwari, Improved Japanese, Snowball-16	Arka Kanti
Cluster bean	Pusa Navbahar, Pusa Sadabahar, Durga Bahar, Pusa Mausumi	Goma Manjari, ARG-80, ARG-13

Table 5: Suitable varieties with resistance to biotic stress

Crops	Disease	Varieties
Tomato	Bacterial wilt	Shakti, Arka Alok, Arka Abha, Utkal Pallavi, Utkal Deepti
	Late blight	TRB-1, TRB-2
	Leaf curl virus	Hisar Anmol, H-86, H-88
Brinjal	Bacterial wilt	Arka Kesav, Arka Nidhi, Surya, Arka Neelkanth, Utkal Tarini
Okra	Yellow vein mosaic	Pusa Sawani, Pusa A-4, Punjab Padmini, Arka Abhay, Azad Kranti
Chilli	Leaf curl virus	Pusa Jwala, Pusa Sadabahar, Punjab Lal, Punjab Surkh
Watermelon	Powdery mildew	Arka Manik
Cowpea	Bacterial blight	Pusa Komal
Crops	Pests	Varieties
Tomato	Fruit borer	Pusa Uphar
	Root knot nematode	Pusa-120, Hisar Lalit
Chilli	Thrips	NP-46A, Chamatkar, Pusa Jwala
	Mites	Kalyanpur Red, Punjab Lal
	Ahids	Kalyanpur Red, Pusa jwala
Okra	Leaf hoppers	Clemson,s Spineless, Siswal
	Shoot & fruit borers	Narnaul Special, Pusa A-4, Red Bhindi
Cabbage	Cabbage butter fly	Green Arc, Red Rock Mammoth

Table 6: Varieties of fruit crops suitable for post harvest utilization

Fruits	Varieties	Purpose
Date palm	Halawy, Barhee, Khalas, Khuneizi, Halawy, Khalas, Sevi	For fresh consumption *Doka* stage *Pind* stage
	Medzool, Shamran, Khadrawy, Chip chap	For dehydration (*Chhuhara*)
Aonla	NA-6, NA -7, Chakaiya, Banarasi, Kanchan, Krishna, NA 10	For preserve For pickle and shreads
Pomegranate	Amlidana	For Anardana
Ber	Umran	For dehydration
	Gola, Seb, Mundia, Banarasi Karaka	For preserve
	Seb, Umran, Mundia	Better shelf life

Table 7: Some Popular Vegetable Hybrids :

Vegetable	Hybrids
Brinjal	Pusa Hybrid 5, Pusa Hybrid 6, Manju, Kalptaru, Azad Hybrid, Neembakar, Arka Navneet, Suphal, Arka Anand
Tomato	Swarna Sampad, Arka Rakshyaka, Rupali, Rashmi, Naveen, NS-815, NS-2530, TH-802, Th-2312, Tolstoi, Minakshi, Arka Vardhan, Pusa Hybrid 2, Karnataka, Vaishali, Mangla, Larica
Chilli	Arka Sweta, Arka Meghna , ARCH-236, CH-1, CH-3, NS-1101, Tejaswani, Agni,
Sweet pepper	Bharat, Kt-1, Indira, Lairo, Heera
Bitter gourd	Pusa Hybrid 1
Bottle gourd	Pusa Hybrid 3, Varad (MGH-4)
Cucumber	Pusa Sanyog, Pant Sankar Khira 1, Malini
Muskmelon	Punjab Hybrid 1, MHC-5, MHC-6
Watermelon	NS-295, Arka Jyoti, Madhur, Milan
Cabbage	Bajrang, Sri Ganesh Gol, Nav Kranti, Manisha, NS-25, Sumit, Konark, BSS-32
Cauliflower	Pusa Hybrid 2, NS-60, NS-66, Punam, Priya, Summer King, Sweta, Pawas, Ageti Himlata

Table 8: Suitable varieties of different fruits crops:

Fruits	Recommended cultivars
Mango	Bombay Green, Himsagar, Zardalu, Krishna Bhog, Langra, Maldah, Dashehari, Chausa, Mallika, Amrapali
Litchi	Shahi, Ajhauli, Rose Scented, Trikolia, Swarna Roopa, China
Aonla	Narendra Aonla-7, Kanchan
Sapota	PKM-1, Kalipatti, DHS-1, Cricket Ball
Jack fruit	Khajva, Swarna Poorti, Swarna Manohar
Guava	Sardar, Allahabad Safeda, Arka Mridula
Custard apple	Balanagar, Arka Sahan
Papaya	Pusa Dwarf, Pusa Nanha, Pusa Delicious, Pusa Majesty, Coorg Honey Dew
Lime	Kagzi, Vikram, Pramalini
Lemon	Assam lemon

Table 9: Year round production technology of some vegetables

Brinjal

Crop	Varieties	Sowing time	Transplanting time	Availability period
Spring-Summer	Pusa Hybrid-6, Pusa Hybrid-9, Pusa Uttam, Pusa Bindu, Pusa Ankur	September-October	November	March

Contd.

Crop	Varieties	Sowing time	Transplanting time	Availability period
Summer	Punjab Bahar	November (seedlings are protected from frost)	February	April-June
Rainy season	Punjab Chamkila, R-34, PPL	February-March	April	June-August
Autumn-Winter	Punjab No.8, Punjab Chamkila, PPL	May-June	June-July	October-December

Tomato

Season	Cultivar	Sowing	Transplanting	Availability period
Early Spring	Pusa Sadabahar Pusa Sheetal	September October	Early November	February - April
Summer crop	Pusa Ruby, Pusa Hybrid-1 Pusa Hybrid-2, Pusa Hybrid-4 Punjab Chhuhara, Punjab Kesari, Punjab Tropic	February	Early March	April -June
Kharif or rainy season crop	Hisar Anmol, H- 36, H-86, H-88	June- July	August	October - January

Sidhu (2008) described some of the technologies for year round production of some vegetables as follows:

Radish

Variety	Sowing times	Availability periods	Maturity period
Pusa Desi	Early August	September	40 days
White -5	Early August	October	60 -70 days
Pusa Reshmi	Early September	October	30-35 days
Punjab Safed	September - October	October - November	45 days
Selection -271	October	November	45 days
Japanese White	Mid September - Mid October	Mid November Mid December	45 days
White Icicle	October - Mid	November - December	30 days

Contd.

Variety	Sowing times	Availability periods	Maturity period
Rapid Red White Tipped	Early December	End of December	25 days
Pusa Himani	Mid December	January – March	40-45 days
Pusa Himani	January	February	40-45 days
Pusa Himani	February	End of March	40-45 days
Pusa Chetki	March	April-October	40-45 days

Cauliflower production in different seasons

Maturity group	Cultivars	Time of sowing	Availability period
Early I to Oct.	Pusa Meghna, Pusa Early Synthetic, DCH-541	End of May	End of Aug.
Early II	Pusa Katki, Pusa Deepali	June	Oct. - Nov.
Mid early	Pusa Hybrid-2, Improved Japanese, P. Sharad	July-Aug.	Nov. to Dec.
Mid-late	Pusa SyntheticPusa Shubhra	End of Aug.	Dec. – Jan.
Late	Pusa Snowball K-1Pusa Snowball K-25	Sept. – Dec.	Jan. – Mar.
Extra late	Kala Patta	Jan.-Feb.	April-May

Table 10: Cultivation of short-duration vegetables for intensive crop rotations and quick return

Crop	Variety	Days to first picking	Crop duration (days)	Av.Yield (q/ha)
Summer Squash	Pusa Alankar	45	90	400
	Austalian Green	45-50	90	350
Radish	Pusa Chetaki	40	50	250
	Pusa Mridula	25-30	45-50	100
	Pusa Himani	50-55	60-65	325
	Japanese White	55-60	70-75	375
Tomato	Pusa Sadabahar	80	100-120	275
Turnip	Pusa Sweti	40-45	50-55	250
	Pusa Chandrima	50-55	60-65	350
	Pusa Swarnima	50-55	60-65	350
Pea	Arkel	55-60	80-85	110
	Pusa Pragati	60-65	85-90	125
Cabbage	Golden Acre	60-65	70-75	250
Knol khol	White Vienna	40-45	50-60	250

Table 11: Intercrops recommended for cultivation:

Crop	Sowing/ planting time	Planting distance (cm)	Fertilization (kg/ha)			Recommended cultivars
			N	P_2O_5	K_2O	
Cowpea	June-July	40 x 15	60	50	50	Arka Garima, Pusa Barsati, Birsa Sweta
French bean (pole)	June-July	40 x 15	50	40	40	Swarna Lata, Birsa Priya
French bean (bush)	August-Sept.	40 x 10	80	50	50	Pant Anupama, Swarna Priya, Arka Komal, Contender
Okra	June-July	40 x 20	120	80	60	Arka Anamika, Arka Abhay, Parbhani Kranti
Elephant foot yam	May	75 x 75	125	50	120	Gajendra, Santragachi
Sweet potato	May-June	60 x 20	75	50	75	Gauri, Sankar, Pusa Safed, Sree Bhadra
Turmeric	May-June	40 x 20	80	60	60	Roma, Suroma, Ranga, Rashmi, Suguna, Sudarshana, Suvarna,
Ginger	May-June	40 x 20	80	60	60	Suprabha, Suruchi, Surbhi
Merigold	June-July	50 x 30	25	25	20	Yellow Drop, Golden Drop, Double Lemon
Gerbera	June	30 x 20	50	40	40	Thalasa, Tara, Sunset
Black gram	July	20 x 10	20	40	20	T-9, Pant U-19, Birsa Urad -1
Pigeon pea	June-July	60 x 25	40	40	20	Asha, UAS-120
Horse gram	August	30 x 10	40	40	20	Madhu

Source: Das *et al.*, (2013)

Role of farm women in livelihood security

Women are the back bone of agricultural work force but much of their work goes unrecognized. In small and marginal farm holdings, women actively participate in almost each and every agricultural activity right from land preparation, weeding, sowing, transplanting, weeding to harvesting and storage of the agricultural produce and secondary agriculture (Grewal *et al.*, 2014). As per census 2011, the work force participation for women at the National level stands at 25.51 % compared with 53.26 % for men. Women play active role in various production and post production activities in horticulture. There are a number of horticultural activities like ornamental nursery, fruits and vegetable nurseries, kitchen gardening, fruit and vegetable processing, vegetable and flower markets in which women are engaged to a considerable extent. Fruit, vegetable, flower also offer much scope for export and additional employment for farmers. Various horticultural based technologies, enterprises and schemes can be promoted for better employment opportunity and as a source of income throughout the year.

Considering the natural resources available in the form of agro wastes, a predominant rural based economy, prevalence of educated youth looks for sustainable vocations and also the increased demands for quality food, there is ample scope for growth of mushroom industry. For rural development, bee keeping can play a vital role as one of the economic activity in comparision to other poverty reduction programmes. Apiculture and Horticulture are interdependent and cannot develop in isolation. Bee keeping is a low input and high return agricultural industry where honey is produced from natural wastes.

Application of precision farming technologies in horticultural crops

A critical analysis of the horticulture sector in India reveals that the productivity of majority of horticultural crops maintains a very low profile when compared with the developed countries. With the passage of time and elevation in the socio-economic status of people the need for horticultural produce is steadily increasing. However, the resource base, particularly land and water, is alarmingly shrinking only because of diversion of these resources more towards housing and industrial sectors. Under such a predicament, it becomes very difficult to expand area under different horticultural crops as advocated in various technical forums. Hence, the only way left out is vertical promotion, which could only be accomplished through increased productivity. Introduction of high yielding varieties and large scale popularization of the concept of precision farming can contribute to the cause of increased productivity.

In this chapter the authors would like to focus the possible role of precision farming which includes components such as, Micro-irrigation, Fertigation, Mulching and Protected Cultivation in increasing the productivity of horticultural crops.

Micro irrigation

The optimum and efficient utilization of water for irrigation using appropriate techniques and practices assumes great significance towards improving water productivity in horticultural crops This could be achieved by adopting micro irrigation and other techniques of improved surface water application. Micro irrigation in horticultural crops could realize 25-70 % of water saving and 20-80 % increases in yield depending on soil, climate, crops and variety. Apart from reducing water consumption, drip method of irrigation also helps to reducing cost of cultivation and improving productivity of crops as compared to the same crops cultivated under flood method of irrigation. The Government of India upscaled centrally sponsored scheme such as the National Mission on Micro irrigation (NMM) and allowed single window approach for all other central sector schemes implementing micro irrigation technologies in the country. The mission is being implemented during the XI plan period for enhancing water use efficiency by adopting drip and sprinkler irrigation systems. The scheme provided assistance at 60% of the system cost for small and marginal farmers and at 50% for general farmers.

The agricultural sector currently consumes over 80 per cent of the available water in the country which continues to be the major water-consuming sector due to the intensification of agriculture. About 90% of the irrigated area in the

Fig. 1: Banana crop under drip irrigation in PFDC farm, OUAT, Bhubaneswar

country comes under traditional surface methods of irrigation with a low water use efficiency (35 to 40 percent) and low field application efficiency (40-50%). Irrigation is essential to the production of most vegetables for good yield and high quality. The enhancement in vegetable productions with lesser quantities of available water is the need of hour. In a bid to save cost and amount of water application, a deliberate reduction in water requirement of vegetable crops should not be more than 20%. In order to have advantage of irrigation under water scarcity, it is advisable to opt for most appropriate precision technology management practices in vegetable crops such as choice of productive land, use of short duration vegetables as well as varieties, maintenance of proper soil fertility and plant health, use of mulches and row covers with drip or micro irrigation, use of pre plant irrigation to ensure rapid emergence etc. Various water management practices in vegetable crops viz. furrow irrigated raised bed planting, ridge and furrow systems, partial root zone drying and alternate furrow irrigation method and pitcher irrigation have the potential to increase the yield grown under excess or scarcity of water. Moreover, grafting induced stress tolerance is also helpful in inducing tolerance to soil-related environmental stresses such as drought, salinity, low soil temperature and flooding and for these suitable tolerant rootstocks may be used. The traits of drought tolerance in certain species are of immense use in developing drought tolerance in cultivated vegetable crops.

Fertigation

Fertigation offers the best solution for intensive and economical crop production, where both water and fertilizers are delivered to growing crops through drip irrigation system. Fertigation provides N, P, K as well as the essential trace elements (Mg, Fe, Zn, Cu, MO and Mn) directly to the active root zone, thus minimizing losses of expensive nutrients, which ultimately helps in improving productivity and quality of farm produce. Moreover, fertigation ensures higher and quality yield along with savings in time and labour which makes fertigation economically profitable. The experiments have clearly demonstrated that through fertigation 40-50 per cent of the nutrients could be saved which is otherwise wasted. Fertigation is ideally suited for hi-tech horticultural production systems since it involves not only the efficient use of the two most precious inputs, i.e. water and nutrients but also exploits the synergism of their simultaneous availability to plants.

For intensive and economical crop production, the best solution for higher productivity is fertigation, where both water and fertilizers are delivered to growing crops through micro-irrigation system. Fertigation ensures higher and quality yield along with savings in the time and labour, which makes it economically profitable. Experiments have proved that the system economizes use of fertilizer and water ranging from 40 to 60 per cent. This is being experienced by a few progressive farmers. Grapes, pomegranate and banana are still beyond the reach of poor farmers. Fertigation is ideally suited for hi-tech horticultural production systems, since it involves not only the efficient use of two most precious inputs, i.e. water and nutrients but also ensures their simultaneous availability to plants. Though micro-irrigation has found widespread use in plantation and horticultural crop production in India (Fig. 1), fertigation is confined to a few high-value crops. Significant yield response coupled with enhanced quality of produce is possible through hi-tech productivity using fertigation. The grape, pomegranate and banana growers in Maharashtra have adopted fertigation to some extent.

Mulching

Mulching is the practice of covering the soil around the plant(s) with organic and /or synthetic materials to make the conditions more favourable for plant growth, development, yield and quality of the produce. The organic materials like dry leaves, dry weeds, dry grass, coconut husk, saw dust, coir dust, moss, compost, paddy straw, paddy husk, sugarcane trash etc. are commonly used for mulching in the plants. Besides these, some inert materials like ash, sand, stones and gravels are also used for mulching purpose. Mulching acts as an insulator for the plant and its roots to protect from extreme temperature fluctuations. Effect of mulching on improvement of phonology, yield and quality of the produce in certain crops through regulation of soil temperature and moisture. The use of organic mulches in the plant is an age old practice. Mulching with organic materials is practiced mainly for controlling weeds, moderating soil temperature and conserving moisture in the soil. But the bulk requirement of organic materials and the scope for reuse as mulch are the major limitations. Sometimes use of organic mulch encourages the breeding and multiplication of harmful pathogens on the soil around the plant. Also the organic mulch materials were found to have inherent weakness and not easily available in large quantities.

As an alternate to organic mulching, plastic mulching is more effective in controlling weed growth, conserving soil moisture and moderating soil temperature and thereby efficiently improves the growth, development, yield and quality of the produce of the plant (Fig. 2 to Fig. 5). Plastic mulch films, however, are easily available, easy to handle, transport and also easy to lay in the field. Plastic mulch is a product used, in a similar fashion to mulch to suppress weeds and conserve water in crop production and landscaping. Certain plastic mulches also act as a barrier to keep methyl bromide both a powerful fumigant and ozone depleter in the soil. The idea of using polyethylene as plastic mulch in plant production was developed in the mid-1950s. Dr. Emery M. Emmert of University of Kentucky was one of the first to recognize the benefits of using LDPE (Low density polyethylene) and HDPE (High density polyethylene) film as mulch in vegetable production. Plastic mulching modifies the radiation of the soil surface and thus suppresses soil water evaporation by creating favourable micro-climatic condition around the plant. The micro-climate created by the use of plastic mulch strongly influence the soil temperature and soil moisture in the root zone, which in turn

Table 12: Recommended area coverage under mulch for different crops

Coverage(%)	Crops recommended
20-25	All creepers grown over pandal system mainly. bottle gourd, snake gourd, bitter gourd, pumpkin, cucumber etc.
40-50	Initial stage of orchard crops.
40-60	Fruit crops and cucurbitaceous crops.
70-80	Papaya, pineapple, vegetables etc.
90-100	Soil solarization

Fig. 2: Plastic mulching in Mango

Fig. 3: Plastic mulching in papaya

Fig. 4: Plastic mulching in Banana

Fig. 5: Plastic mulching in Tomato

Table 13: Response of horticultural crops to plastic mulching

Crop	Thickness of mulch film (micron)	Increase inyield (%)
Chilli	25	50-60
Potato	25	35-40
Cauliflower	25	40-50
Tomato	25	45-50
Capsicum	25	35-40
Okra	25	50-60
Brinjal	25	30-35
Apricot	100	30-35
Peach	100	30-35
Guava	100	25-30
Kinnow	100	45-50
Pomegranate	100	35-40
Strawberry	25	40-50
Arecanut	50	25-30

Source: Research findings of PFDCs, OUAT, Bhubaneswar

Protected structures

The world population is increasing rapidly and cultivable area is decreasing day by day. Productivity of land has become stable. So we need a technology, which gives more production per unit area. Green house, the latest word in Indian agriculture is one such means, where the plant are grown under controlled or partially controlled environment resulting in higher yields than that is possible under open conditions. In extreme cold areas where the temperature is extremely low (-5 °C to -30 °C) during winter season, it is very difficult to grow vegetables. Under such situation, some specific protected structures, called polytrenches, have been proved very useful for vegetable cultivation if there is abundance of sunshine. Similarly, in several parts where biotic stresses during rainy and post rainy season do not allow successful vegetable cultivation due to severe incidence of viruses, protected structures covered with insect proof nets provide a big opportunity of virus free vegetable cultivation even on a commercial scale. It is rather used to protect the plants from the adverse climatic conditions such as wind, cold, precipitation, excessive radiation, extreme temperature, insects and diseases. It is also of vital importance to create an ideal micro climate around the plants.

The crop productivity is influenced by the genetic characteristics of the cultivar, growing environment and management practices. The plant's environment can be specified by five basic factors, namely, light, temperature, relative humidity, carbon dioxide and nutrients. The main purpose of protected cultivation is to create a favourable environment for the sustained growth of plant so as to realize its maximum potential even in adverse climatic conditions. Greenhouses, rain shelters, plastic tunnels, mulches, insect-proof net houses, shade nets etc. are used as protective structures and means depending on the requirements and cost-effectiveness. Besides modifying the plant's environment, these protective structures provide protection against wind, rain and insects.

Table 14: Productivity of different horticultural crops grown under greenhouse and open conditions

Crop		Green-houseyield (kg/m^2)	Open field (kg/m^2)
Leafy Vegetables	Spinach	19.30	6.80
	Amaranth	10.50	4.80
	Fenugreek	2.50	1.40
Vegetable Crops	Okra (var. Indam-62)	2.1	0.72
	Cabbage(cv.Golden Acre)	5.0	2.4
	Tomato(cv.Naveen)	8.8	3.0
Vegetable Nurseries	Tomato	40-95%	30-90%
	Capsicum	35-97%	25-93%
	Chilli	42-96%	25-87%
Grafting, Cutting and Layering	Mango	70-90	60-80
	Pomegranate	60-100	40-60
	Guava	70-90	60-85

Benefits of protected cultivation

Vegetable forcing for domestic consumption and export

During winter the temperature and solar radiations are sub-optimal for growing off season vegetables namely tomato, capsicum, brinjal, cucumber, okra and chilli. In tomato, low temperature and low radiation cause puffiness and blotchy ripening. Hence during extreme conditions of winter season (October-February), these vegetables will be cultivated under polyhouse. The high priced vegetables like asparagus, broccoli, leek, tomato, cucumber and capsicum are most important crops for production around metropolis and big cities during winter season or off-season. Thus during severe winter, it may be useful to grow these vegetables in plastic tunnels as the plants which are protected from cold and frost will manifest faster and better growth resulting in earlier fruiting than the crops grown in the open.

Raising off season nurseries

The cost of hybrid seeds is very high. So, it is necessary that every seed must be germinated. For 100% germination, it requires the controlled conditions. The cucurbits are warm season crops. They are sown in last week of March to

April when night temperature is around 18-20^{0}C. But in polyhouse their seedlings can be raised during December and January in polythene bags. By planting these seedlings during end of February and first week of March in the field, their yield could be taken in 30-45 days in advance than the normal method of direct sowing. This technology fetches the bonus price due to marketing of produce in the off-season.

Similarly, the seedlings of tomato, chilli, capsicum, brinjal, cucumber, cabbage, cauliflower and broccoli can be grown under plastic cover protecting them against frost, severe cold and heavy rains. The environmental condition, particularly increase in temperature inside polyhouse hastens the germination and early growth of warm season vegetable seedlings for raising early crops in spring summer. Vegetable nursery raising under protected conditions is becoming popular throughout the country especially in hilly regions. Management of vegetable nursery in protected structure is easier and early nursery can be raised. Needless to emphasize, this practice eliminates danger of destruction of nurseries by hail storms, heavy rains and protection against biotic and abiotic stresses becomes easier.

Vegetable seed production

Seed production in vegetables is the limiting factor for cultivation of vegetables in North Eastern Hill region of India as well as in other parts of India. The vegetables require specific temperature and other climatic conditions for flowering and fruit setting. Seed production of brinjal, capsicum, cauliflower and broccoli is very difficult in open conditions in those areas due to high rainfall at maturity stage. Hence, a protected environment is essential to reduce such micro climatic condition, Therefore, the seed production of highly remunerative crops namely tomato, capsicum and cucumber is performed under protected environments. The maintenance and purity of different varieties/lines can be achieved by growing them under greenhouse without giving isolation distance particularly in cross-pollinated vegetables namely onion, cauliflower and cabbage. Hence, vegetable production for domestic consumption and export in low and medium cost greenhouse is a technical reality in India. Such production system has not only extended the growing season of vegetables and their availability but also encouraged conservation of different rare vegetables.

Hybrid seed production

During 21st century, protected vegetable production is likely to be commercial practice. In vegetable production hybrids seeds, transgenic, stress resistant varieties, micro propagated transplants, synthetic seeds are likely to replace conventional varieties. Protected environments will be helpful in production of hybrid seeds of cucumber and summer squash by using gynoecious lines. Gibberlic acid is used to maintain such lines followed by selfing. The desired pollen can be used for production of hybrid seed of cucumber. Similarly in summer squash, use of ethephon in inducing female flower at every node would help in the hybrid seed production by using desired pollen parent under protected environments.

Maintenance and multiplication of self incompatible line for hybrid seed production

In case of cauliflower, there is problem of maintaining and multiplication of potential self-incompatible lines for the production of F_1 hybrid seed. Temporary elimination of the self-incompatibility with the use of CO_2 gas has solved this problem. For this purpose, the self-incompatible line is planted in a greenhouse and bees are allowed to pollinate the crop when it is in bloom. Then keeping the greenhouse closed tightly, within 2-6 hours of pollination, it is treated with 2-5% CO_2 gas which allows successful fertilization by temporarily eliminating the self-incompatibility.

Polyhouse for plant propagation

Asparagus, sweet potato, pointed gourd and ivy gourd are sensitive to low temperature. The propagating materials of these vegetables can be well-maintained under polyhouse in winter season before planting their cuttings in early spring-summer season for higher profit.

Higher production per unit area is obtained

The yield may be 10-12 times higher than that of outdoor cultivation depending upon the type of greenhouse, type of crop, environmental control facilities. The U.V. film does not allow harmful U.V. rays to enter the green house thus protecting the crop and inside environment remains under control. The carbon dioxide released by the plants during the night is consumed by the plants itself in the morning. Thus the plants get about 8-10 times more food than the open field condition.

Minimization of irrigation water consumption and wastage of fertilizers

In green house, irrigation is done through pressurized micro irrigation system (drip and sprinkler) which conserves water thus increasing water use efficiency. Fertilizers are mostly applied by fertigation which minimizes the fertilizer consumption due to localized application and thereby reducing leaching losses.

Reduced load of pests and diseases

It is observed that the influence of pests and diseases were comparatively lower (5% and 10%, respectively) under naturally ventilated polyhouse. In summer months, there was less incidences of pests and diseases but when the rain started the incidence of pests and diseases was more under shade house condition It is noted that the pests like thrips, mites and fruit borer were also more under shade house condition than polyhouse. The incidence of pests and diseases under greenhouse were comparatively lower (5% and 10%, respectively) as compared to open field condition (35% and 45%, respectively).

The concept of green house cultivation in the resource poor parts of the world in general and India in particular is gaining momentum. Greenhouse cultivation is being increasingly practiced for production of quality produce in the off season for export. In order to ensure consistency in the quantity and quality of production at reasonable cost, it is necessary to adopt latest technologies of greenhouse production. Though it is widely practiced for commercial crops in developed countries, but is still a growing stage in India. In the global competition, it has become imperative that our produce is competitive, both for domestic market and exports. A hi-tech greenhouse is essentially a proposition for precision farming of horticultural crops because the crop requirement inputs are precisely met. The hi-tech greenhouse crop production results into manifold increase in the input-use efficiencies. Economically viable and technologically feasible greenhouse technology suitable for the Indian agro-climatic and geographical conditions is needed at the earliest. Government efforts in popularizing the greenhouse technology among the farming community of the country are to be strengthened. There is a need for the Government to encourage the farmers by providing timely subsidy for taking up this new technology in a big way.

Horticulture based farming system developed by KVKS

A horticultural based farming system unit was established at K.V.K. Kalahandi (Fig. 6). The existing farming system in the Kanakpur village of Kalahandi district in Odisha was agriculture + dairying, where primary source of income was agriculture enterprise particularly from commodities like paddy. After KVK's intervention the farming systems was transformed to agriculture + horticulture + animal husbandry. The horticulture crop became a primary source of income *i.e* banana, ridge gourd, bitter gourd, cucumber, cowpea, brinjal, tomato, etc grown on in commercial basis which adds significant contribution to their income. Above all the members have shown a positive attitude towards change in the existing farming systems. The Budhia family consists of four brothers with four housewives who all are engaged in farming activity. The eldest brother Murali Budhia was well assisted by younger brothers viz., Prahlad Budhia, Thabir

Budhia and Subal Budhia. Among them Prahlad Budhia is currently leading to take care of all farming operations.

KVK Intervention: Looking at the potential of fruit and vegetable cultivation in the village and his interest during 2012-13 and 2013-14, KVK Scientist advised him to go for developing a small papaya orchard orchard in his 0.4 ha of upland with a spacing of 1.5.m x 1.5 m. Also he adopted drip irrigation in Banana orchard, Sugarcane plantation and also in vegetable crops. Utilization of the interspaces with off season vegetables like tomato, ridge gourd, cowpea, bitter gourd, cucumber etc. in the interspaces of Banana and Papaya were skilfully practiced by him. He was also motivated to include Pisciculture, Milk and paneer preparation by Dairy with cross breed cows (Jersy and Red Sindhi), Cultivation, and Goatery (Jamunapari).

During 2013-14, the tomato Var. Swarna Sampad and Yam var. Orissa Elite were introduced through OFT and FLD programme of KVK. The yield performance of both the crops were excellent (Table-15). The Dean, Extension Education, OUAT, Prof. S.S. Nanda, during his visit to the farms of Mr. Budhia, was very contented while advising him to sustain this Integrated Farming System. From 0.25 ac area, he got an yield of 65 Qntl. in tomato var. Swarna Sampad. Similarly, in yam var. Orissa Elite he has invested only Rs.25/plant with a gross return of Rs. 150/- i.e. per pit Rs. 25/- was invested for digging of pit, application

Fig. 6: Horticulture based Farming system unit at K.V.K., Kalahandi.

Table 15: Cost-benefit analysis of Budhia family, Kanakpur, Bhawanipatna

Sl. No	Enterprises	Area(acre)	Season	Yield(Q)	Cost of cultivation (Rs)	Gross return (Rs)	Profit(Rs)	B:C ratio
1.	Paddy	1.5	Kharif	40	16,000	40,000	24,000	2.5
2.	Banana	0.5	Kharif	500 bunches	27,500	75,000	47,500	2.72
3.	Papaya	0.15	Kharif	79.8	15,000	79,800	64,800	4.65
4.	Sugarcane	0.2	Kharif	20,000(canes)	20,000	52,000	32,000	2.6
5.	Tomato (Var- Swarna Sampad)	0.2	Rabi	45	12,000	45,000	33,000	3.75
6.	Brinjal	0.25	Rabi	20	9,000	32,000	23,000	3.5
7.	Yam (Var. Orissa Elite)	100 plants	Kharif	5 Qntl.	2500	12500	10000	5.0
8.	Maize +cowpea	0.2	Kharif	1000 (maize)5 Qntl	4000	5000-6000	7000	2.75
9.	Cowpea + Beans	0.2	K + R	1511	14,000	40,000	26,000	2.85
10.	Ridge gourd	0.2	Kharif	20	11000	25000	14,000	2.27
11.	Cucumber	0.3	Kharif	20	7,000	20,000	13,000	2.85
12.	Fishery	0.5	Kharif	4.0 q	6,000	18,000	12,000	3.6
13.	Diary	2 nos.	—	8.0 lit/day	5,000	28,800	23,800	5.76
14.	Goat (Jamunapari)	5 nos.		2.0 lit/day				
	Total				1,49,000	4,79,100	3,30,100	3.21

of fertilizer and FYM and from single pit a tuber of size 5-6 kg was obtained which was sold at a price of Rs. 25 per kg. A FLD programme on intercropping of Maize var. MM-1107 with cowpea var. Utkal Manika was introduced during kharif 2012-13 from 0.2 ac, the income generated was quite satisfying. From maize he got a return of Rs. 6000/- while from cowpea he got a gross return of Rs. 5000/-. Besides, proper space utilization he got an additional income of Rs. 6000/- from cowpea. Linkages were facilitated with Horticulture Deptt. With assistance from NHM , drip irrigation system was practied by him for almost all the crops like banana, papaya, sugarcane, maize, cowpea, okra, yam, cucumber etc.

Impact

By seeing his success farmers are shifting from monoculture paddy cultivation to horticulture based farming system (Fig. 7. a, b, c, d, e, f, g and h). Farmers also include new enterprise like dairy and poultry with their paddy-paddy farming system. Income substantially increased with technological intervention in sustainable manner. Many farmers of the district have been motivated by his success and some farmers with average holding size of 2.0 ha. has adopted fruit and vegetable

Fig. 7a.

Fig. 7b.

Fig. 7c.

Fig. 7d.

Fig. 7. a, b, c, d: Horticulture Based Farming, System at Kanakpur village of Kalahandi district

Fig. 7e.

Fig. 7f.

Fig. 7g.

Fig. 7h.

Fig. 7e, f, g, h: Horticulture based farming system developed by Mr. Budhia, (Intercropping of cucurbits in Banana orchard, maize + cow pea intercropping system, Vermicomposting unit utilisting Agri wastes)

based farming model with input assistance like drip irrigation, bore well, weeders, Poly house etc. from ATMA & NHM schemes of the district. KVK has maintained regular liasoning with them.

Horticulture-based farming system for higher income and profit – Success story

Name : Senpal Verma, At- Jurkabadi, Block: Kesinga, Dist: Kalahandi (Odisha)

Mob. No : 9938514100

Description of the technology

Intercropping in Cashew nut and Mango orchard with off season vegetables, Intercropping of turmeric in the Teak plantation & QPM Production in mango and vegetable is presented in the Table- 16 and depicted in the (Fig. 8. a, b, c, d, e, f, g and h).

Horticulture Based Farming System at Jurkhabadi village of Kesinga block (Fig. 8. a, b, c, d, e, f, g and h)

Fig. 8 a. Cashew plantation of Mr. S.Verma

Fig. 8 b. Mango grafts produced by Mr. Verma

Fig. 8 c. Growing of vegetable as intercrops in Cashew nut plantation

Fig. 8 d. Pummelo orchard develooped by Mr.Verma

Dissemination of the technology

1. Capacity building through Training, FLD, OFT, Kisan mela, field visit, KMA and other extension activities by KVK.
2. Method demonstration showcasing all the package of practices
3. Distribution of extension literature on management practices of mango, cashewnut, papaya, cucurbits, banana etc.

Success point

1. Utilization of interspaces in Mango, Cashew orchard and Teak plantation with profitable intercrops.
2. Adoption of high yielding varieties of vegetables and spices as intercrop
3. Drip irrigation and mulching in orchard.
4. Good cooperation and better linkage with the line department officials.

Table. 16: Cost benefit analysis of Sri S. Verma, Jurkhabadi, Kesinga

Crops	Area (Acre)	Yield (Q/acre)	Cost of cultivation (Rs./acre)	Gross return (Rs./acre)	Net return (Rs./acre)	Total net return (Rs./acre)
Mango	8	16	5,000	18,000	13,000	1,04,000
Cashew	3	4	12,000	24,000	12,000	36,000
Vegetable (Tomato, ChilliBrinjal, yam, cauliflower)	4	—	1,20,000	2,00,000	80,000	3,20,000
Intercropping of turmeric in Teak plantation	1	44.0	30,000	66,000	36,000	36,000
Mango grafts	—	40,000 Graftings/Year	4,00,000	8,00,000	4,00,000	4,00,000

Fig. 8e. Vegetable Nursery

Fig. 8 f. Vermicomposting unit

Fig. 8 g. Growing of tomato cv. S.Sampad

Fig. 8h. Growing of Turmeric var. Roma as intercrops in Teak Plantations

Impact

1. Farmers of nearby villagers are now showing much interest to cultivate high value crop due to it market demand.
2. Farmers are now aware about maintenance of old orchard.
3. Farmers are showing interest in drip irrigation and mulching in fruits & vegetable crops.

Conclusion

The agriculture policy of the Government of India immediately after independence was greatly influenced by the needs of resource rich landlords. This neglected the interest of the large chunk of resource poor farmers. And this is the reason for which even after 68 years of Independence three-fourth of our agricultural land remains underutilised and uneconomical. Hence, majority of our small farmers continues to remain below the poverty line. Crop diversification by incorporating pulses, oilseeds, vegetables and other cash crops in a scientific cropping pattern can play an important role in increasing farm incomes and employment to achieve nutritional security. Further, as the average family land holdings at the national level have come down, interventions are further needed to convert such un-economic land holdings into profitable one. Such studies can make a difference to the livelihood as well as food & nutritional security to the people. As such, the focus of extension functionaries should shift to farming system diversification. One of the main channels through which diversification towards high-value crops can reduce poverty is via the participation of small farmers. However, although smallholders have the benefits of proportionally larger labour pools, this may be offset by constraints such as lack of access to credit. Thus, there is continued debate as to whether smallholders can successfully diversify into the high-value sector. Smallholders show more participation in high-value fruit and vegetable production compared to larger farms. The high capital intensity and greater gestation lags in fruits seem to be deterrents for small farmers having a minimal capital base (physical and human) and a low appetite for the riskier fruit market. Knowledge and skill development are important attributes of successful horti-entrepreneurship for small farm holders. As long as production of horticultural crops remaining relatively more profitable as compared to alternatives, it is feasible to promote crop diversification towards horticultural crops. One of the main channels through which diversification towards high-value crops can reduce poverty is via the participation of small farmers. Given the high labour endowments in India and the preponderance of smallholders, the share of resources allocated to high-value agriculture continues to be relatively small, although it is increasing over time. Conditional on supporting infrastructure and institutions, smallholders have an advantage when adopting labour-intensive crops such as vegetables. The bias towards vegetables rather than fruits clearly points to the role of enabling factors in transforming the potential advantages of the smallholders (such as larger families) into realized crop choices that favour high-value products.

Under such a predicament well thought of Integrated Farming System involving the concepts of crop diversification and precision farming, taking the need and resource base of micro units of the major Agro Ecological Zones can play an important role in enhancing the farm income and employment to achieve the expected livelihood security of the resource poor farming community. This in due course would increase the contribution of Agri Horticultural sector to the National GDP.

References

Chadha, K.L., Kalia, P and Singh, S.K. (2014). *In : Horticulture for Inclusive growth.* Paper presented in the 6th Indian Horticulture Congress, held at Tamil Nadu Agricultural University, Coimbatore from 6-9th November, 2014.

Chand, R., Raju, S.S. and Pandey, L.M. (2008). Progress and Potential of Horticulture in India. *Indian J. Agric. Econ.*, 63.

Chand, R., Raju, S.S. and Chauhan, S. (2014). Agriculture diversification towards Horticulture : Trends and Prospects. *In : Horticulture for Inclusive growth,* pp. 1 to 8, Edited by K.L. Chadha et al.

Das, B., Singh, R. and Kumar, S. (2013). In : Fruit based multitier cropping system for uplands. L.Patra and S.C.Swain Edited "Sustainable Horticulture: Development and Opportunities", Mangalam Publishers and Distributors, New Delhi.

Ghosh, S.P. (2014). Horticulture for Livelihood security. *Plenary Lecture delivered in the 6th Indian Horticulture Congress, held at Tamil Nadu Agricultural University, Coimbatore* from 6-9th November, 2014.

Gangwar, B. (2014). *Diversification for sustainable Farming Systems.* Paper presented in the 8th National Conference of KVKs, held at Bangalore from 23-26 October, 2013 on the theme sustainable intensification of small holder farms.

Grewal, N., Moharana, G. And Babu, N. (2014). Role of Farm women in livelihood security through Horticulture base farming system. *Horticulture for Inclusive growth, pp. 817-824.* Paper presented in the 6th Indian Horticulture Congress, held at Tamil Nadu Agricultural University, Coimbatore from 6-9th November, 2014.

More, T.A. and Sharma, B. (2008) . *Recent Technologies for Developing Arid and Semi-Arid Horticulture in India.* Paper presented in the 3rd Indian Horticulture Congress, at OUAT, Bhubaneswar.

Peter, K.V., P.G.Sadhan Kumar & S. Nirmala Devi (2008). *Advances in Vegetable production.* Paper presented in the 3rd Indian Horticulture Congress, at OUAT, Bhubaneswar.

Patra, D. (2012). The marketing strategy of economic development in Kalahandi. *Odisha Review*, September, 2012.

Sidhu, A.S., (2008). Technologies for improving year round production of vegetables in India. Paper presented in the 3rd Indian Horticulture Congress, at OUAT, Bhubaneswar.

Singh, H.P. (2008). *Horticultural Research in India- Impact and Initiative.* Paper presented in the 3rd Indian Horticulture Congress, at OUAT, Bhubaneswar.

Singh, I. S. (1996). *Agro-forestry for salt affected Lands.* NDAUT, Faizabad pp 9-20.

Integrated Farming System Practices: Challenges & Opportunities, pp 107-158

New India Publishing Agency, New Delhi, India

5

Pisciculture in Pond Based Integrated Farming System

R.N. Mishra

Today the world is facing a variety of interrelated problems such as food production, energy use, pollution, unemployment and urbanization, mainly because the developmental strategies contravene the basic philosophy of ecology. The world ecosystem is becoming increasingly more unstable through an ever growing dependence on finite resources of fossil fuels. Food production system in developed countries, both terrestrial agriculture and fisheries are energy intensive and are thus, not entirely suitable for developing countries. An alternative strategy to an increasing reliance on fossil for agriculture development should depend on locally available sources of renewable energy and resources. A significant effect could be made through recycling organic wastes into fish. Organic wastes can be recycled in ecologically sound pond based systems fueled by solar energy. The reuse of wastes would safeguard the environment from pollution and lead to a reduction in diseases of insanitation. Recycling of animal wastes is the cheapest way of protein production by reducing cost of production. Fish farming in Europe, Israel and USA is energy intensive, since more farmers use commercially prepared feed formulations which resulted in cost to feed often exceeding 50% of the total operating costs.

In recycling of waste through fish culture, fish is considered to have a great potential to change raw protein into high energetic edible protein. The rationale

College of Fisheries, Rangeilunda, Berhampur, Orissa University of Agriculture & Technology, Bhubaneswar, Odisha, INDIA behind raising fish on organic manure is that about 72.98% of the nitrogen, 62.87% of phosphorus and 82.92% of potassium in the feed rations fed to animals are recovered in their excreta. In most of the world, cereals and legumes account for most of the proteins and calories consumed, but fish produced cheaply by recycling wastes would provide a low energy way to produce an animal product and diversify the diet. Fish, being the cheap source of animal protein is now widely accepted to meet the persistent and widespread malnutrition in developing countries.

Aquaculture is the fastest growing food production sector in the world with annual growth in excess of 10% over the last two decades. Much of this development has occurred in Asia, which also has the greatest variety of cultured species and systems. Asia is also perceived as the 'home' of aquaculture, as aquaculture has a long history in several regions and knowledge of traditional systems and is most widespread.

Rural poverty is a continued crisis in Asia. Each year millions more children are added to the farming households without much hope of a better livelihood and also millions of hectares of natural resource base become degraded. At this juncture, small scale farming systems must provide a reasonable rural livelihood, a clean conserved environment, and adequate food, fuel and fibre products. One option for sustainable development in farming is small scale pond based integrated farming system. The diversification that comes from integrating crops, vegetables, livestock, and fish imparts stability in production, efficiency in resource use and conservation of environment. Pond based integrated farming involving aquaculture is the concurrent or sequential linkage between two or more activities of which at least one is aquaculture. Benefits of integration are synergistic rather than additive. The integration of aquaculture with livestock and crop farming offers great efficiency in resource utilization, reduces risk of diversifying crop, and provides additional food and income. This system involves recycling of waste or by-products of one farming system which in turn serves as an input for another system and efficient utilization of available farming space for maximum production. Thus, pond based integrated farming system are considered to be sustainable.

As per Brundtland Report (WCED, 1987) "*Sustainable development is that which meets the needs of the present without compromising the ability of future generations to meet their own need*". The interpretation and measurement of sustainability have become focal points of rural development. In a small holder farmer's world, key parameters of sustainability have been identified as high levels of species diversity, nutrient cycling, capacity (total production) and economic efficiency (Dalsgaard *et al.,* 1995). At the micro level, watershed, community, farm, plot and pond may be used as a basis for assessing sustainability.

For centuries, small scale farmers of India have sustained themselves by practicing different kinds of crop diversification. About 80% of India's population lives in rural areas at subsistence or near subsistence level. These rural folk are greatly undernourished and need a large supplement of animal protein in their diet along with new sources of gainful employment. Integrated fish farming is well developed in China, Hungary, Germany and Malaysia, and is accepted as a sustainable form of aquaculture and major contributor of farmed fish.

Fresh water aquaculture in India is organic based and derives inputs from agriculture and animal husbandry. The fresh water aquaculture resources in India comprise 2.25 million ha of ponds and tanks, 1.3 million ha of beels and derelict waters, 2.09 million ha of lakes and reservoirs and also 0.12 million km of irrigation canals and channels and 2.3 million ha of paddy fields. Besides, India possesses the largest bovine population of over 307 million cattle heads, along with 181 million sheep and goats, 16 million pigs and over 150 million poultry and other livestock. India being an agrarian economy produces large quantities of plant and animal residues to the tune of over 322 and 1000 million metric tonnes respectively annually. In addition activities like mushroom cultivation, rabbit, sericulture and apiculture, fodder, aquatic plants etc provide huge quantities of organic material for aquaculture. Agro-based industries like food processing plants also produce effluents which could be recycled for aquaculture in addition to domestic sewage and biogas slurry to the extent of over 4000 million litres per day.

Aquaculture and agriculture complements and supplements each other by balancing the economy of natural resources. Pond embankments could be used for growing vegetables, plants, fruit trees, fodder etc which are fertilized with pond silt rich in plant nutrients and used as feed by grass carp and as human foods (vegetables and fruits). The main potential linkages between aquaculture and livestock concern use of nutrients like nitrogen and phosphorus, particularly reuse of livestock manures for higher fish production by stimulating natural food webs. Cattle, sheep, goat and pigs could be raised on pond dykes or in the vicinity. Poultry farming and aquaculture could be effectively combined, since poultry litter serves as a good fertilizer for fish culture. Duck raising with fish culture has a special advantage in that they feed on snails which serve as vectors of certain infectious diseases. Processing wastes through organisms viz. earthworms that feed on them and concentrate nutrients to produce 'live feeds' is an alternative approach raising fish needing high levels of dietary animal protein. The optimal uses of such a wide variety of wastes help to control pollution through recycling along with enhancing fish production. About 40 to 50 kg of organic manure can produce 1 kg of fish.

India has to sustain 16% of the world's population on 2.4% of the global land area. It has to feed its burgeoning population using 3% and 5% of global farm land and water resources respectively. Hence, its dependence on aquatic resources for production of additional food is obvious and shall become more and more obligatory. Fish supplies about 30% of the total animal protein in the diet of Asian population, 20% in Africa and 10% in Latin America. The rationale of aquaculture has socio-economic and marketing advantages for small scale farmers in developing countries like India. Besides, high fecundity (up to 1 million eggs) and faster growth coefficient of fishes are the advantages to offer considerable scope for increasing production and achieving nutritional security to a great extent.

Continued growth in aquaculture is only possible if the relative availability of its primary inputs namely land, water and feeds can be increased. The quantum of water required to produce one kilo of food is as high as 4000 litres for rice and 900 liters for wheat, whereas it is as low as 400 liters for fish. In many developing countries, suitable land and water is already used for agriculture and therefore, further aquaculture development will depend on profitability of aquatic products or as an integrated enterprise with the existing one.

1.1 Ecosystem of integrated fish farming

Integrated fish farming system works in following way:

a) Trapping of solar energy and production of organic matter by primary producers.

b) Utilization of primary producers by photographs or tertiary consumers.

c) Decomposition of primary producers and photographs by saprotrophs or osmotrophs.

d) Release of nutrients for producers.

The animal wastes in water body enter into the food chain in three different ways

i. Feed

Certain bottom feeders like *Cyprinus carpio* and *Cirrhinus mrigala* directly utilize the organic particles which are generally coated with bacteria along with other material.

ii. Autotrophic production

Some of the decomposed portion of waste products provides nutrients for the micro-flora (autotrophs), while non-mineralised portion provides food base for bacteria and protozoa (heterotrophs). Temperature, light, micro and macro flora, inorganic nutrients, carbon, phosphorous and nitrogen are the basic inputs required for photosynthesis process.

iii. Heterotrophic production

Micro fauna (zooplankton) feed on small manure particles coated with bacteria. In the process, bacteria are digested while rest is excreted. In this heterotrophic production system, micro fauna (protozoans and zooplanktons) are produced finally shortening food chain. This system of production is not linked with the process of photosynthesis.

1.2 Pattern of energy flow:

There are two basic patterns of energy flow under integrated farming system. They are (a) flow of energy through farm animals, and (b) flow of energy through pond eco-system.

a) Flow of energy through farm animals

i) When food is provided to the farm animal, the animal derives its energy from a part of the feed because of assimilation of food and the rest of the feed is released to the system in the form of faecal matter which can be represented as

   Energy of feed (E) = Energy assimilated (A) + Energy released through Faeces and Urine (F+U)

ii) Further, a part of the assimilated energy (A) is used by the animal for metabolic activity and ultimately lost through respiration (R) and the rest part of the assimilated energy is stored and utilized by the animal for growth (G) which is expressed as

   Energy assimilated (A) = Energy for Respiration (R) + Energy for growth (G)

iii) The unused energy released through faecal matter includes the energy of faeces (F) and energy of urine (U).

   Thus, the energy of feed of animals (E) = Energy for growth (G) + Energy for respiration (R) + Energy released through faeces and urine (F+U).

b) Flow of energy through pond eco-system

Pond ecosystem derives energy mainly by the ways like energy input through allochthonus sources (animal excreta as main input of energy and nutrient), through primary producers (energy fixed by photosynthetic organisms), through chemosynthesis (oxidation of inorganic substances to derive energy which is very less or not significant) and through heterotrophs or decomposers (oxidation of organic matters consumed by heterotrophs).

Hence, it is assumed that major quantity of energy in a pond ecosystem under integrated farming comes from allochthonus and autochthonus source. Thus, the energy available at the lowest trophic level under integrated farming system can be expressed as

Energy available in the pond = Energy fixed by the producers + Energy input as animal excreta + Energy input by chemosynthesis + Energy input by heterotrophs

In pond based integrated farming system, the available energy at the lowest trophic level flow to the next higher trophic levels through either grazing food chain i.e. herbivores to carnivores or, through detritus food chain i.e. dead and decayed organic matters are consumed by detritivores followed by predators. There are gradual reduction in the energy level from lower trophic level to higher trophic levels. If the food chain involves more trophic levels, then it reduces the net harvest of fish. Therefore, in any integrated farming system, the main aim is maximum utilization of energy with a minimum loss.

1.3 Important integrated fish farming systems

Pond based integrated farming systems include two major systems:

A. Aquaculture – agriculture integration		B. Aquaculture – livestock integration	
a	Rice – fish integrated system	a	Cattle – fish integrated system
b	Horticulture – fish integrated system	b	Pig – fish integrated system
c	Mushroom – fish integrated system	c	Goat – fish integrated system
d	Seri – fish integrated system	d	Rabbit – fish integrated system
e	Fodder – fish integrated system	e	Duck – fish integrated system
f	Aquatic plant – fish integrated system	f	Poultry – fish integrated system

1.4 Fish culture

India is a carp country where in Indian Major Carps (catla, rohu and mrigal) are grown together under polyculture system or along with three exotic carps (silver carp, grass carp and common carp) as composite carp culture. These six species are selected because of their compatibility for habitat preference and food to utilize all the ecological niches of the culture system. While catla and silver carp are basically surface feeders showing preference for zooplankton and phytoplankton respectively, mrigal and common carp are omnivorous bottom feeders. Rohu is a column feeder and grass carp shows high preference for submerged vegetation. These carps which feed at the base of food chain especially, phyto- and zooplankton, detritus and aquatic weeds, utilize natural productivity with greater output. These carps are also reared under mixed culture system incorporating fresh water prawn (*Macrobrachium rosenbergii*, *Macrobrachium malcolmsonii*) or magur (*Clarias batrachus*).

The standard package of practice for carp culture comprises of a) pond preparation, b) liming, c) fertilizer application, d) stocking of fish seed, e) feeding and feed management, f) water quality management, and g) health management. The management practice involves pre-stocking, stocking and post-stocking operations.

i) Pre-stocking management: This includes the following practices.

a) Control of aquatic weeds

b) Application of lime

c) Eradication of predatory and weed fish

d) control of aquatic insects

e) Pond fertilization

ii) Stocking management: This includes

a) Stocking density

b) Ratio of different species

c) Acclimatisation/conditioning of fish seed

iii) Post-stocking management: This includes the following practices.

a) Pond fertilization

b) Supplementary feed and feeding

c) Health management

d) Harvesting

1.4.1 Control of aquatic weeds

The earthen ponds are often infested with marginal, floating, submerged emergent aquatic weeds and algal blooms. These weeds cause many problems in the pond: (i) absorb major portion of the available nutrients retarding the growth of plankton, (ii) prevent light penetration, thereby reducing photosynthesis, (iii) cause deficiency of dissolved oxygen during night, (iv) cause higher diurnal pH fluctuation, (v) harbor aquatic insects and predators etc. Hence, complete removal of weeds from the culture pond is a pre-requisite.

The aquatic weeds can be controlled by employing manual, mechanical, chemical and biological methods. Manual method is preferred for smaller nursery ponds. For submerged weeds, weed cutters may be employed. Stocking of weed eating fishes like grass carp, common carp, gourami (*Osphronemus gouramy*) and silver barb (*Puntius gonionotus*) is effective for grow out ponds. Common carp feeds on the bud, tender leaf and shoots of the aquatic plant and also uproots the germinating plants through its nibbling habit and thus prevents their establishment.

A wide range of weedicides is available for control of heavy infestation of aquatic weeds.

a) **Floating weeds**: Water hyacinth (*Eichhornia crassipes*) is one of the most important floating weed. Depending on its degree of infestation, they are categorized into three groups, viz. small, medium and big, based on their wet weight per unit area. The recommended doses of the herbicide 2–4-D (2,4- Dichlorophenoxy acetic acid) are 2,7 and 12 kg/ha for small (13 kg/m^2), medium (23 kg/m^2) and big (35 kg/m^2). Addition of a detergent (0.2 % concentration) to the aqueous solution gives better results. The dilution for better coverage has been estimated at 400 l/ha. Water lettuce (*Pistia stratiotes*) another floating weed often causes a serious problem in fish ponds which can be controlled with 0.1–0.2 kg of paraquat/ha. Salvinia forms a thick surface mat in ponds and can be conveniently controlled by the application of foliar spray of paraquat at the rate of 1 kg/ha. Usually it takes 30–40 days for the weeds to be killed and settled in the pond. Smaller floating weeds, e.g. Spirodella, Lemna and Azolla can also be cleared with 0.1 kg/ha of paraquat.

b) **Emergent weeds:** Water lily, lotus, and floating heart can be cleared by spraying the herbicide 2–4-D at the rate of 8–10 kg/ha with detergent (0.25%). The chemical is diluted at the rate of 300 l/ha and sprayed through a foot pump sprayer.

c) **Marginal weeds:** Ipomea, Jussiaea, etc., could be controlled by spraying the herbicide 2–4-D at the rate of 8 kg/ha.

d) **Submerged weeds:** Ottelia, Vallisneria, Hydrilla, Najas, Potamogeton and Ceratophyllum can be controlled by paraquat at the rate of 3–4 ppm within two weeks. It can also be controlled by application of anhydrous ammonia at the rate of 15–20 ppm.

e) **Algal blooms and mats:** Microcystis bloom is cleared with 0.3 to 0.5 ppm of Diuron. Simazine also clears the bloom in 16–20 days and the rate of application is 0.3–0.5 ppm. Both the chemicals do not have harmful effect on fish.

1.4.2 Application of lime

The first step in fertilization is the application of lime. Liming raises the soil pH to a desirable level (near neutral) for establishing a strong buffer system, stimulates microbial decomposition of organic matter which favours mineralization of nitrogen and other nutrients from organic matter. It reduces toxicity of harmful compounds including disinfecting the environment. Treatment of pond bottom with Calcium Oxide (CaO) or Calcium Hydroxide {$Ca(OH)_2$} will destroy parasites and other undesirable organisms. Slightly acidic to neutral soil with pH 6.5 to 7.0 is considered productive.

Various kinds of limes are available in the market. Based on the soil pH, amount of lime required for correction of soil is given below.

Table 1: Amount of lime required for correction of soil based on its pH

Soil pH	Amount of Lime required (x100kg)/ha						
	Pure $CaCO_3$	Agric. Lime	Calcite $CaCO_3$	Dolomite $CaMg(CO_3)_2$	Slaked lime $Ca(OH)_2$	Quicklime CaO	Shell powder
6.5	2.5	2.8	2.8	2.8	4.2	2.3	3.2
6.0	5.0	5.5	5.6	5.7	8.5	4.6	6.4
5.5	7.5	8.3	8.4	8.5	12.7	6.9	9.6
5.0	10.0	11.1	11.1	11.3	17.0	9.2	12.8

$CaCO_3$: Calcium Carbonates $CaMg(CO_3)_2$: Calcium Magnesium Carbonate

However, generally Agricultural lime ($CaCO_3$) is extensively used and applied either to the pond soil while dry or broadcast over the water surface in a single application or in several equal installments. Calcite and dolomite are used for treatment and pH correction of pond water during culture period. Alkaline pH in pond water is corrected through agricultural gypsum ($CaSO_4$) @250-500kg/ha or Alum [$Al_2(SO_4)_3$] @ 25-50kg/ha.

1.4.3 Eradication of predatory and weed fishes

Predatory fish prey upon the spawn, fry and fingerlings of carps and the weed fish compete with carp for food, space and oxygen. Therefore, predatory and weed fish should be completely eradicated from nursery, rearing and stocking ponds before these ponds are stocked with fish seed. Absolute removal of these unwanted predatory fishes (Channa spp., *Clarias batrachus, Heteropneustes fossilis, Pangacius, Mystus* sp. *Ompok spp*., *Wallago attu* and *Glossogobius giuris* etc.) and weed fishes (*Puntius sp*., *Oxygaster sp., Ambassis sp., Amblypharyngodon mola. Colisa sp., Rasbora sp., Aplocheilus sp., Laubuca sp., Esomus danricus,* etc.) from the pond is highly essential. Followings are different methods to eradicate predatory and weed fishes from the pond.

a) Mahua oilcake is used @2.0-2.5 t/ha-m of water for total eradication of predatory and weed fishes. It should be applied at least 3 weeks prior to fish seed stocking for detoxification of the poison from water.

b) Derris root powder is effective for shallow water ponds with 1.0 to 1.5 m depth, especially in hot sunny days with temperature more than 25°C. Its application @15-20ppm is sufficient to kill all weed fishes, tadpole and air breathing fishes. However, its poisonous effect persists for 4-12 days. So, fish seeds should be released after about one month of application of derris root powder.

c) Commercial grade bleaching powder with 30% chlorine is applied @350kg/ha-m to kill all fishes in pond.

d) Alternatively 100kg urea/ha-m followed by 175kg bleaching powder/ha-m after 18-24hr is found equally effective.

When bleaching powder is used, subsequent application of lime is reduced. When mahua oil cake is used, subsequent application of raw cow dung is reduced.

1.4.4 Control of aquatic insects

Pond ecosystem harbours number of aquatic insects like back swimmer (*Anisops bouvieri*), pond skater (*Gerris*), lesser water boatman (*Corixa*), water scorpion (*Nepa*), diving beetle (*Cybister*) etc. which find their way from the adjacent water body and increase their population tremendously just after fertilization of the pond. They not only compete with the carp seed for food, but also cause extensive damage, often killing them through devouring or pricking and sucking the body fluid resulting in poor seed survival. These need to be controlled before releasing the seed to the pond for culture.

a) Small pond: In case of small ponds like nurseries and rearing ponds, repeated netting with suitable small mesh size is best preferred.

b) Large ponds: In stocking ponds, simple and effective method of eradication of aquatic insect is through application of soap oil emulsion, i.e. 18kg/ha of cheap soap and 56kg of vegetable oil applied at 1-2 days before stocking of spawn. This is very effective during calm weather. Kerosene @100-200 litres or diesel @75 litres/ha can be used as an alternative to vegetable oil and detergent powder @2-3kg/ha substitutes soap cake.

1.4.5 Pond fertilization

The Indian major carps and exotic carps at their early stages are planktivorous, with zooplankton as the preferred natural food. Sustained zooplankton population in a pond depends on a good phytoplankton population base, which is further ensured through adequate availability of major nutrients like nitrogen, phosphorus and carbon, besides certain micronutrients in water. When these are at low levels in both pond sediment and water, such nutrients need to be supplied through application of organic and inorganic fertilizers. Decomposition of organic manures in the pond bottom leads to slow release of nutrients to the over lying water column and helps in long term maintenance of rich plankton population. In contrast, inorganic fertilizers dissociate into elemental form, which are readily available for utilization by phytoplankton.

Combined use of manures (Raw cow dung / poultry manure) and chemical fertilizer give good result for sustaining natural productivity in pond. Raw cow dung (RCD) is applied @10 t/ha in phased manuring. When mahua oil cake (MOC) is used, RCD will be reduced to 5 t/ha. Poultry manure is 2-3 times rich in nitrogen content than RCD. So, it is used @3-4 t/ha. Phased manuring with a mixture of 750kg/ha groundnut oil cake, 200kg/ha RCD and 50kg/ha single super phosphate are effective in production of desired plankton in a fish pond. Half of it made into thick paste by addition of sufficient water, are applied as basal dose 2-3 days prior to stocking and remaining is applied later in 2-3 split doses depending on plankton production of pond.

1.4.6 Nursery pond management:

- In the nursery pond, 3-4 day old spawn (5-6 mm) will grow to fry of 25-30mm during 15 to 20 days.
- **Size of the pond** : Earthen ponds-0.02 to 0.1ha with average water depth of 1.0-1.5 m. Concrete tanks- 50-100 m^2 provided with a soil base of 15-20cm
- **Stocking Density**: 3-5 millions/ha (Earthen nurseries) or, 10-15 million/ha (concrete tanks)
- **Survivality**: Approx. 50%.

Supplementary feed and feeding:

- **Feed:** Groundnut oil cake and polished rice bran at equal proportions are ground and powdered and vitamin mineral mixture of 0.5% is added to it and is used as supplementary feed.
- **Quantity:** 6kg/million/day for first 5 days and 12kg/million/day for subsequent days.
- **Method of feeding**: Broadcasting in equal splits during morning and evening.

1.4.7 Rearing pond management

- Fry of 20-25mm size are raised further for 2-3 months in the rearing ponds to get fingerlings of 80-100 mm size.
- **Pond size**: 0.05 to 0.2 ha with 1.2 to 1.5 m water depth
- **Stocking density:** 0.2 to 0.3 million fry/ha
- **Ratio:** In a poly culture system comprising of only Indian major carps, Catla:Rohu:Mrigal at 1:1:1 or 3:4:3 ratio can be stocked. In a composite fish culture system, Catla:Rohu:Mrigal:Silver carp:Grass carp:Common carp may be stocked at equal proportions.

Supplementary feed and feeding

- **Feed:** Groundnut/ mustard oil cake and rice bran/ wheat bran at 1:1 ratio by weight are best preferred supplementary feed. Other ingredients such as fish meal, soyabean flour @ 5% of total mass and vitamin-mineral mixture @0.5% level are added to improve food quality. Feed is provided @8-10% of initial biomass of fry/day during the 1st month, followed by 6-8% of the standing biomass during subsequent two months. Daily ration is provided in 2 equal installments during morning and evening hours. A part of ration is broadcasted and the rest is placed in feeding trays. Duckweeds like spirodella, lemna, azolla, wolffia are provided when grass carp is stocked as a component.

1.4.8 Grow out pond management

- The fingerlings are grown in the grow out/ stocking pond to table size within a period of 9 months.
- **Pond size:** Extensive and Semi-intensive method: 1-5 ha

  Intensive method: 0.4 – 1.0 ha
- **Pond shape:** Rectangular (Maximum pond width should be 80 m)
- **Depth of pond:** Desirable pond depth is 1.5 – 2,0 m

A. Low input system:

- Stocking density : 4000/ha
- No supplementary feeding
- Daily application of RCD @35-40 kg/ha or, Biogas slurry @80-120kg/ha
- Fish production: 1-2 t/ha/year

B. Medium input system

- Stocking density : 4000-6000/ha,
- Periodic fertilization and regular feeding with cake-bran mixture coupled with water quality and fish health monitoring
- Fish production: 3.0-6.5 t/ha/year

C. High input system

- Higher stocking density of 15000-20000/ha with higher feed inputs
- Water exchange and aeration
- Pond size : 0.4 to 1.0 ha
- Single stocking- multiple harvest, multiple stocking- multiple harvesting or rotational type of culture practice followed
- Fish production: 15 t/ha/year
- **Pond fertilization**: It is done with RCD @10 t/ha or Poultry manure @4-8 t/ha and Single super phosphate @50kg/ha /year. Basal dose is 25-30 % which should be applied a fortnight before stocking and remaining to be used daily in equal amounts
- **Stocking density:** 10000 fingerling/ha (30-40 % silver carp and catla, 30-35% rohu, 30-40% common carp and mrigal and 5-10 % grass carp).

Supplementary feed and feeding

- Groundnut/ mustard oil cake and rice bran/ wheat bran at 1:1 ratio by weight. Other ingredients such as fish meal, soybean flour @ 5 % of total mass and vitamin-mineral mixture @0.5 % level are added to improve food quality.
- Feed is provided @3-5 % of initial biomass of total body weight of stocking material/day initially and subsequently at sliding scale from 3 to 1%. Daily ration is provided in 2 equal installments during morning and evening hours.
- Feeding is provided either by hanging in perforated gunny bags or in trays.
- Grass carps are provided with aquatic vegetation like duckweeds, viz. lemna, spirodella, and wolffia, Hydrilla, Najas and ceratophyllum at periodical intervals.

Health management

- Application of Potash ($KMnO_4$) @5ppm in each month during winter season is advocated to prevent occurrence of disease to fish.

- Besides, Carbonate lime : turmeric powder at10:1 ratio @200kg/ha may be applied in three split doses during winter months.
- Further, CIFAX @1.0 L/ha-m of water mixed with 400 L of water may be sprayed as a precautionary measure to avoid any fish disease.

Harvesting: Carps are harvested after a culture period of 9-10 months during which they reach a marketable size of 0.8 to 1 kg. Fishes of about 6-8 t/ha can be harvested from the system.

Fig. 1: Pond filling

Fig. 2: Application of raw cow dung

Fig. 3: Supplementary feeds

Fig. 4: Application of $KMnO_4$

Fig. 5: Application of Lime

Fig. 6: A harvest of fish

1.5 Aquaculture – agriculture integration:

1.5.1 Rice-Fish integrated system:

Paddy is the dominant cereal crop in Asia and rice is the staple food for over 1.6 billion people in the world. Over 90% of the paddy is produced in Asia. This single rice crop is virtually the sole livelihood for most rural farmers. The practice of collecting wild, naturally occurring fish for food from rice field is probably as old as rice cultivation itself. Fish culture in rice fields was introduced into Southeast Asia from India about 1,500 years ago. The problems of food supplies during the Second World War gave an impetus for extensive fish culture in rice fields.

Rice-fish culture plays an important role in rural economy of Southeast Asia. Fish culture can be used in conjunction with rice cultivation to increase productivity. Paddy-fish integration provides a net annual income which accounts for several fold increase over traditional practice. It also facilitates crop diversification, thereby reducing investment risk and also generating year round employment opportunities in farm.

Paddy-cum-fish culture needs modification of rice plot, digging of peripheral/ one side trench, construction of dykes, pond refuge, sowing improved varieties of rice, manuring, stocking of fish at 10,000/ha and finally feeding of stocked fish with rice bran and oil cakes at 2-3% of body weight. Harvesting is done when fish attains marketable size. Vegetables, fruits-bearing plants, papaya and banana, can be grown on fallow land/dyke of pond for an additional production. Such system provides a net income of about Rs35,000 to 40,000/ ha/year which is much higher over traditional practice.

Advantages

1. It intensifies land and water use.
2. It contributes to increased and diversified food supply.
3. Stocking of fish in paddy fields effectively uses the available food and space.
4. It controls weeds when herbivorous fish is stocked within one week of transplanting paddy.
5. When stocked at the start of paddy production, fish feeds on competing phytoplankton.
6. The activity of feeding fish and their excretion improves paddy fertility.
7. It minimizes additional labour requirement.

8. Fish acts as a biological control of paddy pests namely stem borers, plant and leaf hoppers.
9. Higher level of water management for fish improves conditions for plant crop.

Fish culture in paddy fields may be concurrent with rice culture or in rotation. In concurrent type, fish is reared simultaneously with the growing of rice crop. The fish helps to increase the rice production by its residues and providing aeration to the plants. It also helps to control pests, weeds, insects and mollusks from the paddy fields. In rotational culture, fish may be cropped either after a single annual rice crop or between harvest and next replanting. The plant residues (e.g. rice stubble, split grain) benefit the fish. Also the fish residues benefit crop production.

a) Field design of rice-fish plots

The rice fields which retain water for a fairly long duration and free from flooding are generally suitable for rice-fish integration. Some modification of rice-fish plot is required to make the system more profitable. Clay soil is suitable. Two different designs are generally accepted for the practice.

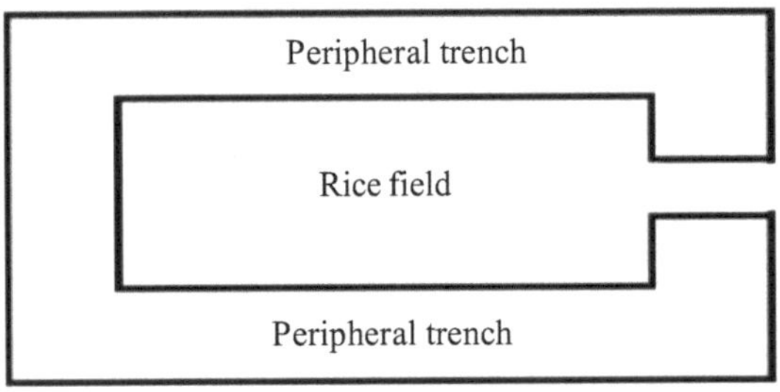

Fig. 7: Rice – Fish plot Design I

Design I: A peripheral trench is excavated around the rice growing area (width 3.5 - 4.0 m, depth 1.5 m) which is blocked at one place and connected to the main land for easy access for farmers and agricultural appliances to the rice plot. The rice plot may range from one acre to one hectare or more and preferably be rectangular or even square. A dyke is constructed all around.

For a 1 ha plot area required for dykes, trenches, pond refuge and field will be: Dykes 2000 sq. m. (20%), Trenches and pond refuge 1300 sq. m. (13%) and Field 6700 sq. m. (67%).

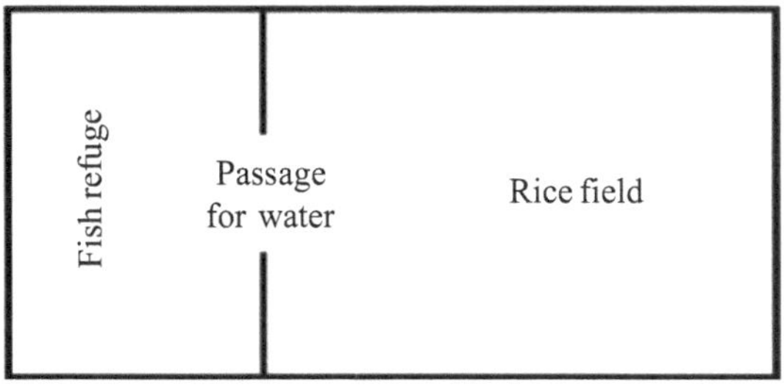

Fig. 8: Rice – Fish plot Design II

Design II: Approximately one third area of the plot at low lying side of its width was dug up to 1 m for fish cultivation. The soil, thus excavated was used for making the inside bund leaving space of 1.0 m at the centre for water to enter the rice field and strengthening the bund of whole plot. The inlets and outlets for all the plots were made and cement pipes were arranged with 40 mm mesh, so that miscellaneous fishes will not enter into the ponds. The rice cultivated area was left free so that when water level increases, it enters into the rice field through the passage at the centre of the inner bund and about 20 to 30 cm of water was always allowed in the rice field, so that the fishes will enter into the rice field and can eat the insects available there.

Modifications in size may be made as per availability of land.

b) Culture methods

Improved varieties of rice like Panidhan, Tulasi, CR 260-77 are cultivated in season which have tolerance to submergence and pest attacks. Fertilization schedule includes 40 kg N and 20 to 30 kg each of P_2O_5 and K_2O/ha at the time of seeding, besides FYM at 5 to 10 t/ha.

Fish and Prawn

Catla, rohu, mrigal and common carp in combination with freshwater giant prawn are stocked in equal proportions @ 10,000 individual/ha. These are fed with rice bran and mustard/ groundnut oil cake @ 2-3% of the total body weight. Manuring schedule includes application of cow manure at 10 t/ha/yr, while liming is done @ 200 to 500 kg/ha based on soil pH. Harvesting is generally done periodically along with receding water levels.

c) A Case study

Comparative field experiments were conducted on bulk rice farming and rice-fish farming during the kharif season of 2009-10 at farmer's field in Puri district of Odisha to evaluate the economics of the system. Approximately one third

Table 2: Comparative statement of On-Farm Research showing economics of rice and fish in bulk rice culture and rice-fish culture systems in Puri district of Odisha during 2009-10

Parameters	Experiment Plot No. I			Experiment Plot No. II			Experiment Plot No. III		
	Only Rice	Rice– Fish system		Only Rice	Rice – Fish system		Only Rice	Rice – Fish system	
		Rice	Fish		Rice	Fish		Rice	Fish
Area (m^2)	340	320		240	266		300	321	
Production (q)	0.765	0.564	0.185	0.636	0.554	0.127	0.621	0.551	0.163
Yield (t/ha)	2.25	2.82	1.69	2.65	3.26	1.51	2.07	2.55	1.55
Income (Rs.)	436	1151		363	880		354	1039	
Expenditure (Rs.)	259	450		195	282		210	385	
Net Income (Rs.)	177	701		168	598		144	654	
Net Profit (Rs./ha)	5206	22500		7000	22481		4800	20374	
C:B Ratio	1:0.68	1:1.56		1:0.86	1:2.12		1:0.69	1:1.70	

area of the plot at low lying side of its width was dug up to 1 m for fish cultivation. Long duration rice variety, Sarala was cultivated in rice fields where as Indian Major carps along with common carp and fresh water prawn, *Macrobrachium rosenbergii* were cultured in pond refuges designed especially for rice-fish system. The stocking density of catla, rohu, mrigal, common carp and prawn as 10,000 nos/ha was maintained at a ratio of 10:20:20:20:30. The carp fingerlings and juveniles of prawn survived up to a level of 61% and 38.7% respectively within a mean culture period of 165 days. The production, productivity, net income and C:B ratio for both only rice cultivation and rice-fish cultivation in three experimental plots were presented. The C: B ratio for only rice farming ranged from 1: 0.68 to 0.86 whereas for rice-fish farming it ranged from 1: 1.56 to 2.12. Thus, integrated rice-fish culture system was found to be an efficient and lucrative approach in terms of yield and economic output (Table - 2).

Fig. 9: Management of rice-fish farming

Fig. 10: Fish and prawn catch from rice cum fish farming

1.5.1.1 Horticulture on the dykes

After the harvest of rice certain crops which require lesser amount of water like water melon, groundnut, vegetables, cow pea, etc. can be grown in the field. Top of the bund which is 10% of the pond area is utilised for growing vegetables and fruit bearing plants. Rice-fish system results in 168% intensity of cropping in field and 400% on bunds as compared to 52% in the case of traditional monocropping of rice. Rice-fish system provides a net annual income of around Rs.75,000/ha in the first year which accounts for about six folds income over farmer's traditional practices and three folds over the improved monocropped rice. This system encourages synergism between rice and fish leading to increase in grain yield by 5-15% and straw yield by 5-9%. It facilitates crop diversification, thereby reducing investment risk. It promotes gainful linkage between rice, fish, prawn, vegetables, fruit crops and other resulting in better resource utilization as well as conservation of the ecosystem. It generates year-round employment in the farm.

Dilapidated or narrow dykes of the farm can be used for the cultivation of black gram. In addition to extra crops, it biologically controls the weeds which are heavily infested on dykes. The cultivation provides 0.80 t pulses, 1.8 t dry pod and 3.64 t straw per hectare in 100 days. Cultivation of black gram requires only broadcasting of seeds after cutting weeds during monsoon season. Irrigation, fertilization or post-sowing care is not required. An income of Rs 10,000 is achieved in 100 days of cultivation and saving of recurring cost towards periodical de-weeding. The foliage of black gram is utilised by grass carp feed. In mixed culture practice of grass carp, rohu, catla, mrigal, silver carp and common carp in 50:15:15:10:5:5 ratio at a density of 7,500/ha yielded fish to the tune of 4,000 to 5,000kg/ha of fish fed with dyke grown green fodder. This integration system provides 25% higher return when compared to extensive method of aquaculture only.

1.5.1.2 Azolla in the rice field

The cultivation of azolla in rice-fish fields were initiated in Philippines in the mid 1980's. *Azolla,* is a nitrogenous biofertiliser on the water surface in fish ponds producing nitrogen, phosphorus and potassium by trapping the solar energy. Initial focus was given to assess the effectiveness of fresh azolla supplement for giant tilapia (*Oreochromis nilotica*). Fish species grown concurrently with low land rice, feeds heavily on fresh or processed *azolla*. When about 30% of the inorganic nitrogen fertilizer for rice was substituted with *azolla*, it resulted in 10-15% increase in net income. In polyculture pond of 1 ha water spread area; the nutrient requirement met through application of 40 t/ha/yr *azolla* provides over 100 kg of nitrogen, 25 kg of phosphorus and 90 kg of potassium in addition to 1,500kg of organic matter. Due to high protein content (13.0 to 30.0%), the feed formulations using *azolla* meal as the major component was found to be as effective as commercial feeds in aquaculture system.

Fig. 11: Azolla Culture in Rice – Fish plot

1.5.1.3 Fish seed production in rice field

Production of fry, fingerlings and yearlings are all practised in rice fields. Fry and fingerlings require a much shorter culture time and are more suitable when rice production methods require regular draining of the paddy. Also where rice production is almost year round, a crop of fry or fingerlings are quickly taken in between. Supplementary feeds are given especially when the culture period is limited. Such feeding allows faster individual growth to ensure that the required size fish seed are produced.

Fig. 12: Fish seed production in rice field

1.5.2 Horticulture – Fish integrated systems

Integrated vegetable production and fish farming could both improve household nutrition and increase socio-economic status of the rural farmers. In India, floriculture trade is also blooming up providing excellent avenues for employment generation and as a whole for earning foreign exchange. These crops can be increased by bringing more area under cultivation. Ponds are well suited for this purpose. The top, inner and outer dykes of ponds as well as adjoining areas can be best utilised for horticulture crops. These crops are fertilized by the pond silt and fertile pond water is used for watering. The success of the system depends on the selection of plants. They should be of 1.Dwarf variety, 2.Less shady, 3.Evergreen, 4.Seasonal, and 5.Highly remunerative.

Dwarf variety of fruit bearing plants like mango, banana, papaya, coconut, lime can be grown around the pond. This will not obstruct the sunlight to the water bodies and also the pond will be free of dry leaves. Pineapple, ginger, turmeric, chilly can be grown as intercrops. Pond dykes are used for growing

Fig. 13: Horticultural crops on pond dyke

vegetables solo as well as intercrops. During summer season, brinjal, tomato, chilli, gourds, cucumber, melons, ladies finger and during winter months peas, beans, cabbage, cauliflower, carrot, beet, radish, turnip, spinach, ethic etc. can be raised. Pond silt and pond water is used for providing nutrient for these crops.

Flower bearing plants like tuberose, rose, jasmine, gladiolus, marigold, cassandra, chrysanthemum can be grown on the pond dykes. These flowers have tremendous market potential in the cities which provides additional income as well as employment to the farmers.

a) Farming practices

Farming practices are carried out on broad dykes which can stand ploughing and irrigation. Ideal management involves utilisation of the middle portion of the dyke covering about 2/3 of the total area for intensive vegetable cultivation and the rest on the area along the length of the periphery through papaya cultivation keeping sufficient space on either side or netting operation. Semi intensive farming is done where the dykes are not good. Crops of longer duration like beans, ridge gourd okra, papaya, tomato, brinjal, mustard and chilli are suitable for such dykes. Narrow dykes are suitable for cultivating sponge gourd, sweet gourd, bottle gourd, citrus and papaya. Where the dykes are shaded ginger ad turmeric can be cultivated.

b) Fish culture practices

Large quantities of leaves of cauliflower, cabbage, turnip and radish are available from the horticulture source at the farm site. These are fed to the fishes like Grass carp which is one of the ideal fish for this purpose. A monoculture of grass carp with a stocking density of 1000/ha will give a production to the tune of 2000 kg/ha/yr. During summer, amaranth and water bind weeds are fed to the grass carp cutting fortnightly. If possible, common carp can also be added. Grass carp is a voracious feeder. Only part of the intake food is digested, while the rest released as faecal matter which serves as a good feed for the common carp. This results in additional production without involving any cost towards feed. In mixed culture of grass carp along with rohu, catla, mrigal in the ratio of 50:15:20:15 at a density of 5000 fish per hectare results in yield of 3000 kg/ha/yr.

Table 3: Calendar of activities for fish-horticulture farming (Mixed fish farming of catla, rohu, mrigal with grass carp and horticultural crops of papaya, banana and vegetables)

Month	Activities to be undertaken
June-July	Pond preparationDyke preparation and planting of papaya and banana and vegetables
August	Stocking of fishApplication of inorganic fertilizers to crops
September	Pest control, if necessary
October	Harvesting of vegetables, inorganic fertilizer application
November	Harvesting of vegetables
December	Harvesting of vegetablesHarvesting of Papaya
January	Preparation of dyke for second crop of vegetablesHarvesting of PapayaPlantation of second crop of vegetables
February	Partial harvesting of fishHarvesting of papaya and banana
March	Harvesting of papaya and banana
April	Harvesting of papaya and banana and vegetables
May	Harvesting of papaya and banana and vegetablesFinal harvesting of fish

c) Economic viability of the system

Fig. 14: Horticulture- fish integrated system dyke

In horticulture-fish system, when mixed farming is done with three IMCs and three Exotic carps in 1 ha water spread area along with vegetables, papaya and banana on the dyke, about 3000kg fish can be harvested along with 1500kg papaya, 60 to 80 bunches of banana and 1500 kg green vegetables resulting a B:C ratio of 2.4. Thus, integrated agri-horti-aquaculture practice, though labour intensive, generates about 20-25% higher returns compared to aquaculture alone besides generating employment opportunity round the year.

1.5.3 Mushroom - Fish integrated system

Mushrooms are fleshy fungi and are the most preferred food item now-a-days in the affluent society. Mushrooms are rich source of protein with 60-70 per cent digestibility. These contain all essential amino acids and are good source of vitamin B, C, D and K and contain appreciable quantities of niacin, pantothenic acid, riboflavin, nicotinic acid and minerals like calcium, phosphorus and potash. It has been established that 100-200g (dry weight) of mushrooms are sufficient to maintain nutritional balance in human being. Since the mushroom cultivation requires high degree of humidity, it would be possible to extend its cultivation in the vicinity of fish ponds.

It has been observed that the paddy straw after growth of mushroom is much enriched in proteins, inorganic nutrients and other organic matter. This supplemented nutritive value of used paddy straw after the harvest of mushrooms can be well utilized for cattle feed which has beneficial effects on enhanced milk yield. In turn the excreta of cattle are recycled in fish pond fertilization enhancing pond productivity through detritus food chain. Compost of mushroom beds known as spent mushroom substrate (SMS) contains 1.9% nitrogen, 0.4% phosphorus and 2.4% potash which could be used in aquaculture and agriculture.

In such mushroom-fish integration system, composite fish culture can be managed utilizing the compost of mushroom beds as fertilizer to the system. Thus, it fetches about 15-20% additional return compared to pisciculture alone and provides ample opportunities for the youths.

Cultivation of edible mushrooms

In most parts of India, three types of mushrooms are being cultivated commercially for consumption purpose. They are European button/ white button (*Agaricus bisporus*), paddy straw (*Volvoriella* species) and oyster mushrooms (*Pleurotus* species). Standard techniques for their cultivation were adopted by small growers in India.

i) Paddy straw (*Volvoriella* species) mushroom

Fig. 15: Paddy straw mushroom cultivation

Large parts of India have temperature conditions favourable for growing this mushroom either throughout the year or during a part of the year. Paddy straw, the basal substrate is also available in plenty. The optimum temperature for its cultivation is between 25^{0}C and 45^{0}C. In general, the optimum season for growing paddy straw mushroom is from April to October in the states of Punjab, Haryana, Uttar Pradesh, Madhya Pradesh, West Bengal, Odisha, Andhra Pradesh, Tamil Nadu and Maharashtra. A number of crops can be raised during the favourable season since one crop cycle takes about 30-45 days. An average yield of 3 kg per bed (made from 32 kg paddy straw with 200 g of gram powder) is normally obtained within a period of 45 days. The cost of production is about Rs 30/- per kg and the sale price fluctuates between Rs.80 to 120 per kg.

ii) Oyster (*Pleurotus* species) mushrooms

Fig. 16: Oyster mushroom cultivation

These mushrooms are popularly known as, 'hingiri'. In India, cultivation is done on different substrates viz paddy straw, saw dust, maize stalks etc. Optimum temperature for growth of *Pleurotus* species is between 25^{0}C to 35^{0}C. The method of cultivation involves the use of dried paddy straw chopped into 1-2 cm long bits and soaked in water overnight. Excess water is drained off and horse gram powder (@8g/kg) and spawn (@30 g/kg) are added and mixed with wet straw in alternating layers. The polythene bags with perforations are filled with this substrate and kept in a room at 21-35^{0}C with sufficient light and ventilation. The mycelial growth takes about 11-14 days to penetrate the substances in the bags. After this period, the polythene bags are cut open on the sides without disturbing the bed which becomes quite compact during this period. Water is sprayed over it twice a day. In a few days, the crop is matured for harvest. Since the oyster mushrooms can grow on a wide variety of agricultural wastes, which are easily available and cheap, the cost of production is lower than of others. The yields per unit area of the substrate are also good, the cost of production being about Rs 20/- per kg of mushrooms. The price for market ranges between Rs 60/- to Rs 80/-per kg in Odisha.

1.5.4 Seri - Fish integrated system

Sericulture is an agro-industry and plays an important role in rural economy of India. This includes mulberry cultivation, sericulture, and silk extraction. Fish farming is done using silk worm faeces, silk worm pupae and waste water. In this system mulberry is the producer, silk worm is the first consumer and fish are the second consumers ingesting the silk worm faeces directly. In the fish ponds having this integration the energy flow occurs in the following patterns:

- Silk worm's faeces are directly consumed by the fishes and part of the detritus is filtered by the filter feeding fishes.
- Inorganic nutrients in the silk worm faeces are utilized by phytoplankton and heterotrophic bacteria and these phytoplankton and bacteria are, in turn, consumed by filter feeding fish, either directly or indirectly.
- Leftover feeds and fish faeces are decomposed by hydro-microbes, releasing inorganic nutrients and then, the same process occurs as above.
- At the same time pond silt which is composed of all kinds of sediments return to the pond dyke and a new cycle begins.

Energy passes through the complex food-web of the dyke-pond system and undergoes a series of exchanges as it flows among the sub-systems. This energy enters the system via three pathways:

- Absorption by the dyke crops converts energy into chemical energy during photosynthesis.
- Absorption by phytoplankton in the pond and conversion to chemical energy via photosynthesis, and
- Direct input with the pond of chemical energy stores in plant material and waste production used, as fish feed and pond fertilizer respectively.

In general 75 percent of the mulberry leaves supplied is consumed by the silk worms. Together with silk worm excrement, the remaining 25 percent of unconsumed leaf debris is dumped into the pond. Mulberry dykes yield leaves at the rate of 30 t/ha/yr. When fed to silk worms, 16.2 t of waste is produced in which the energy store is 66% of that supplied to the silk worms. The energy intake by the fish accounts for only 32% of the total energy input to the pond. About 72% of this intake energy is absorbed and the remainder is with fish excrement and in the process of respiration. Based on these facts of energy exchange, a two way energy exchange system exists between the dyke and the pond. Energy enters the pond via materials grown on the dykes and then fed to the fish and in this mulberry-dyke system via silkworm excrement. This is then returned to the dyke in the form of silt.

In 1 ha mulberry dykes-pond system 50% of the area is dyke and 50% water body. Of the former, 0.45 ha is planted with mulberry and 0.05 ha is under crop. During winter season vegetables are inter-planted with the mulberry. A production of 30 t/ha mulberry leaves, 21 t/ha silk worm cocoons, 225 t/ha crop and 3.75 t/ ha vegetables is achieved. Waste of the vegetables which account for 50% is fed to the fish while rest 50% of total produce is consumed by the human beings. Approximately 16 t of waste is produced per ha mulberry which is put in the pond. Mulberry leaves are harvested for feeding silk worm 7-8 times a year. Mulberry grows best at a temperature of 12^{0}C to 25^{0}C.

a) Mulberry cultivation - technical management

i. Nursery bed preparation

The nursery bed is prepared 3.5 months before transplanting and soil is treated with a pesticides (BHC) @ 45 kg/ha) before broadcasting the mulberry seed @ 20 kg/ha. Sprouting occurs within a week. Seedlings remain for 3 months in the beds. Periodically these are fertilised using urea @ 38 kg/ha. Any pest problem is controlled by the Roger or DDVP (Dichlorvos).

ii. Transplantation of seedlings

Dykes are prepared before the transplantation of seedlings which include activities like tilling, sun drying, breaking-up of soil, making rows about 60 cm apart and separated by a shallow drainage ditch and application of manure. Planting is done at about 1.25 lakh plants/ha. These are periodically watered, using pond water.

iii. Management practices

De-weeding, pest eradication and manuring are the main management practices. Urea and salt are main features. The use of slurry is also very common. Other management practices include the removal of buds which give birth to the new plants making each plant bushy.

iv. Leaf harvesting

Leaves are harvested several times a year as per the requirement. These provide good nutrition for silk worms. Generally the dykes are divided into several sections to ensure daily supply of fresh leaves for silk worm. Generally the leaves production goes down after 5-6 years. At this juncture plants can be uprooted and new crop should be planted to maintain the productivity.

b) Silk worm culture

The optimal temperature and humidity for silk worm culture ranges between 15-32^{0}C and 50-90% respectively. Silk worm rearing trays are cleaned and disinfected and sheds are prepared. The basic unit of calculation in silk worm rearing is the sheet of paper (16 x 21 cm) on which the moths are induced to lay eggs. Each sheet carries 18000-20000 eggs of which approximately 90% are viable.

Fig. 17: Silk worm culture

The worms produced by one sheet of eggs require the leaves from 0.067 ha of mulberry dike to produce cocoons. Eggs are hatched in the rearing shed with sufficient space for ventilation. Feeding rates increase sharply during the rearing period. Successful worm-rearing demands supply of fresh leaves. Waste material is removed daily from the rearing trays. At the end of the rearing period, the worms become ready to spin cocoons. At this stage, they are removed from the rearing trays and placed in separate compartments which completes within 48 hours. They are then removed and dried in the field.

1.5.5 Fodder – Fish integrated system

In case of fodder-fish integrated system, plants of terrestrial or aquatic origin may be grown specifically for use as fish feed. Farmers use various areas of their farms, including fields, small plots of unused land, pond dykes and drained ponds for cultivation of aquatic plants using available water resources.

Advantages

1. Fodder plants are used as a direct feed for herbivorous macrophyte feeders in pond fish culture.
2. Fodder plants when added to ponds before stocking of fish, vegetation decomposes and acts as a fertilizer or green manure.
3. Fodder plants are used as organic fertilizers during preparation of pond prior to filling.
4. The soft materials of crops of grains and legumes can be fed directly to cattle.
5. Grass species, easily produced on the farm serve as low cost supplemental feeds for fish.

6. These are used to stabilize pond embankments.
7. Fodder plants serve as an additional source of income.

a) Technical management

Grasses are cultivated on the wide dykes exclusively as fodder for herbivorous fish. Elephant or Napier grass (*Pennisetum spp.*), cassava (*Manihot spp.*) and Sudan grasses (*Sorghum sudanenese*) can be grown with pond sediments and water. A variety of vegetables particularly the cabbage family, maize, sorghum and sweet potato are grown for feeding the fish.

Napier grass and cassava are propagated by vegetative means. Fencing should be provided around the area to prevent grazing of animals. Fertilizer/ compost is to be applied every month. Napier can be harvested by cutting at 7 cm from the ground from 6-8 weeks after planting and then cutting regularly every 2-4 weeks. Cassava is harvested by cutting 0.5m from the ground, 8 weeks after planting, then regularly after every 4 weeks. Legumes are harvested through cutting 8-12 months after planting, then regularly after every 8-12 weeks, 0.3m from the ground.

b) Integration with fish farming

Leaves of these fodder crops are used as feeds. However, for cassava, the tuber can also be used as human feed. The leaves are chopped in small pieces before feeding to fry/ hatchlings. For big fish, the leaves are simply placed in the pond. Generally, grass carp, big head carp, tilapia and prawn may be selected for fish culture who can take the fodder leaves directly as feed. In this system, about 3-4 partial harvests can be done before final harvest. For prawn, harvesting is done after 6-7 months and for fish, after 10-12 months. Survival rate is about 70-90% for fish and about 30% for prawns.

1.5.6 Aquatic plant-fish integrated system

In India, trapa (*Trapa bispinosa)* and *Makhana* (*Euryale ferox*) are two seasonal, aquatic cash crops which are grown extensively in Madhya Pradesh and Bihar, respectively either in concurrent or rotation method. While the environment is not congenial for Indian carps, common carp goes well with trapa and air breathing fish like magur with makhana.

a) Concurrent trapa – fish culture

- Trapa seedlings are transplanted in May/ June in a perennial pond. These plants make use of the available organic matter for their growth.

- Common carp yearlings are stocked at 800 nos (50g each) in 0.4 ha water spread area in September/ October.
- Trapa fruits ripen in winter and are harvested from November to January. A production of 3-4 tons of fruits can be obtained from that pond.
- Fish are harvested in April/ May when they attained 750-1000g weight. A total of 5-6 quintals of fish are harvested from one acre pond.

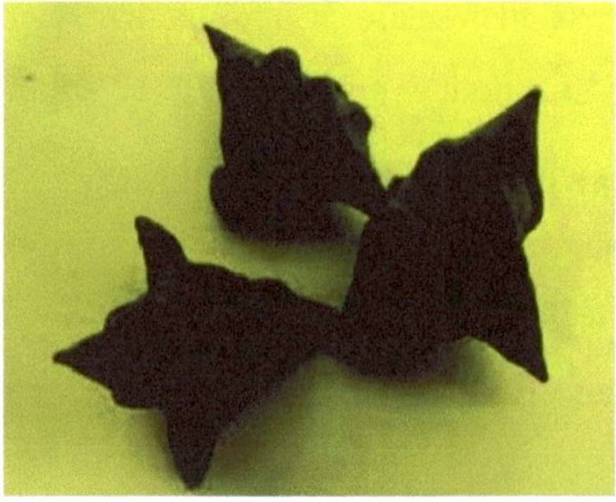

Fig. 18: Makhana

b) Rotational makhana – fish culture:

- The seed of makhana sprout in February and the leaves cover the pond fully in May/June.
- The plants start fruiting in August and burst in October, scattering the seeds at the pond bottom which are collected by scanning the bottom mud.
- Air breathing fish (*Clarias batrachus*) are stocked with 1200 (8-10g) nos in 0.4 ha pond in November and harvested in April; when about 500kg of fish can be obtained.

1.6 Aquaculture-livestock integration

1.6.1 Cattle-fish integrated system:

1. Fish farming using cattle dung is one of the common practices all over the world.
2. A healthy cow weighing about 400-450 kg excretes over 4,000 – 5,000 kg dung, 3,500 – 4,000 L urine in a year.
3. The manure particles of cattle sink at a rate of 4.3 cm/min which provides sufficient time for fish to consume edible portions available in the dung.
4. Manuring of fish pond with cow dung results in increase of natural food organisms, detritus and bacteria. The faeces and urine are extremely beneficial for filter feeding and omnivorous fishes. Therefore, silver carp and catla are usually the major species cultivated with assorted omnivorous fish (common carp) and herbivorous fish (15-20 percent).

5. The optimal output of herbivorous fish in fish-cow integration should be around 12 percent of the total output of the pond.
6. A unit of 5-6 cows can provide adequate manure for 1 ha of pond. If 0.024 kg of fresh cow manure is applied to 1 m^3 of water every day, inorganic N and P will be 0.897 and 0.024 mg/L, respectively. These levels are close to the inorganic N (0.97-2.06 mg/L) levels in high-yield fish ponds. The N/P ratio will be 36.9.
7. The conversion factor of cow manure is 3.15 in dry weight or 21 in wet weight at an average weekly manuring rate of 0.17 kg/m^3 in filtering and omnivorous fish farming. In silver carp it is 3.3 in dry weight or 26 in wet weight at the same manure input.
8. The susceptibility of cow manure not only enables the fish to obtain more feed but also reduces oxygen consumption caused by manure stacking and avoids the formation of harmful gases.
9. The BOD of cow manure is lower than that of other livestock manures because the cow forage had already been decomposed by micro-organism in the rumina. The BOD of 1 kg of cow manure in 5 days is 20.6 g,
10. In addition to 9,000litre of milk, about 3,000 to 4,000 kg fish/ha/year can also be harvested with such integration.

Table 4: Composition of excreta of cattle

Excreta of cattle	Moisture, %	Organic material, %	Nitrogen, %	P_2O_5, %	K_2O, %
Milk cow dung	85	11.4	0.36	0.32	0.20
Cow dung	80-85	14.6	0.30-0.45	0.15-0.25	0.05-0.15
Cow urine	92-95	2.3	0.60-1.20	trace	1.30-1.40

Source: Agricultural Handbook

a) Technical management

Animal sheds can be built close to the fish ponds to simplify the handling of the manure. The faeces and urine may be collected separately. If the floor is higher than the pond dyke, a manuring ditch can be dug to collect the faeces and urine together and the mixture can be flushed directly into the fish ponds. This method saves time and labour. The area of the fish pond should match with the number livestock and waste food, the species ratio and target output of fish, etc. The frequency of manuring depends on the conversion of the manure, which changes seasonally, and the fluctuation of food organisms in the fish ponds.

Cows feed mainly on grass and during the grass-growing season (about 7 months), an adult cow requires 9000-11000 kg grass. Around 3000 kg of this grass, however, is leftover. That period of time is the highest ingestion season of herbivorous fish. Therefore, this waste fodder can be utilised as fish feed to the grass carp. The food conversion factor of terrestrial wild grass is 40-50. In addition, the matted grass in the cow shed can be used as compost for the pond. The leftover fine fodder for cows can also be used as fish feed. Dairy cows are fed with a mixture of roughages, concentrates and supplements in quantities that will fulfill their nutritional requirements which are determined by animal weight, productivity level and butter fat percentage in the milk. Water must be available at all times because of the high water requirements for maintenance and growth.

Economics of cattle-fish integration system

i. Capital investment		**Amount (Rs)**
1	Cost of 5 nos of cross bred cows @30,000/cow	150000
2	Construction of cow shed (5x50 sq.ft/cow) including inlet outlet structures	100000
3	Other equipments	30000
	Total capital investment	**280000**
ii. Costs		
a)	**Fixed cost**	
1	Depreciation @10% (except for cow)	13000
2	Interest on capital @15%	42000
3	Insurance @6% on value of cows	9000
4	Wages to permanent labour for 12 months @Rs5000/month/labour	60000
	Total fixed costs	**124000**
b) Variable costs		
1	Pond preparation, fish seed, feed and fertilizer cost for 1 ha pond area	75000
2	Cost of cattle feed (4kg/cow/day for 300 days+2kg/cow/day for 65 days)	80000
3	Cost of fodder grass15kg/cow/day	28000
4	Veterinary, electricity and miscellaneous expenses	12000
	Total variable costs	**195000**
	Total costs (Fixed + variable)	**319000**
c) Net returns		
1.	Sale of Fish (5000kg @ Rs80/kg)	400000
2	Sale of Milk (10litres/cow/day x 300 days x 5 xRs20/liter)	300000
	Total returns	**700000**
	Net returns	**381000**

Fig. 19: Cattle farming by a marginal farmer

1.6.1.1 Biogas slurry in fish culture

The fish yield in a pond can be more than doubled if the cow dung is first fed to a biogas plant and the digested slurry then used instead of raw cow dung. After pre-stocking management of pond, fingerlings of IMC and Exotic carps together can be stocked, ponds need to be fertilized daily with 30litres of biogas slurry for an acre of water spread area. The slurry is rich in nitrogen and phosphorus, and is free from toxic gases which are produced when cow dung decomposes in pond (Tripathy and Sharma, 2005). At the same time, the gas is used in kitchen and for lighting the house. The slurry is not applied on a cloudy day or when fish come to the surface to gulp air.

In this system, the surface feeders reach to about 1 kg individual body weight in 6 months. All marketable fish are then harvested every 2 months and replenished with an equal number of fingerlings. A total of 4 to 5 tons of fish can be obtained using biogas slurry from an acre pond area.

Advantages

- Savings on inorganic fertilizers and feed.
- Environment friendly, no oxygen demand
- Savings on fuel and electricity
- Cooking with biogas removes drudgery of women for fuel wood collection and helps in keeping the kitchen clean and environment clean.

Limitation

- Slurry/ gas production is poor during cloudy days or when temperatures are low.

1.6.2 Pig – Fish integrated system

Pig cum fish culture has been practiced for many years in various parts of Asia. It is the most prevailing integrated fish farm model practiced throughout China. Pig-fish farming in backyard is most successful in Vietnam and Philippines. Fish farmers can reduce the production cost of fish to an extent of 60% by reducing the feed and fertilizers with pig manure.

a) Advantages of pig-fish farming

1. Pig-fish farming maximizes land use by integrating two farm enterprises in the same area.
2. The left over residues of kitchen, aquatic plants, agriculture by-products and wastes are used as feed for pigs and pig excreta are in turn used as organic manure for fish ponds.
3. Per unit area animal protein production in pig-fish integration is very high when compared to other integrated fish farming practices.
4. Aquatic weeds if any in the pond are utilized as roughage to pigs.
5. The pig dung acts as an excellent fertilizer for ponds for raising biological productivity which in turn helps in increasing fish production.
6. Some of the fishes feed directly on the pig excreta which contains 70% digestible food for the fish.
7. Pond water can be used for cleaning the pig sheds and for bathing pigs.
8. The pond dykes provide space for erection of pig housing units.
9. Pig-fish integration provides additional income and also acts as a cheap source of animal protein.

In India, pig-fish farming has a special significance as it can improve the socio-economic status of many of the weaker rural communities especially the tribal communities which traditionally rear pigs. Pig rearing is the fattening of piglets (2-3 months old) to a marketable size (60-100 kg) over a feeding period of 5-6 months. As pigs attain slaughter size within 5-6 months and fish raising of Indian

Fig. 20: Pig rearing adjacent to pond

exotic carps is done for 10-12 months, two lots of pigs can be raised along with one lot of fish. The pigs are marketed at a mean weight of approx. 80-150 kg. The pigs reared for breeding deliver 6-12 piglets at a time. Their age at first maturity ranges from 6-8 months. The most important varieties of pig like (a) White Yorkshire, (b) Hampshire, and (c) Landrace are widely used for integration with fish culture.

b) Pig housing

The pig's house is collectively called as 'pigsty'. It has one or more pens. In each pen, one or more pigs may be kept. Proper hygienic conditions must be maintained in the pigsty. The pens should have adequate floor space (1-1.5 m^2 per pig), facilities for feeding and providing drinking water. Floor of the pigsty should be cement concrete and height 2-2.5m from floor level. Pigsty should be covered with thatched roof or RCC roof. Over heating should not be allowed to occur in the pigsty. The pigsties are built mostly at the pond site and at times even over pond. The washing of pens containing dung and urine is channelized into ponds.

Fig. 21: Pigsty adjacent to pond

i) Feed and feeding of pigs

Feed plays an important role for pig production. Quality of ration of pig determines the rate of growth, general resistance to diseases and parasites and vigour of litter. A complete diet consists of protein (more than 20%), carbohydrate, fats, minerals, vitamins and clean water. They are also fed with chopped grasses or aquatic plants such as azolla, duckweed, water hyacinth etc. once in a week to prevent mineral deficiency. Pigs should be fed twice a day Feeding occurred at a rate of approx. 3.5 – 7 % body weight per day.

ii) Pig excreta as manure

- Each fully grown pig voids between 500 to 600 kg dung in a year and excreta released by 40-45 pigs is adequate to fertilize 1 ha water area under polyculture of fish during the year.

- Pig manure is fine in texture and contains major inorganic nutrient components (N,P,K) in addition to trace elements like Ca, Cu, Zn, Fe and Mg.
- It has been reported that a pig excretes about 1000kg of faeces and 1200 kg of urine in the fish culturing period of 8 months i.e. from piglet to an adult size. Pig excretes about 10-20% of its body weight.
- Pig manure is used as fertilizer in fish ponds in raw or fermented form which is done by composting. However, fresh pig manure application to a pond gives maximum yields than pig compost.

Table 5: Chemical composition of pig manure and urine

Sl. No.	Components	Faeces (%)	Urine (%)
1.	Moisture	70-77	90-95
2.	Dry matter	23-30	5-10
3.	Organic matter	15	2.54
4.	Inorganic matter		
	a. Total carbon	2.72-4.00	-
	b. Total nitrogen	0.60	0.40
	c. Ammonia nitrogen	0.24-0.27	-
	d. Phosphorus	0.50	0.10
	e. Potassium	0.40	0.70
5.	C/N ratio	14:1	-
6.	pH	6.6-6.8	7.6

c) Fish culture

The application of pig dung in ponds provides a nutrient base for dense bloom of phytoplankton, which in turn forms a base for intense zooplankton development. Thus, polyculture of IMCs and exotic carps is undertaken in fish cum pig farming ponds. Pond is stocked @ 8000-8500 fingerlings/ha and a species ratio of 40% surface feeder, 20% column feeder and 20-30% bottom feeders and 10-20% macro vegetation feeders (grass carp) is preferred for high fish yield. About 200 kg lime should be applied to the pond water in a year. Due to abundance of natural food in the pig- fish integrated pond, the fish attains marketable size within 8-10 months and about 6000kg fish can be harvested from the pond with B:C ratio of 1.6. Thus, the system of pig-fish culture is quite significant in improving the socio-economic status of poor farmers.

d) Economics of pig-fish integrated farming system

i. Capital investment	Amount (Rs)
1. Cost of construction of pigsty 35nos x @16 sq. ft/pig	168000
2. Construction of inlet outlet structures	12000
3. Other equipments including pump set	30000
Total capital investment	**210000**
ii. Costs	
a) Fixed cost	
1. Depreciation @10%	21000
2. Interest on capital @15%	31500
3. Insurance @6% on value of pigs	1680
4. Wages to permanent labour for 12 months @Rs5000/month/labour	60000
Total fixed costs	**114180**
b) Variable costs	
1. Pond preparation, fish seed, feed and fertilizer cost for 1 ha pond area	75000
2. Cost of piglets (8kg size 35 nos/cycle x 2 cycles ((@Rs400/piglet)	28000
3. Cost of pig mash ((@400 kg/animal x 35 animals x 2 cycles @Rs1/kg)	28000
4. Veterinary, electricity and miscellaneous expenses	14000
Total variable costs	**145000**
Total costs (Fixed + variable)	**259180**
c) Net returns	
1. Sale of Fish (5000kg @ Rs80/kg)	400000
2. Sale of pig (4200 kg @Rs60/kg)	252000
Total returns	**652000**
Net returns	**392820**

1.6.3 Goat – Fish integrated system

Fig. 22: Goat rearing

Goat and sheep are found to play a significant role in many types of small farm system in Asia, Africa and Latin America. In Asia, over 60% of the sheep and goat population is raised on farms and India has 25% of the world's goat population. Goat farming is an age-old practice but its integration with fish culture has not been explored. Goats not only provide meat and milk but also a good amount of manure. Annual production of manure from a goat s around 1.5 – 2.0 t.

If animal manure is not properly used, it causes pollution of water and environment. It has been observed that 40-50 kg of animal manure produce 1 kg of fish. Animal manure and green fodder can totally replace the commercial feed for fish farming achieving a similar fish production.

Fig. 23: Goat house

There are several well known Indian breeds of goats apart from local non-descripts scattered throughout the country. The breeds are described in 5 regions.

1. Himalayan region: Chamba, Gaddi, Kashmiri, Pashmina, Chegu
2. Northern region: Jamnapari, Beetal, Barbari
3. Central Region: Marwari, Mehsana, Zelwadi, Berri, Kathiwari, Sirohi, Jhakrana
4. Southern region: Surti, Deccani, Osmanabadi, Malabari
5. Eastern region: Bengal, Assam hilly breed, Ganjam

a) Distribution of fibre, meat and milk type breed

Fibre type	Meat type	Milk type
Himalayan, Chegu	Bengal, Assam hill goat,Deccani, Osmanabadi,Jhakrana, Sirohi	Jamanapari, Beetal, Barbari, Marwari, Mehsana, Kutchi, Surti, Malbari

a) Technical management

- The goat is a versatile animal. Goats can be kept with little expense on undulating lands with an inexpensive shelter.
- Goat housing should be dry, comfortable, safe and protected from excessive heat. Adequate space, proper ventilation, sanitation, drainage, sufficient light should be provided while constructing a house.
- Goats are selective feeders and relish, napier grass, cowpea, soybean, cabbage, cauliflower leaves, lettuce, leaves of shrubs are well consumed by the goats.
- Goat excreta is very good organic fertilizer and contains 60% organic carbon, 2.7% nitrogen, 1.78% phosphorus and 2.88% potassium. Its solid excreta are extremely rich in nutrients.

- Goat droppings (about 1.0 cm size, pellet shape) are coated with mucus that helps in bacterial aggregation resulting in formation of organic detritus in the water.
- An adult goat weighing about 20kg discharges 300 – 400 g excreta on a daily basis. For manuring 1 ha of water area, a herd of 50-60 goats will be needed.
- It has been observed that rohu and mrigal grow well when pond is manured with goat excreta.
- This integration can produce 3.5 to 4.0 tones/ha/year of fish without supplementary feed or fertilizer in addition to goat meat which has a ready market throughout the country.

1.6.4 Rabbit – Fish integrated system

Rabbit is considered as a pet animal by the common citizens in India. It plays an important role as a non-conventional meat animal for hilly, tropical rain forests, roughage, legumes and horticulturally rich areas. But currently rabbit has emerged as an alternate meat source for the future. Rabbit meat has been regarded as a dieticians choice for the health-conscious meat consumers. Rabbit meat is low in fat content. Rabbits have the highest reproduction rate and can attain the growth rate comparable to modern broiler chicken. Rabbit fur skin which is an important by-product supplements the farmers' income.

Fig. 24: Rearing of rabbits

a) Breed, varieties and strains

Approximately 60 individual breeds and varieties of rabbits are recognised world over today. Some of the popular breeds are as follows:

Meat type: Soviet Chinchilla, Grey Giant, White Giant and New Zealand White

Wool type: Russian Angora and German Angora

b) Technology of operation

- Rabbits may be reared in three ways a. Cage system b. Hutch system, and c. Floor system.
- Cage system is used when rabbits are reared in semi-commercial and commercial scale. Hutch system is generally used for breeding and maturity purpose.
- Atmospheric temperature (15-20^{0}C), proper ventilation, 5% higher RH than outside, effective lighting for 16hrs a day and kutcha floor are preferred for the shed of rabbit.
- Rabbit is a monogastic animal but presence of microflora in the hind gut (caecum) and habit of coprophagy makes it capable of consuming a variety of feed. Concentrate feed includes grain such as oats, barley, maize etc.(energy sources) and protein supplements namely soyabean meal, groundnut cake, seasame meal etc. table wastes, kitchen wastes can be used to feed rabbits. Common salt needs to be added at about 0.5% of whole diet. Availability of adequate fresh and clean water should be assured to rabbits.
- Strict sanitation practices help in disease control.

c) Rabbit excreta - A new potential aquaculture manurial input

The potential of rabbit excreta was evaluated through studies on its composition and effects on hydrobiological conditions, in comparison with the traditional organic manure, cow dung. Rabbit excreta contain 50% organic matter, 2% nitrogen, 1.33% phosphate and 1.2% potassium. The high nitrogenous rabbit excreta (10 times higher than that of cow dung) was found to be releasing nutrients gradually, sustaining high plankton production over a long period of time. Rabbit manure has a greater value as a direct food for fish. The pellet size and semi-floating nature of rabbit manure encourages the enthusiastic and direct ingestion by fish. It is evident from the above facts that rabbit excreta, low in moisture and high in nitrogen content, can very well be used for sustained plankton production and hence, rabbits can be efficiently integrated with fish.

1.6.5 Duck – Fish integrated system

- Duck-fish integration is the most common integration, mainly practised in China, Hungary, East Germany, Poland, Russia and India. It utilises the mutually beneficial biological relationship between fish and duck.
- In varied affluent countries, duck meat is a delicacy, and ducks are mainly reared for table purposes.

Fig. 25: Duck cum fish culture

- Duck eggs are an important source of protein food in India. These are very cheap to produce and can play an important role in balancing the diet of the Indian people. Consumption as well as production of duck eggs in India is mostly done by socially weaker sections of the community.
- The production of duck eggs is about 400 million/year which is 5% of the total egg output in the country.

a) Importance of duck rearing

- Rearing of ducks is limited to watershed regions. It is very popular among villagers as a profitable back-yard enterprise as average egg production from ducks is higher than local fowls.
- They have great foraging capacity.
- Maintain egg productivity of 300 eggs/bird/annum for almost up to the age of 2 to 3 years. Energy level is higher in duck eggs than in hen egg.
- Eggs are bigger in size and because of thicker and strong shell, transportation is easier and breakage during transit is lesser. Cholesterol level in duck egg is less as compared to the eggs.
- Ducks feed on a large variety of organisms like snails, flies, earthworms, insects, etc, that are vector of diseases. Ducks may serve as effective biological control of a number of human and animal diseases.
- Ducks keep the water clean by controlling potato beetles, grass hoppers and other aquatic fauna. They feed on green algae and weeds thereby helping in the control of unwanted plants.
- Ducks are quite hardy, easily brooded and resistant to the common diseases. They need less attention and area easily manageable.

- Cannibalism behaviour is not usually encountered in ducks.
- Ducks do not require extensive housing.

b) Varieties of ducks

Ducks are of several kinds as the egg type, the meat type and the ornament type. In India, mainly ducks of egg-laying type are reared.

i) Ducks for eggs

The famous Indian duck breeds for eggs are: (a) Sylhat mate, (b) Nageswari, (c) Indian Runner (d) *Khaki Campbell*. *Khaki campbell* is recommended for integration with fish culture. As they are prolific layers they can be reared economically. These ducks start laying eggs when they are 3 months old. The annual average egg production is about 300 eggs or more.

ii) Ducks for meat

The ducks for meat include White Pekin, Muscovy, Buff and Swedish breeds. White Pekin is most popular bird from taste point of view. It grows very fast weighing between 3.5- 4.0 kg.

c) Housing and management

A comfortable house should provide adequate accommodation, be reasonably cool in summer and sufficiently warm during winter. It should provide adequate fresh air and sunshine and always remain dry. The house should give adequate protection against sudden changes and extreme temperatures, to the birds from their natural enemies like jackals, foxes, dogs, cats, rats, snakes, kites and crows. In Centralised enclosure housing system, a relatively large duck shed is constructed in the vicinity of the fish ponds with a cemented area of dry and wet runs outside. Average stocking rate is about 4 ducks/m^2. The dry and wet runs must be cleaned daily.

The fish-pond housing system is most common. The dykes of grow-out ponds are partly fenced to form a dry run and part of the water area of a corner of the pond is fenced with used material to form a wet run.

Fig. 26: Duck house on fish pond

d) Care of ducklings

The brooder house should be prepared with dry litter with provision for water to drink by dipping its beak in water. The birds need heat for first 8-11 days depending on the weather. For the first 3 weeks starter feed containing 17% protein will be enough for proper growth of ducklings. Ducks are voracious eaters and foragers. Apart from compound feeds snails, fingerlings, earthworms, insects and other vegetative forms should be provided to them when reared in ponds. It reduces the feed cost. Ducks have difficulty in swallowing dry mash. Ducks are very much susceptible to aflatoxin produced by fungus, *Aspergillus flavus*, in groundnut oil cake, maize and rice polish. Duck can tolerate up to 0.03 ppm aflatoxin. Annual consumption of feed is about 50-60 kg per duck. It requires about 3 kg of feed to produce a dozen eggs and 3.22 kg feed to produce 1 kg of broiler duck. The layer requires 200-210 grams of food per day. Ducks should be vaccinated in time for prevention of viral diseases.

e) Fish culture

Duck cum fish culture in ponds provides a nutrient base for phytoplankton, which in turn forms a base for intense zooplankton development. Polyculture of IMCs and exotic carps is generally undertaken in fish cum duck farming ponds. Pond is stocked @ 8000-10000 fingerlings/ha and a species ratio of 40% surface feeder, 20% column feeder and 20-30% bottom feeders and 10-20% macro vegetation feeders (grass carp) is preferred for high fish yield. Lime should be applied to the pond water based on water pH. The fish attains marketable size within 8-10 months and about 6000kg fish can be harvested from 1 ha pond in a year.

f) Benefits of duck-fish integration

A fish pond is a semi-closed biological system. In fish ponds, there are many aquatic animals and plants, most of which are natural food organisms of fish, some are detrimental to fish but can be utilized by ducks. Fish ponds provide ducks with an excellent, essentially disease free environment. Ducks consume juvenile frogs, tadpoles, and dragonfly larvae, thus eradicating many predators of fry and fingerlings. Furthermore, the protein content of these natural food organisms of duck is high. Therefore, duck raising in fish ponds reduces the demand for protein in duck feeds. For ducks raised in pens, the digestible protein content in the feed must be 16-20 percent; for ducks raised in fish ponds, the digestible protein content can reduce to 13-14 percent. This can save 200-300 g available protein for each duck, equivalent to 2-3 percent the duck feed. Duck droppings directly into the pond, providing carbon, nitrogen, and phosphorus go and stimulating the growth of natural food organisms. This direct fertilization

Fig. 27: Ducks in fish pond

of the pond has two merits: 1) there is no loss in the availability of manure, and 2) it becomes more homogeneous and avoids any heaping of duck droppings. For these reasons, raising ducks on fish ponds promotes fish growth, increase fish yields and eliminates pollution problems that might be caused by excreta in duck pen.

The quality and quantity of duck excreta, however, are dependent on species, feeds applied, culturing management, and climatic conditions. The stocking rate of ducks is generally 300-500 ducks/ha and each duck produces about 7 kg of dropping during the 36 day fattening period. If 500 ducks are raised, 3500 kg of excreta would be produced in that period. The moisture content of duck excreta is 56.6 %; organic substance, 25.2 %; C, 10 %; P2O5, 1.4 %; N, 1 %; K2O, 0.62 %; Ca, 1.8 %; and others. Each egg-laying duck annually produces 7.5-10 kg (dry weight) of excreta (equivalent to 70 kg wet weight).

Duck feeds are fully utilised in fish-duck integration. Ducks lose 10-20 percent (23-30 g/day) of their feed being undigested. This feed can directly be consumed by fish. Fish-duck integration also promotes the recycling of nutrients in the pond ecosystem. In shallow areas, a duck dips its head to the pond bottom and turns the silt to search for benthos. By virtue of this digging action, nutritional elements locked in the pond humus will be released. Ducks also act as pond aerators. Their swimming, playing and chasing disturb the surface of the pond and aerates the water.

Duck raising in fish ponds has three advantages over raising ducks in pens. (i) The feed efficiency and body weight of each duck improved. The higher feed efficiency also implies that the waste feeds are utilised by the fish. (ii) The food conversion factor in fish-duck integration was reduced from 3.84 to 2.64. (iii) The survival rate is increased by 3.5 percent because fish ponds provide a clean environment for the ducks.

It was believed that if fish and ducks were raised in the same pond, the ducks would eat the small fish. Hence, when the fish juveniles attain 5 g or more, the ducks are to be released into the pond. Fishes above 5 g are able to escape from the duck.

Table 6: Daily amount (g) of duck excreta and split input into fish ponds

Time(days)	Excreta		Feed	
	Wet wt	Dry wt.	Applied	Split
21-30	127	68	194	29 (15)
31-40	248	74	227	34 (15)
41-50	420	73	248	37 (15)

(Values in parenthesis are the percentages of feed split)

g) Economics of duck-fish integration system:

i. Capital investment	**Amount (Rs)**
1. Cost of construction of duck pen 200nos x @1 sq. ft/duck	20000
2. Construction of inlet outlet structures	10000
3. Other pump set	20000
Total capital investment	**50000**
ii. Costs	
a) Fixed cost	
1. Depreciation @10%	5000
2. Interest on capital @15%	7500
3. Insurance premium @Rs1.50/bird	300
4. Wages to permanent labour for 12 months @Rs5000/month/labour	60000
Total fixed costs	**72800**
b) Variable costs	
1. Pond preparation, fish seed, feed and fertilizer cost for 1 ha pond area	75000
2. Cost of ducks 200 nos ((@Rs100/duck)	20000
3. Cost of duck feed ((@30 kg/duck x 200 ducks @Rs10/kg)	60000
4. Veterinary, electricity and miscellaneous expenses @ Rs10/duck x 200	2000
Total variable costs	**157000**
Total costs (Fixed + variable)	**229800**
c) Net returns	
1. Sale of Fish (5000kg @ Rs80/kg)	400000
2. Sale of duck egg (180 nos x 120 eggs/bird @ Rs 3/egg)	64800
3. Sale of culled ducks (180 nos x 1.5kg/bird @Rs60/kg)	16200
Total returns	**481000**
Net returns	**251200**

An accurate economic analysis of fish-duck integration is impossible because of the variations in production costs, duck yields in different countries. even in the same district, fish and duck species, stocking densities, quality and efficiency of feeds, rearing management techniques and varied climatic conditions. From the viewpoint of input - output relationships fish-duck integration is the best model providing a B:C ratio of 1.8.

1.6.6 Poultry – Fish integrated system

Fish-cum-poultry farming is practiced in many countries of the world and especially in Asia. It is not only an efficient way of recycling farm wastes but also produces high economic returns. Poultry manure is a complete fertilizer. The integration of aquaculture with poultry results in a more efficient use of resources. In this system, the costs associated with fish culture are reduced by about 70%.

Generally, chicken raising for meat (broiler) and eggs (layers) can be integrated with fish culture which is common in India. The poultry excreta is recycled to fertilise the fish ponds and also it is used as a fish feed ingredient (dietary protein source) because it contains an undefined quantity of uneaten chicken feed crumbles. The poultry droppings of fully built up poultry litter is recycled in the polyculture fish ponds which results in production of 4,500-5000 kg fish/ha/ year. Broiler production gives good and immediate returns to the farmers.

The most important factor a farmer should consider before taking up broiler production is to investigate the market conditions, where the product will be sold. There should be steady demand for his chickens, so that all the stock could be disposed of immediately when they are ready for market. Success in broiler production depends mainly on the efficiency of the farmer, his experience, aptitude and ability in the management of the flock.

a) Chicken varieties

Chickens are grown for their egg and meat. Accordingly, they are classified as egg-type chicken and meat-type chicken. Besides, dual purpose birds for both egg and meat have been developed.

i) **Breeds of Egg-type chicken (Layers):** BV-300, Leghorn, Hyline, Babcock

ii) **Breeds of meat-type chicken (Broilers):** White synthetic male line (WSML), White synthetic dam line (SDL), Coloured synthetic male line (CSML), Coloured synthetic female line (CSFL).

iii) **Breeds of dual purpose:** Gramshree, Kalinga brown, Vanaraja, Giriraja.

b) Poultry farming and management

- Two distinct housing structures for poultry birds such as caged system (for egg laying operation) and deep litter or floor system (for broilers, layers and breeders) are generally used in India.
- With floor operations, absorbent litter materials like paddy husk, saw dust, sugarcane bagasse, rice bran etc need to be placed at a thickness of 4-6inch.

- Chicken can be raised in cages over or adjacent to the ponds. Each broiler bird requires 1 sq.ft floor space and each layer bird 2 sq. ft. floor space area.
- Egg production commences at the age of 22 weeks upto 70 weeks. The birds lay 200-250 eggs/bird/year.
- Chicken integrated with fish culture are generally fed commercial rations which are highly balanced and has high protein content for growth and production. For broilers, starter mash is fed for 1-4 weeks and finisher mash for 5-8 weeks and it is given as much as they can consume. For chicken layers, feed is given @ 80-110g/bird/day for the first 16 weeks and 110-120g/bird/day from 17th week onwards. Water should be provided at all times.
- Proper vaccination and prophylactic measures against diseases are needed for better economic returns.
- Eggs are collected twice a day, i.e. morning and evening. Layer birds are discarded after 18 months of rearing as eggs production goes down. Marketing of boilers starts after 5-6 weeks of rearing, during which birds weigh 1.4 to 1.6 kg.

Fig.28: Broiler chicken

Fig.29: Dual purpose "Banaraja" bird

c) Poultry manure

Each bird produces approx. 40g of manure per day. The conversion coefficient for chicken manure is 6.8. These manures can be converted into quality fish food by stimulating microbial activity in the water column and at the pond bottom and realizing the nutrients and minerals originally bound in relatively indigestible form. These nutrients and minerals provide the substrates for photosynthetic (autotrophic) and microbial production, which can be utilized by fish. The total protein content of chicken manure is as high as 10-20%. About 80% of the

Table 7: Nutritional status of poultry manure

Sl. No.	Type of poultry manure	Crude protein (%)	Nitrogen (%)	Phosphorus (%)	Potash (%)	pH	Crude fat (%)	Crude fiber (%)	Organic Carbon
1.	Fresh poultry dropping	10	1.4-1.6	1.5	0.8-0.9	6.9	11.98	9.59	26% on dry weight basis
2.	Deep litter	19	3	2	2	7.8	-	-	23% on wet weight basis

manure represents undigested foodstuff with 25% dry matter content, which can be directly used by fish as feed.

d) Integration with fish farming

- The feed waste of animals and animal excreta are utilised to increase the biological productivity of water. Probably supplemental feed and fertilisers may not be needed in such a system and the cost on inputs, therefore, may be reduced.
- The ponds are prepared and stocked in the same way as in duck-fish farming. The built up poultry litter removed from the poultry sheds is suitable and is applied to the pond in daily doses @ 40 - 50 kg/ha/day every morning after sunrise. The application of litter is deferred on the days when algal bloom appears in the pond.
- It has been estimated that 25-30 birds can produce one tonne of deep litter fertiliser in a year time. As such 500-600 birds are adequate to produce manuring for a hectare of water area under fish polyculture.
- In India, scientists have developed economically viable system for integrating fish culture and poultry farming without use of supplementary feed or fertilizer. This type of farming system produces 4500 to 5000kg fish, more than 70,000 eggs and about 1250 kg of poultry meat on an annual basis using 500 to 600 birds per ha of pond area.

e) Economics of poultry-fish integration system

i. Capital investment	**Amount(Rs)**
1. Cost of construction of poultry shed 500nos x @1 sq. ft/duck	100000
2. Construction of inlet outlet structures	10000
Electrical installations and overhead water provision	20000
3. Other pump set	20000
Total capital investment	**150000**
ii. Costs	
a) Fixed cost	
1. Depreciation @10%	15000
2. Interest on capital @15%	22500
3. Insurance premium 500 nos x @Rs2.00 /bird	1000
4. Wages to permanent labour for 12 months @Rs5000/month/labour	60000
Total fixed costs	**98500**
b) Variable costs	
1. Pond preparation, fish seed, feed and fertilizer cost for 1 ha pond area	75000
2. Cost of layer birds 500 nos (@Rs100/bird)	50000
3. Cost of poultry feed (@40 kg/bird x 500 ducks @Rs12/kg)	240000

Contd.

i. Capital investment	Amount(Rs)
4. Veterinary, electricity and miscellaneous expenses @ Rs10/bird x 500	5000
Total variable costs	**370000**
Total costs (Fixed + variable)	**468500**
c) Net returns	
1. Sale of Fish (5000kg @ Rs80/kg)	400000
2. Sale of egg (450 nos x 250 eggs/bird @ Rs 3/egg)	337500
3. Sale of culled ducks (450 nos x 2kg/bird @Rs80/kg)	72000
Total returns	**809500**
Net returns	**341000**

1.7 Multi-utilisation systems

In multi-utilisation system, more than two farming systems can be undertaken in the pond area to utilize the pond water, peripheral areas etc resulting in extra income from the land area. As the pond bottom is rich in organic nutrients it reduces the application of chemical fertilizers and also results in higher and good quality product yield. Where ponds get dried up in winter months, farmers grow short term oil crops with low water irrigation that resulted in better utilization of land area with more crop yield. The excess nutrient in soil trap by these green plants resulted in better fish production in the next season. Various multi-utilisation system includes a. Rice-fish-azolla system, b. Rice-fish-duck system, c. Rice-fish-poultry system, d. Rice-fish-goat farming system, e. Rice-fish-pig farming, fish-horticultural crops-livestock system.

1.8 Nutrient dynamics in integrated fish farming systems

Nutrient dynamics in any aquatic ecosystem deals with the study of major plant nutrients (carbon, nitrogen, potassium and phosphorus) that stimulate phytoplankton growth. The source of available nutrients is of two types (i) internal - autochthonous, and (ii) external – autochthonous. Potassium is available in high quantities in many aquatic environments. To maintain adequate phytoplankton as natural food to support desirable fish yield, fertilization is imperative. Generally, total phytoplankton production is directly proportional to the initial concentration of limiting nutrients. Hence, optimal pond fertilization is necessary to provide a balanced supply of major nutrients.

Nitrogen, phosphorus and occasionally carbon are the most common limiting nutrients to be maintained. Algae are capable of continuously utilizing the nutrients above this concentration and deposit the surplus nutrients in their cells without concomitant growth, which is termed "luxury consumption". This phenomenon is particularly well known for phosphorus uptake. The standing algal crop in fish ponds is continuously consumed by filter feeding organisms and fish like herbivores in the pond ecosystem. Thus, the amount of nutrients required is only to compensate for algal growth and the losses through grazing and sinking. Furthermore, only a fraction of the nutrients is ingested/ assimilated by fish and other organisms and a large portion is recycled back to ponds as waste. Besides, the nitrogen, phosphorus ratio (N:P) is an important consideration in pond fertilization and the typical N:P ratio in algal biomass is roughly 10:1. Excess nitrogen inputs may lead to ammonia toxicity to fish. When the appropriate N:P ratio in a pond is not known, a surplus of phosphorus is safer than a surplus of nitrogen. The ammonia concentration in pond water should not exceed 0.5ppm.

Animal manures are generally low in N:P ratio (<3). Such a low ratio is not optimal for phytoplankton growth. Thus, fish pond can be fertilized at a reduced manure rate and supplemented with inorganic nitrogen fertilizer to provide adequate nitrogen and avoid oxygen depletion. Sedimentation of particulate organic matter and detritus formed in the water column are the major sources of organic matter in fish pond sediment. thus, sediment or mud in fish pond is regarded to be an "energy sink" in the fish pond ecosystem. Bioturbation by benthivorous and detritivorous fishes facilitates the penetration of dissolved oxygen into the sediment pore water and enhances the mineralization of organic phosphorus in the sediment.thus, the applied fertilizer and manures along with the derived nutrients in any aquatic ecosystem undergo complex dynamic processes.

1.9 Conclusion

In the present situation, the farmers are based on traditional knowledge and experience without proper planning, shortfall of latest technology and management techniques. For the practice of scientific integrated fish farming, the farmers need specialized training, introduction of high yield varieties, multi-cropping knowledge and innovation in farming to extract maximum output from the small land holding. In addition, marketing of produce, proper utilization of internal resources, mechanized farm machinery, climatic change and update knowledge of world trade criteria are some of the challenges faced by farmers. To overcome these problems, integrated fish farming is the only alternative to enhance production and productivity from existing agricultural land and water. A multidisciplinary approach is needed, including technological, economic, social and political aspects. Strategy should also be developed to trap the potential water bodied in rural areas for introduction of pond based integrated farming for socio-economic development of the farmers.

References

Dalsgaard, J. Lightfoot, C. and Cristensen, V. 1995. Towards qualification of ecological sustainability in farming systems analysis. Ecological Engineering. 4: 181-189.

Diver S. 2006. Aquaponics–Integration of Hydroponics with Aquaculture, ATTRA Publication IP163.

International Fund for Agricultural Development. 2008. Improving Crop-Livestock Productivity through Efficient Nutrient Management in Mixed Farming Systems of Semi-arid West Africa. http://www.ifad.org/lrkm/tags/384.htm.

International Fund for Agricultural Development. Undated. Community Approach to the Development of Integrated Crop/Livestock Production in the Low Rainfall Area. http://www.ifad.org/lrkm/tans/3.htm

Kumar, K. and Ayyappan, S. 1998. Current practices in integrated aquaculture in India, In: integrated aquaculture in Eastern India. FAO, working paper No. 5, pp.25.

Little, D.C. and Edwards, P. 2003. Integrated livestock-fish farming systems. FAO, Rome. PP. 166.

Preston, T.R. and Murgueitio, E.1992. Sustainable Intensive Livestock Systems for the Humid Tropics. World Animal Review 72:2-8.

Singh H. 2002. A highly integrated fish farming system in Punjab: A case study. (Abstract No.4, pp 48) Proceedings of Second Indian Fisheries Science Congress held at Bhopal during 23-25 October 2002 organized by Indian Society of Fisheries Professionals, Mumbai, India.

Singh, C. S. 2002. Fisheries based integrated farming system. Fishing chimes, Vol. 22(1), April 2002 issue, pp 146.

Singh, J.P., Gangwar, B., Pandey, D.K. and Kochewad, S.A. 2011. Integrated Farming System Model for Small Farm Holders of Western Plain Zone of Uttar Pradesh. PDFSR Bulletin No. 05, pp. 58. Project Directorate for Farming Systems Research, Modipuram, Meerut, India.

Singh, J.P., Gangwar. B. and Pandey, D.K. 2011. Nutrient Management in Farming System Perspective. Indian J. Fert., Vol. 7(11), pp. 16-21.

Tripathy, S.D. and Sharma, B.K. 2005. Biogas slurry in fish culture In: Integrated agriculture-aquaculture A primer, FAO Series Technical Paper No. 407,pp.128-130.

Verma. A.M. 1994. Integrated fish farming with Makhana (*Eurale ferox*) Fishing chimes, 14 (2):13

World Commission on Environment and Development (WCED). 1987. Our Common Future. Oxford. Oxford University Press, P. 43.

http://www.vuatkerala.org/static/eng/advisory/fisheries/culture_fisheries/integrated_farming/introduction.htm

Integrated Farming System Practices: Challenges & Opportunities, pp 159-182

New India Publishing Agency, New Delhi, India

6

Livestock Based Integrated Farming Systems

R.K.Swain

Introduction

Agriculture accounts for a large share of Gross Domestic Product (GDP), exports, and employs more than 60% of the work force in India (FAO, 2001). The failure or success of agriculture determines the economic growth of the country. As human population grows so does the demand for food and, therefore, demand for land to grow food. Ironically, the demand to convert agricultural land to commercial and residential developments is also very high due to population growth and urbanization. This is reducing the amount of land available for agricultural purpose. This scenario is being noticed in most of the developing countries like India. Against this background, agriculture will be expected to meet food, feed, and fiber demands of a world population that is anticipated to grow from approximately 6 billion in 1999 to 11 billion by 2050 (USDA, 2010). Small and fragmented land holdings do not allow farmers to have independent farm resources like draught animals, tractors, bore wells, tube wells and other sophisticated farm machineries for various cultural operations. While crop production is the primary farming enterprise in the country, livestock raising serves as an auxiliary activity to crop production. About 90% of livestock population are in the hands of the smallholders where adequate supply of high quality forage is a common problem. Integrated farming system (IFS) models

College of Veterinary Science &Animal Husbandry, Orissa University of Agriculture & Technology, Bhubaneswar-751003, Odisha, India developed in different parts of the country involving dairy, duckery, poultry, horticulture, apiary, pisciculture and plantation crops viz. coconut, cocoa, nutmeg, banana, pineapple *etc.* along with crops, have been found to increase net profit significantly as compared to cropping alone.

Livestock based integrated farming system

An integrated livestock based farming system represents a key solution for boosting agricultural production on an overall basis and safeguarding the environment through prudent and efficient resource use. In crop-livestock systems, often referred to as mixed farming systems (Sere and Steinfeld, 1996), livestock and crops are produced within a co-ordinated framework. In many mixed systems, the waste products of one component serve as a resource or input for the other: manure from livestock is used to raise soil fertility in order to enhance crop production, while crop residues and byproducts are used as feed for the animals. The increasing pressure on land and the growing demand for livestock products makes it more and more important to ensure the effective use of feed resources, including crop residues. The necessity of integrating livestock and crop husbandry is becoming more pronounced due to deterioration in soil fertility, high cost of inorganic fertilizers and scarcity of fodder for livestock particularly during the dry season in many marginal areas with fragile ecosystems.

Agriculture and livestock production is severely affected by frequent droughts, inadequate rainfall, extreme temperature, poor quality of land and groundwater, high population growth and poor infrastructure. Faced with low productivity and high uncertainty in crop production, rural people are heavily dependent for their livelihood on common property resources based on livestock rearing. Integrated farming refers to agricultural systems that integrate livestock and crop production. Moreover, the system help poor small farmers, who have very small land holding for crop production and a few heads of livestock to diversify farm production, increase cash income, improve quality and quantity of food produced and exploitation of unutilized resources. Population growth, urbanization and income growth are fuelling a substantial increase in the demand for food of animal origin, while also aggravating the competition between crops and livestock (increasing cropping areas and reducing rangelands). There is a need to effective linkages and complementarities of various components to develop holistic farming system. Different farming systems have been evolved independently and being practiced by the farmers without any rationale for utilizing the wastes and residues arising out of cropping/animals and other associated enterprises at farm resulting wastage of resources. To fulfill the basic needs of household including food (cereal, pulses, oilseeds, milk, fruit, honey, fish meat, etc.) for human, feed and fodder for animals and fuel & fibre for general use warrant an attention about integrated farming system.

Integrated crop-livestock farming systems

An integrated crop-livestock farming system represents a key solution for enhancing livestock production and safeguarding the environment through prudent and efficient resource use.

Crop-Livestock Integrated Farming System refers to an agricultural system that is characterized by the systematic production of livestock and crops on the same farm. In crop-livestock integrated farming systems the most visible feature is the synergy between crops and livestock. At one level, animals gain from crops produced on the farm. For example, crops provide animals with fodder from grass, legumes and crop residues. On the other hand, crop farming takes advantage of the animals on the farm to improve the environment in which crop production takes place. The animals provide draught power in crop production where the practice of animal traction is popular and their dung (or waste matter) is used as manure to improve soil fertility. Animals can also be used in weed control when they graze under trees and on stubble (Fig. 1).

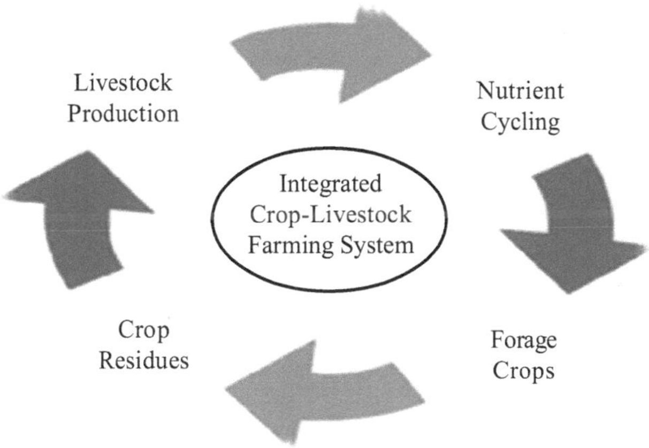

Fig. 1: Integrated Crop–Livestock farming system: Key aspects

Crop–animal interactions

Various interactions exist between crop and animals in a mixed farming system are discussed in Table 1. Draught animal power has a long history of use in smallholder farming systems in Asia. Large ruminants provide power for land preparation, soil conservation practices and haulage. Draught animal power is especially important in farming systems that are isolated from infrastructure and have a high land to population ratio. Various crops and animal interaction exists in various parts of Asia (Table 2). In northeast Thailand, draught animal power is particularly associated with the production of rain-fed lowland rice and mixed upland crops/lowland rice.

Table 1: Main crop–animal interactions in mixed farming systems

S.N.	Crop production	Animal production
1	Crops provide a range of residues and by-products that can be utilized by ruminants and non-ruminants.	Large ruminants provide power for operations such as land preparation and for soil conservation practices
2	Native pastures, improved pastures and cover-crops growing under perennial treecrops can provide grazing for ruminants	Both ruminants and non-ruminants provide manure for the maintenance and improvement of soil fertility. In many farming systems it is the only source of nutrients for cropping. Manure can be applied to the land or, as in Southeast Asia, to the water which is applied to vegetables whose residues are used by non-ruminants.
3.	Cropping systems such as alley-cropping can provide tree forage for ruminants	The sale of animal products and the hiring out of draught animals can provide cash for the purchase of fertilizers and pesticides used in crop production.Animals grazing vegetation under tree crops can control weeds and reduce the use of herbicides in farming systemsAnimals provide entry-points for the introduction of improved forages into cropping systems. Herbaceous forages can be sown under annual and perennial crops and shrubs or trees established as hedgerows in agroforestry-based cropping systems.

Table 2: Some examples of crop–animal interactions in Asia

Country	Crop–animal interactions
Indonesia	Use of manure in rice/maize/grain legume systems in upland Java.Introduction of forages in crops for use by cattle on Bali.Use of cattle for draught power in rice in southern Sumatera
Cambodia	Use of buffalo for draught power in rice in Siem Reap Province.Use of rice straw by buffalo in Siem Reap Province.Use of manure for rice production in Siem Reap Province
Malaysia	Use of large and small ruminants for weed control and manure application under rubber and oil palm.Introduction of forages under rubber and oil palm for ruminants.
Myanmar	Use of cattle for land preparation in rice production in Bago Division.Utilisation of rice straw by cattle in Bago Division.
Philippines	Use of small ruminants for weed control under coconut in southern Luzon.Use of manure from cattle feedlots for pineapple production in Northern Mindanao.Use of ducks in rice to control golden snails (rice pests) in southern Luzon

Contd.

Country	Crop–animal interactions
Thailand	Utilisation of rice straw by cattle and buffalo in the northeast Province.Use of manure from stall-fed large ruminants for rice production in the northeast Province.Use of buffalo for draught power in rice in the northeast Province.
Vietnam	Use of buffalo for draught power in rice in Song Be Province.Utilisation by buffalo of crop residues in Song Be Province.Use of weeds by ducks in ponds fertilised by pig manure in central and north eastern areas.
Bangladesh	Use of buffalo for draught power in rice Use of rice straw by cattle and buffalo Use of manure from large ruminants for rice
Bhutan	Use of rice straw by cattle in the western lowlands.Use of manure from cattle for cropping throughout the country.Use of cattle for draught power in the lowlands.
India	Use of manure from small ruminants folded on arable land in Gujarat and Rajasthan states.Use of sorghum residues by cattle in Andhra Pradesh.Use of cattle for draught power in rice–wheat systems on the Gangetic plains.
Pakistan	Introduction of forages in irrigated cropping systems in Sindh and Punjab provinces.Use of crop residues by buffalo and cattle in the Barani areas of Sindh and Punjab provinces.Use of manure from large ruminants for cropping in Sindh and Punjab Provinces
Srilanka	Use of buffalo for land preparation in rice production in the wet and intermediate zones.Utilisation of rice straw by cattle in the irrigated dry zone.Use of cattle for weed control and manure application under coconut in the intermediate zone.

In Indonesia, draught animal power is used in lowland rice-farming systems, medium altitude mixed garden/rice systems, low-altitude rice/medium-altitude mixed garden systems, upland maize systems, and soyabean production in the transmigration areas. Both buffalo and cattle are used for draught purposes, although buffalo are the most important draught animals in Southeast Asia

In Bangladesh, some 80–85% of land preparation is carried out with large ruminants despite an increasing interest in mechanization (Devendra *et al.*, 1997). In Bhutan, cattle, mithun and yak are used for draught purposes. In India, cattle, buffalo, equines, camels and yak are important for draught power in the different agro-ecological zones. The production of draught bullocks is still an important aspect of cattle rearing in India and it is estimated that there are some 70 million working animals. The number of working bovines per hectare of net sown land in India is about 0.6 (Vaidyanathan, 1988). In addition to cattle, buffalo and yak, Baruwal sheep and Sinahal goats are used in Nepal for haulage in the mountain regions. These small ruminants can carry weights of 10–20 kg depending on body size. In Pakistan, cattle are the main draught animals. In Sri Lanka, both cattle and buffalo are used, with 90% of the swamp buffalo providing power predominantly for cultivation in rice.

Draught animals can assist farmers to improve soil tillage and introduce soil conservation practices such as terracing, ridging and the broad-bed/furrow system; operations that are unlikely to be undertaken with hand-cultivation. Improved tillage requires extra power, for which resources of hand-labour are presently inadequate. Animals can provide this power. The lower compaction resulting from land preparation using animal traction, compared with tractor ploughing, also reduces the erosion hazard. In Asia, hillsides have been leveled into terraces for rice fields initially, and then re-leveled using draught animal power annually, to ensure spread of water. Without such a system, erosion of rice fields would make farming unsustainable within a few years. Heavy textured soils, such as vertisols, have a high water-storage capacity and become waterlogged during much of the growing season. To tap the inherent fertility of these soils, draught animal power can be harnessed to shape broad-beds and improve surface drainage

Animal feeds from crops

Crop production provides a range of residues and agro-industrial by-products (AIBP) that can be utilised by ruminants and non-ruminants. These include cereal straws (e.g. rice and maize), sugarcane tops, grain legume haulms (e.g. groundnut and cowpea), root crop tops and vines (e.g. cassava and sweet potato), oilseed cakes and meals (e.g. oil palm kernel cake, cottonseed cake and copra cake), rice bran and bagasse. In Asia, rice straw is the principal fibrous residue fed to over 90% of the ruminants. Devendra *et al.* (1997) has calculated that 30.4% of rice straw is used for feed in Asia.

Crop-livestock systems in India

The existence of interactive mixed crop-livestock systems is a typical characteristic of the Indian agrarian economy. Crop-livestock systems evolve in response to human population pressure, the opportunity cost of land, labour, capital, and terms of trade between outputs from the crop and livestock sectors. This is in addition to the impact of Government policies on input-output prices, investment, subsidies, trade, environment, etc., which plays a vital role. While livestock provide manure and draught power, generate income, and control weeds in plantation, crops furnish feed in the form of residues and allow for integration of forages.

The livestock farming system in rainfed agriculture are complex and generally based on traditional socio-economic considerations. An understanding of production factors and processors that effect animal production is pre-requisite for livestock integration. The productivity of livestock in farming systems in rainfed agriculture can be improved by increased fodder production as an intercrop with cereals, relay and alley cropping, forage production on bunds, improving the feeding value of stover by chopping, soaking with water, urea treatment, strategic supplementation of concentrate, urea molasses mineral block for enhanced utilization, improvement in productivity of grasses. Quantum and distribution decides the effective growing season and it becomes critical in selecting cropping systems for a given reason. An increased in 15-25% milk yield was observed by cultivation of legumes in degraded lands, establishment of fodder banks in areas where surplus fodder is available, artificial insemination with semen of approved bulls, removing low-grade animals through castration and adoption of preventive measure like vaccination and de-worming through health camps (Mishra, 2002).

For centuries, the use of cattle for milk and draught purposes has been nearly universal. While the buffalo is an important species maintained mainly for its milk, sheep, goat, pig, and poultry are raised for their meat. The bulk of livestock production takes place in rural areas, although specialized peri-urban dairy and poultry enterprises are also emerging. Crop residues and by-products are the major sources of feed. There are significant regional variations in the composition of livestock population and feed availability. For instance, buffaloes are often found in more well-endowed and irrigated regions, while cattle, sheep, and goat are predominant in rainfed areas. The buffalo's ability to adapt to a wide range of climates has made it popular even in its non-traditional breeding tracts. As a result, its population has been increasing at about 2% a year, while that of cattle is heading towards stabilization. The ambit of meat production is gradually shifting from small ruminants to large ruminants and monogastrics. In fact, pig and poultry populations have grown at about 4% a year during the last two decades. The distribution of livestock asset value is more equitable than land. In numbers, marginal and small landholders (<2.0 ha) comprise 63% of rural households but account for only 34% of the arable land. In contrast, they account for 67% of the bovines, 65% of the ovines, 70% of the pigs, and 75% of the poultry. Livestock in such households not only serves the purpose of augmenting income, employment, and food security, but also acts as a storehouse of capital and an insurance against crop shocks. Besides, livestock enterprises being women-oriented, they promote gender equity.

According to vision 2050, Indian Veterinary research Institute (IVRI) the value of output from livestock and fisheries sectors together at current prices is about Rs. 4,08,386 crores during 2009-10, which is about 29.7% of the value of output of Rs. 13,76,561 crores from total agriculture and allied sectors. Livestock in total contributes 3.93% (Rs. 2,41,177 crores) of national GDP and 22.14% to the agricultural GDP. The contribution of milk in national economy is higher (Rs. 2,28,809 crores) than paddy (Rs. 1,35,307 crores), wheat (Rs. 1,03,226 crores) and sugar cane (Rs. 37,366 crores). Animal husbandry sector provides self employment opportunities and about 6.7% of work force in rural areas is engaged in this sector. All these figures speak about the importance of livestock sector

Advantages of crop livestock integration

In an integrated system, livestock and crops are produced within a coordinated framework. The waste products of one component serve as a resource for the other. For example, manure is used to enhance crop production and nutritive value of crop residues feed the animals, thus contributing to improved animal nutrition and productivity. The result of this cyclical combination is the mixed farming system, which exists in many forms and represents the largest category of livestock systems in the world in terms of animal numbers, productivity and the number of people it services (Kathiresan, 2009).

Animals play key roles in the functioning of the farm, and not only because they provide livestock products (meat, milk, eggs, wool and hides) or can be converted into prompt cash in times of need. Animals transform plant energy into useful work. Animal power is used for ploughing, transport and in activities such as milling, logging, road construction, marketing, and water lifting for irrigation.

Animals also provide manure and other types of animal waste. Excreta have two crucial roles in the overall sustainability of the system:

(a) Improving nutrient cycling: Excreta contain several nutrients including nitrogen, phosphorus, potassium and organic matter, which are important for maintaining soil fertility and increase production.

(b) Providing energy: Excreta is the basis for the production of biogas and energy for household use (e.g. cooking, lighting) or for rural industries (e.g. powering mills and water pumps). Fuel in the form of biogas or dung cakes can replace charcoal and wood.

The overall crop livestock integration have following advantages:

1. **Productivity**: Increase economic yield per unit area - per unit time by virtue of intensification of crop canopy, crop rotation and allied enterprises.
2. **Profitability**: The system, as a whole provides opportunity to make use of produce/ waste material of one component as input on the other component at the least cost.
3. **Potentiality**: In Integrated Farming System, organic supplementation through effective utilisation of by-products of linked components as a measure is possible and this will certainly provide opportunity to promote soil health.

4. **Balanced food:** In Integrated Farming System, components of different nature are linked enabling to produce different sources of nutrition, namely, protein, carbohydrates, fats, minerals, vitamins, etc from the same unit area. It will provide opportunity to mitigate malnutrition of the farmers.
5. **Pollution:** In crop based activity, some of the organics are left as waste materials which in turn pollute the environment on decomposition. Application of huge quantity of fertilizers, pesticides, weedicides, insecticides, etc pollute soil, water and air. Much of the wastes could be converted/ recycled to some other forms of economic/ecological/social value, under the Integrated Farming System.
6. Integrated Farming System provide opportunities as crop insurance cover as income are obtained from different farm produces round the year.
7. Technology Infusion (R&D) integrated with indigenous/Traditional knowledge.
8. Mitigating energy crisis.
9. Climate change Programme from the perspective of adaptive & mitigation
10. Mitigating the wood, Fodder crisis, etc
11. Avoid degradation of land resources.
12. Provide opportunities for Agri-oriented industries, tourism and related tourism based activities, etc
13. Mitigating rural- urban exodus

Demand for animal food

Sustained economic growth, increasing urbanization, and a shift in diets in favour of high protein foods are fuelling the growth in demand for animal foods. Demand for animal food is more income elastic. India has a huge livestock population comprising of 290 million bovines, 170 million small ruminants, 13 million pigs, and 310 million poultry birds, producing 121.8 million ton of milk, 4.9 million ton of meat, and 63024 million no. of eggs (BAHS, 2012). Since 1970, there has been a consistent rise in the production of milk (4.7%) and egg (5.58 %) (BAHS, 2012). It is projected that by 2020, the demand for milk will be 131- 158 million ton, for meat 10-14 million ton, and for eggs 3.6 - 4.8 million ton in India. So integrated livestock based integrated farming can fulfill the requirement in current scenario. In a crop and livestock integration various sources flow is presented in fig.1

Factors linking crops and livestock in a crop and livestock integrated system

Integrated crop-livestock farming systems continue to dominate in most developing countries. The physical and financial stability of crop-livestock production systems arises from the complementary interactions between components of the production systems (Fig. 2.).

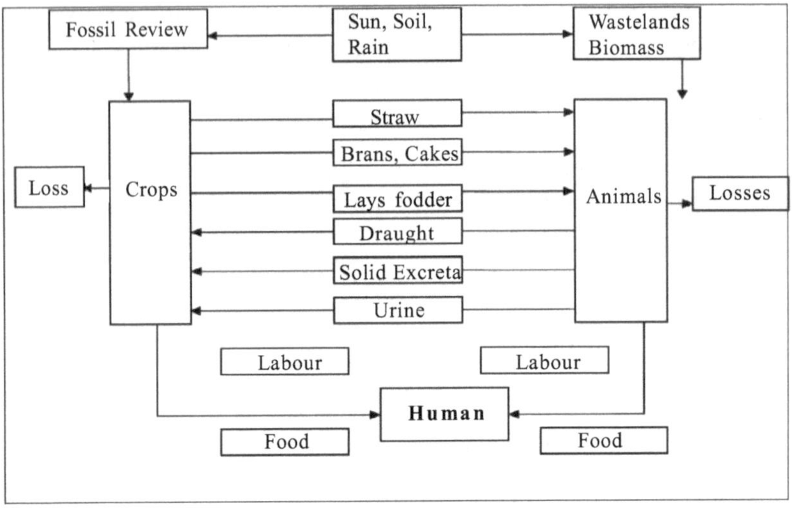

Fig. 2: Outline of different resource flows in crop livestock system (FAO, 2001)

Income linkages

Poor soil fertility and low and erratic rainfall remained as major limitations to crop production in a number of agro ecosystems (Pingali, 1993). When faced with such conditions, many households are diversifying into livestock with the intension of reducing risk providing insurance in the case of crop failure. Also, livestock are used as a source of liquidity and investment in the absence of savings and credit institutions. In relation to crop production, income from the sale of livestock can be used to improve crop production by providing investment capital needed to enhance productivity. In addition, income earned from livestock production may increase the demand, and hence profitability of food production (Hopkins and Reardon, 1993). Income obtained from the sale of livestock provides benefits to crop production both at household level and at macro level. At household level, income influences crop production directly by allowing households to invest on inputs such as fertilizer, hired labour, and carts. It also has an indirect influence by allowing poor households to improve the nutritional status, and thereby productivity of their own labour. At macro levels, when traded, livestock can serve as a source of export revenues, therefore they become a catalyst for economic growth. Such growth has an influence in stimulating the demand for locally produced food staples as well as other higher-valued non-staple products (Powell *et al.,* 2004). This provides an opportunity for farmers to benefit from overall economic growth. Finally, the income derived from livestock can provide the capital needed to initiate remunerative non-agricultural activities, which in turn provide the cash needed to finance crop production activities (Hopkins and Reardon, 1993).

Use of animal power in crop production

Animal power is used to assist farmers in the production, harvesting, processing, and marketing of crops. This reduces the amount of labour that is required by the farmers and helps to intensify production (Pearson *et al.*, 1998). Farmers owning draught animals tend to cultivate larger pieces of land and reap higher yields. This pattern is attributed to the ability of farmers with draught power to expand cultivated area due to labour-saving ability to cultivate in time (Sumberg and Gilbert, 1992). Thus, the use of animal power improves the timeliness of planting and therefore, increases yields in areas where growing seasons are short. Nonetheless, there is a danger when the use of animal power results in the cultivation of less suitable marginal land. Animals can also be used as a form of transport, usually together with a cart. Animal transport can reduce post harvest losses from pests by allowing timely removal of crops from the fields. Animal transport can also be used to move crop produce to the market, increasing the chances of selling crops at desired prices.

Nutrient cycling

The integration of livestock into cropping systems converts some crop residues into animal products such as meat and milk. Additional nutrients may also be introduced when animals are fed on concentrates and forages. Part of these feed nutrients return back to the fields in the form of excreta.

Crop residue

In crop-livestock farming systems, crop residues are used as livestock feed and sometimes provide income through their sale. Farmers use various methods to feed crop residues to their livestock, such as animals having open access to residues left on harvested fields, harvest and removal of stalks, storage of residue for feed or harvest of crop thinning from fields for selective feeding before harvest of main residue (Powell *et al.,* 2004).

Manure for crop production

The availability of manure for cropping is influenced by livestock types, numbers, the location of livestock during the time when manure is needed; and the efficiency of manure collection by the farmer. The chemical composition of dung produced from different species of animals are presented in Table.3. As argued by Powell *et al* (2004), the N and P content in manure of grazing cattle is up to three times greater during the wet season and crop residue–grazing period than during the dry season

Table 3: Chemical composition of dung produced from different species of animals

Type of animals	Quantity excreted (kg) /d	Water (%)	Nitrogen (%)	Phosphorus (%)	Potassium (%)
Cattle	15-20	82.40	0.30	0.18	0.18
Buffalo	15-20	81.10	0.26	0.18	0.17
Sheep/Goat	1.20	61.90	0.70	0.51	0.29
Poultry	0.11	54.70	1.46	1.17	0.62
Pig	5.0	80.70	0.51	0.46	0.43

Advantages of crop- dairy integrated farming system

In an integrated system, livestock and crops are produced within a coordinated system. Much of this coordination derives from the inherent interdependence that is exploited under integrated crop-livestock systems (Timsina, 1991). In such a system, waste products of one component serve as a resource for the other. If this system is well managed, it can lead to an increase in productivity on both crops and livestock. Animals transform plant energy into useful work for example animal power is used for ploughing and transport and also provide manure. On the other hand, crops provide a valuable, low-cost feed resource for animal production, and are the major source of nutrients for livestock in developing countries (IFAD, 2009; Powell *et al,* 2004). According to IFAD (2009) and Powell *et al.* (2004) the benefits of crop-livestock integration include:

- Retrieval and maintenance of soil productive capacity
- Product diversification and higher yields and quality at less cost.

It helps to increase profits by reducing production costs where farmers can use fertilizer from livestock operations, especially with high fertilizer prices. It provides diversification of income sources, guaranteeing a buffer against trade, price and climate fluctuations (Van Keulen and Schiere, 2004; IFAD, 2009).

- Reduction of crop pests (less pesticide use and better soil erosion control). It helps to improve and conserve the productive capacity of soils, with physical, chemical and biological soil recuperation. It results in greater soil water storage capacity, mainly because of biological aeration and the increase in the level of organic matter.
- Social, through the reduction of rural-urban migration and the creation of new job opportunities in rural areas (IFAD, 2009).

Constraints of crop and livestock integration

According to IFAD (2009), the main drawbacks associated with integrated farming systems include.

- Nutritional values of crop residues are generally low in digestibility and protein content.
- Crop residues are primarily soil regenerators, but most of the time they are either disregarded or misapplied.
- Intensive recycling can cause nutrient losses.
- If manure nutrient use efficiencies are not improved or properly applied, the import of nutrients from feed can not be optimised.
- Farmers prefer to use chemical fertilizer instead of manure because it acts faster and easier to use.
- Resource investments are required to improve intake and digestibility of crop residues.

Table 4: Opportunities and constraints related to crop-livestock integration

Opportunities for crop–livestock integration	Constraints that militate against fostering crop– livestock
Livestock enterprises are more lucrative than crop farming so it is advantageous to integrate livestock into farm activities	Competition for resources such as land, labour, capital, management, and water by the crop and livestock sectors
Intensification of agriculture which is currently occurring in most farming systems favours crop–livestock integration.	Land use and tenure policies that inhibit livestock mobility and limit farmers' access to manure and livestock access to feed
Poor soil fertility, unavailability or increases in prices of fertilizers, and labour shortages, have forced farmers to rely on manure and traction	Since manure is bulky and is required in large quantities, high labour and transportation costs may be involved
Many indigenous, emerging, and developed technologies are available to support sustainable crop–livestock integration. These include improved cereal and grain legume varieties, cropping systems, weed and nutrient management strategies, the eradication of most livestock diseases, and the development of modeling and all-year-round feed packages for animals	Smallholder farmers are reluctant to grow improved forages; this is related to the whole systems approach

Contd.

Opportunities for crop–livestock integration	Constraints that militate against fostering crop– livestock
There is scope for improving the efficiency of the integration by diversifying the use of animals. For instance, the use of cows for traction will also provide milk and manure. Farmers can also crop in the wet season and engage in livestock enterprises in the dry season.	Lack of research on holistic approaches; this requires in depth knowledge of integrated crop–livestock systems. Wrong targeting of crop–livestock integrated systems.

Role of technology to improve dairy based integrated farming system

The sustainability of number-driven growth in livestock output would be constrained by declining per capita land availability and feed and fodder scarcity, implying higher prices. For livestock production to be competitive, growth in production must result from yield-increase and/or cost reducing technologies. Numerous technologies are available but not much progress has been made in their application on farmers' fields. Reproduction technologies such as artificial insemination of nondescript indigenous animals with semen of exotic high yielding breeds is widely acknowledged for its contribution to improve animal productivity. However, its adoption has remained limited and sporadic. Currently, only 10% of female cattle, 5% of sheep, and 15% of pigs are crossbred. Considerable progress has however been made in the poultry sector.

An impressive growth in rice and wheat production has augmented the availability of crop residues. The availability of straw from coarse cereals has remained stagnant due to a decline in area, resulting in a rise in their real prices. Dry matter availability from grazing lands and common property resources has declined due to encroachment, overgrazing, and lack of management. Cultivated forages account for only 4% of the total cropped area. Though the availability of concentrate feed has improved, its use is limited to large landholders and commercial peri-urban systems. Improved animal nutrition technologies such as urea treatment of straws, urea molasses mineral blocks, and bypass protein have not gained wide acceptance. Diseases pose a major constraint to improve livestock productivity. Foot and mouth disease (FMD), hemorrhagic septicemia, black quarter, etc. occur frequently, causing enormous economic losses. Though there has been an uncontrolled growth in veterinary infrastructure in terms of hardware, manpower and animal coverage, but the delivery system remains poor. There are about 7500 livestock units per veterinarian (Birthal, 2002). Technology is the key for the future growth of mixed crop livestock systems. The livestock sector offers several technologies, but they have yet to gain wider acceptance. For instance, crossbreeding technology in India has largely targeted cattle and despite better performance in select pockets, its wide spread adoption is constrained for various reasons:

- Mechanization of agriculture diminished the utility of male cattle as a source of draught power.
- Cattle being sacred to the local population, its slaughtering is banned creating problems of disposal of surplus male cattle.
- There has been a considerable decline in the performance of crossbreds due to inadequate institutional support, particularly the delivery of breeding and health services.
- The availability of feed and fodder is inadequate. The option of improving the productivity of indigenous cattle and buffalo through better feeding and management could be explored. There is also considerable scope to improve the productivity of short-generation animals such as sheep and goat through crossbreeding. However, lack of institutional support and deteriorating grazing lands discourage small scale producers from rearing crossbred animals. These issues need to be addressed through research and policy interventions.

A huge livestock population not commensurate with available feed resources is an impediment to improve productivity. Options such as optimizing livestock numbers to match available feed resources and/or improving feed availability through breeding of dual-purpose crop varieties with better digestibility coefficient, improving the cost-effectiveness of existing nutrition technologies, and bringing more land under fodder crops need to be explored. Conservation and management of common grazing lands through legislative means (protection from encroachments and ban on acquisition and redistribution of common lands under rural development programmes) and enhancing community participation in the management of common lands would help improve poor households' access to grazing resources at little cost. Optimizing livestock population does not seem a practical step in the short run given the low local demand for cattle and buffalo meat, the socio political environment, and the restrictive trade barriers related to sanitary and phyto sanitary norms. On the policy front, there is a need to arrest the diminishing synergy between crops and livestock. Policy-induced price distortions for food crops in both input and output markets have acted as deterrents to crop diversification. The livestock sector receives meager policy support in terms of investment and subsidies. The authors roughly estimate that public spending on this sector comprises just about 5% of the value of its output. There is a need to improve the delivery system of livestock services by redefining the role of veterinarians as agents of technology transfer, and improve the supply of medicines and equipment. Disease prevention should get precedence over curative treatments. Vertical integration through cooperatives has proved successful in improving dairy farmers' access to markets. The model has, however, not always succeeded due to lack of investment, mismanagement, and politicization of societies. Existing cooperative legislation is restrictive, and continues to inhibit the potential and growth of the cooperative movement.

Livestock and the environment

Mixed crop-livestock systems are environment friendly. By recycling huge amounts of crop residues, they make land available for growing food crops. These systems provide dung as manure and domestic fuel, in addition to draught power, and protect the environment from an overuse of chemical fertilizers and petro-fuels. Disruptions in crop-livestock linkages could cause severe damage to agro ecological systems and consequently to agricultural productivity. Kumar *and* Deogarh (2003) observed the degradation of fertile soils in the rice-wheat-dominated Indo- Gangetic plain of India due to excessive use of chemical fertilizers and overexploitation of groundwater, accompanied by a decline in the use of dung manure.

Economics of dairy based integrated farming system

In a dairy based integrated farming system income sources are mainly from crop, milk, manure *etc*. An income and expenditure from a small dairy based integrated farming system is presented in Table 5.

Table 5: Economics of milk production in dairy based integrated farming system

Breed	Jersey 5 (3 Lactating+2 dry)
Dry Fodder	18kg/ day
Green Fodder	75 kg/day
Concentrates	15 kg/day
Average Milk Yield	9,000 lit/year (3,000 lit /animal/ year)
Milk Production Cost	Rs.15/litre
Milk Sales Price	Rs.25 /litre
Gross income	Rs.2,25,000/year
Expenditure	Rs.1,35,000/year
Net income	Rs.90,000/year

Integrated crop-livestock-fish farming systems

The objectives of an integrated farming system incorporating aquaculture are to provide green manure for the cropping subsystem, pond fertilization and fish feed supplementation. In the parts of tropical Asia where rainfall exceeds 1,000 to 1,200mm/year, cropping systems are usually based on monocrop of rice. Rice is grown at the peak of the rains because rice is the only crop that tolerates flooding. However, it may be possible to plant upland crops such as maize, mung bean, cowpeas, and sweet potato in mixed cropping systems at the end of the rains to utilize residual moisture (Beets, 1982).

There is also the question of whether the crop is grown primarily for human, livestock or fish feed. Fodder crops, particularly legumes could be introduced into existing cropping systems as intercrops, relay crops, or sequential crops without upsetting the regular cropping system (Javier, 1978). Aquatic grasses could be grown in rice fields as a fodder crop as an alternative to rice. There are several other possibilities with upland crops. More maize than normal could be sown and the excess thinned-out for fodder. Fast growing legumes such as soybean, mung bean and pigeon pea could be intercropped with maize. It is possible to sow annual *Stylosanthes* on rice field to provide seed which can be broadcast into the stubble after harvest and grazed with volunteer weeds (Perkins *et al.,* 1986). There is a vast array of trees and shrubs that serve as animal fodder, many of which fix nitrogen; for example *Leucaena leucocephala* (Brewbaker, 1986). The greatest potential for integrated farming systems with fish probably lies with mixed farms having crop and livestock subsystems because livestock manure is a useful pond input. Farms that have only livestock fall into two distinct categories:

1. Grazing farming systems in semi-arid and arid areas, with little potential for aquaculture because of water constraints.
2. Feedlots: Intensive feedlot livestock farming may be conveniently integrated with fish but it is a capital-intensive operation because of the purchase of feed and has limited relevance for small-scale farmers.

Most potential feed resources on the farm are in the form of crop residues such as rice straw and maize stover which have low digestibility (Javier, 1978). These are probably best to feed ruminants (rather than used directly as fishpond inputs) and the ruminant manure can be used as a pond input (Javier, 1978).

The highest fish yields from integrated farming systems have been reported from ponds receiving feedlot livestock manure. The livestock had received high quality feed and their manure therefore had a high nutrient content. Since livestock are a key component in a productive integrated farming system with fish, a major research effort is required to develop technology for increasing the quality and quantity of livestock feed produced on small-scale farms so as to increase livestock production directly and fish production indirectly in integrated farming systems. Ruminants can process fodder which is indigestible in human but research is needed to identify strategies to upgrade the quality of their manure as a pond input because ruminants grazed on rough pasture and/or stover have manure with a low nutrient content. Small ruminants (sheep and goats) are normally considered to be animals of arid areas but they are important in certain areas in the humid tropics. Small ruminants are sometimes stall-fed using the "cut and carry" system. Free ranging ruminants, both large and small, are often paddocked at night which also facilitates manure collection as a pond input. The collection of nitrogen-rich urine (as well as manure) as a pond input has important potential. For stall-fed ruminants, this resource is usually wasted, but trials are now beginning on the use of cattle urine as a pond input in India. A possible on-farm interaction between the various subsystems in a crop-livestock-fish integrated farming system has been given below (Fig. 3).

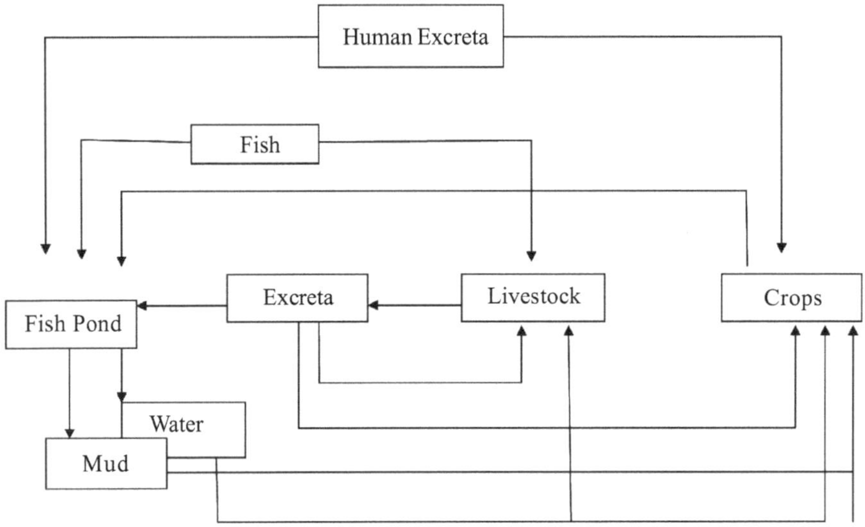

Fig. 3: Scheme of possible on farm interactions between various subsystems in a crop-livestock-fish integrated system

Goat based integrated livestock farming

Goats are reared all around the world for a variety of reasons, and across a variety of climates. Intensive, high-yielding dairy goats are mostly a feature of Europe and North America. In Afghanistan cashmere goats are kept on rangelands for meat and fibre. In Africa, subsistence farmers keep dual-purpose goats on smallholdings for milk and meat. Here the Savannah goats cope with the heat & well protected by their black skin. Rangeland goats are found in Australia, exported for meat in significant volumes, while in New Zealand feral goats are run alongside sheep and beef herds to manage weeds and scrub in pasture (Pasha, 1991). The goat is a versatile animal. It is known as the poor man's cow. Goats can be kept with little expense on undulating lands with an inexpensive shelter. In addition to good quality chevon and milk, shoes, clothes and bags were prepared from their hide and fibre. It is very much helpful in integrated farming system as follows:

Better pasture quality

Research shows that grazing goats encourages increased clover percentage in the long term. Grazing goats on pasture before lambs also grooms the pasture, setting up better quality pasture and allowing better growth rates in lambs (Kumar and Jain, 2002).

Cheaper weed control

One of the roles of 'meat' goats is the control of weeds such as blackberry, gorse, broom and thistles. Grazing by goats can control many weeds as they are capable of browsing on and controlling many spiny and noxious weeds. Under rainfed upland conditions, integrating goat rearing, especially for grazing during the off-season complimented weed control by 45% and addition of goat manure to fields grazed upon by goats contributed for 31% weed control during the cropping season (Kathiresan, 2009).

Monitoring shows running goats helps boost other stock

Goats are useful to boost other stocks. They bring in an income from meat. This is despite eating a diet of weeds and poor quality pasture. Secondly, they improve pasture quality for other stock, particularly sheep. Thirdly, they save money for weed control costs.

Goats as 'sustainable' livestock

Rearing of goats is a low-input and low-impact enterprise compare to other livestock. They are more sustainable than other livestock species due to production of high-quality protein rich meat. Goat causes less environmental degradation by producing lower ammonia than other livestock. They are largely kept extensively around the world in low-input systems, with a comparatively low expenditure. Goats also serve as a livelihood for people in marginal areas, where you can't farm anything else.

Crop- Goat integration

The goat production is concentrated to smallholder farmers in India. The goat manure is applied to the crops as a supplemental fertiliser. One of the factors, which limits goat production, is inadequate supply of good quality feeds during the whole year, especially in the dry season. The boundaries of the farms should be planted with different leguminous shrubs and trees, such as subabool (*Leucaena leucocephala*), babool (*Acacia arabica*), neem (*Azadirachta indica*), ber (*Ziziphus mauritiana*), tamarind (*Tamarindus indica*) and pipal (*Ficus religiosa*) *etc.* The green leaves and dried form of the plants were very much relished by goats. The goat sheds were constructed with slatted floors so that the manure could be collected from beneath the sheds.

Integrated Fish with Goats

Goat farming is an age-old practice but its integration with fish culture has not been explored. Goats not only provide meat and milk but also a good amount of manure. The production of manure from a goat is around 1.5 - 2 tonnes per year. If animal manure is not properly used, it causes pollution of water and environment. It has been observed that 40-50 kg of animal manure produce 1 kg of fish (Kumar and Ayyapan, 1998) . Animal manure and green fodder can totally replace the commercial feed for fish farming achieving a similar fish production (Fig. 4 (a) and 4 (b)).

Sheep based integrated farming system

Raising of small ruminants is now getting popular among smallholders due to the relatively lower capital investment and risk as compared to cattle. Sheep is small in size and can be easily attended by women and children. They have early maturity, shorter reproductive cycle and can be fed with different forages including weeds. All farmer-co-operators practise semi confinement system where they allow the sheep on roadsides, rice fields, fallow lands and even in park in the morning and get them back in the afternoon (Walsh and Feinstein, 2003). During the rice growing period, farmers feed the sheep with rice straw and weeds from rice field. These weeds include *Echinocloa crusgalli*, *Cyperus rotundus*, *Oplisimus composites, Ipomea aquatica* and *Brachiaria mutica.* Various benefits obtained from sheep based integrated farming systems are:

- In rice-based farming system, raising of large ruminants is not commonly practiced as feed supply is limited and fear of crop destruction and source of stock pose problems. Nevertheless, sheep production has great potential as mutton is becoming a popular dish in India
- Raising sheep in a lowland rainfed farming system requires very less time.
- Saves the cost of labour in weeding the rice paddies and bunds during the rice cropping period.
- Sheep are less destructive to crops than goats.

Overall benefits of livestock based integrated farming system

- It improves and conserves the productive capacities of soils, with physical, chemical and biological soil recuperation. Animals play an important role in harvesting and relocating nutrients, significantly improving soil fertility and crop yields.

- It reduces crop pests
- It requires less pesticide.
- It reduces rural to urban migration by creating new job opportunities in rural areas.
- It is quick, efficient and economically viable because grain crops can be produced in four to six months, and pasture formation after cropping is rapid and inexpensive.
- It helps to increase profits by reducing production costs.
- It results in greater soil water storage capacity, mainly because of biological aeration and increase the level of organic matter.
- It provides diversified income sources, guaranteeing a buffer against trade, price and climate fluctuations

One key advantage of crop-livestock production systems is that livestock can be fed on crop residues and other products that would otherwise pose a major waste disposal problem. For example, livestock can be fed on straw, damaged fruits, grains and household wastes. Integration of livestock and crop allow nutrients to be recycled more effectively on the farm. Manure itself is a valuable fertilizer containing nitrogen, phosphorus and potassium. Adding manure to the soil not only fertilizes, it but also improved its structures and water retention capacity. In Colombia sheep are sometimes used to control weeds in sugarcane. Draught animal power is widely used for cultivation, transportation, water lifting and powering food processing equipment. Using draught animal reduces the need for foreign exchange to buy expensive tractors and fuel. Cow dung is highly valued for being used for cooking in many countries.

Challenges of integrated farming system

- Develop strategies and promote crop livestock synergies and interactions that aim to
  - (a) Integrate crops and livestock effectively with careful land use;
  - (b) Raise the productivity of specific mixed crop-livestock systems;
  - (c) Facilitate expansion of food production; and
  - (d) Simultaneously safeguard the environment with prudent and efficient use of natural resources.
- Devise measures (for instance, facilitating large-scale dissemination of bio-digesters) to implement a more efficient use of biomass, reducing

pressures on natural resources; and develop a sustainable livestock manure management system to control nutrient losses and environment pollution.

Opportunities of livestock based integrated farming system

- Intensification of agriculture which is currently occurring in most farming systems favours crop– livestock integration.
- Poor soil fertility, unavailability or increase in prices of fertilizers, and labour shortages, have forced farmers to rely on alternatives such as manure and traction.
- Farmers can grow crop in the wet season and engage in livestock enterprises in the dry season.
- Livestock enterprises are more lucrative than crop farming so it is advantageous to integrate livestock into farm activities.
- Many indigenous, emerging, and developed technologies are available to support sustainable crop–livestock integration. These include improved cereal and grain legume varieties, cropping systems, weed and nutrient management strategies, the eradication of most livestock diseases, and the development of modelling and all-year-round feed packages for animals.

Conclusion

India has a considerable livestock, poultry population and crop wastes. All efforts have to be mobilised to reclaim the resources and to put them to use effectively. Suitable technology has to be developed for the treatment of wastes and their all round effective utilisation, so that, it can help into reducing the poverty and malnutrition and strengthen environmental sustainability. The increase in demand for livestock products presents opportunities for small farmers who can increase livestock production and benefit from related income. However, in terms of environmental impact, the growing number of livestock and the increase in livestock processing can have a negative impact on natural resources unless actions are taken to identify farming practices that are economically and ecologically sustainable. The highly improved integrated crop-livestock system can guarantee more sustainable production and therefore constitutes a valid new approach.

Fig. 4(a). Goat based integrated farming system

Fig. 4(b). Goat and fish integrated farming system

Reference

BAHS, Basic Animal Husbandry Statistics. 2012. Government of India, Ministry Of Agriculture, Department Of Animal Husbandry, Dairying And Fishries, Krishi Bhawan, New Delhi.

Beets, W.C. 1982. Multiple cropping and total farming systems. Gower publishing Company Limited.156p

Brewbaker,J.L.1986 .Leguminous trees and shrubs for South-East Asia and South Pacific, p 43-50.In G.J.Blair, D.A.Ivory, and T.R.Evans(eds.)Forages in South-East Asian and South Pacific agriculture. ACIAR Proceedings No. 12. Australian Centre for International Agriculture Research, Canberra, Australia.

Devendra, C., Thomas, D., Jabbar, M.A., Kudo, H., 1997. Improvement of Livestock Production in rainfed Agro-ecological Zones of South-East Asia. ILRI, Nairobi, Kenya

FAO, 2001. *Farming Systems and Poverty: Improving Farmers' livelihoods in a changing World.* Food and Agriculture organization of the United Nations, Rome pp 412.

FAO, 2001. Food and Agriculture Organization of the United Nations. Mixed Crop-Livestock Farming: A Review of Traditional Technologies based on Literature and Field Experience. Animal Production and Health Papers 152. Rome: FAO.

Hopkins, J., & Reardon, T. 1993. Agricultural price policy reform impacts and food aid targeting in Niger. Int. Food Policy Res. Inst. Rep. to USAID/Niger; Grant no. 683–0263- G-IN-9022–00; NEPRP 683–0263. Int. Food Policy Res. Inst., Washington, DC

IFAD, 2009. Integrated Crop-Livestock Farming Systems, Communities of Practice for pro-poor livestock and Fisheries/aquaculture development, Rome.

Javier, E.Q. 1978. Integration of fodder production with intensive cropping systems in Southeast Asia, p. 19-36. In Feeding stuffs for livestock in South East Asia. Malaysian Society of Animal Production, Serdang, Selangor, Malaysia.

Jayanthi, C. 2002. Sustainable farming system and lowland farming of Tamil Nadu. IFS Adhoc Scheme. Completion Report.

Kathiresan, R.M. 2009. Integrated farm management for linking environment. *Indian Journal of Agronomy*, 54(1).

Kumar, K. and Ayyapan, S. 1998. Integrated Aquaculture in Eastern India, Integrated Aquaculture Research Planning Workshop, Purulia, India.

Kumar, S and Deoghare, P.R. 2003. Goat production system and livelihood security of rural landless households. *Indian Journal of Small Ruminants*, 9 (1): 19-24

Kumar,S. and Jain, D.K. 2002. Interactions and changes in farming systems in semiarid parts of India: Some issues in sustainability. *Agricultural Economics Research Review*, 15 (2): 217-230.

Mishra, A.K. 2002. Integration of livestock in land use diversification. (In): Summer school on "Land Use Diversification in rainfed Agro-Ecosystem during 15 April to 5 May, 2002, Central Research Institute for Dryland Agriculture, CRIDA.

Pasha, A.S. 1991. Sustainability and viability of small and marginal farmers: Animal husbandry and common property resources. *Economic and Political Weekly*,26 (13): A27-A30.

Pearson, R.A., Nengomasha, E.M. & Krecek, R.C. 1998. The challenges in using donkeys for work in Africa. p. 190–198. *In* P. Starkey and P. Kaumbutho (ed.) Meeting the 102 challenges of animal traction. Proc. ATNESA Workshop, Ngong Hills, Kenya. 4–8 Dec.1995. Intermediate Technol. Publ., London.

Perkins, J., R.J. Petheram, R. Rachman and A. Semali. 1986. Introduction and management prospects for forages in Southeast Asia and the South Pacific, p. 15-23. *In* G.J. Blair, D.A. Ivory and T.R. Evans (eds.) Forages in Southeast Asian and South Pacific agriculture. ACIAR Proceedings No. 12. Australian Centre for International Agricultural Research, Canberra, Australia.

Pingali, P.L. 1993. Crop–livestock systems for tomorrow's Asia: From integration to specialization. Proc. Int. Workshop on Crop–Livestock Interactions, Khan Kane, Thailand. 27 Sept.–1 Oct. 1993. IRRI, Los Banos, the Phillipines.

Powell, J.M., Pearson, R.A & Hiernaux, P.H. 2004. Crop-Livestock Interactions in the West African Dry lands. *Agronomy Journal*, 96: 469-483.

Sere, C. and Steinfeld, H. 1996. World Livestock Production Systems. Current status, issues and trends. FAO Animal Production and Health Paper No. 127. FAO, Rome.

Sumberg, J. and Gilbert. E. 1992. Agricultural mechanisation in the Gambia: Drought, donkeys and minimum tillage. *Afr. Livest. Res.* 1:1–10.

Timsina, J., Singh, S.B. and Timsina, D. 1991. Integration of crop, animal and tree in rice-based farming systems of hills and Terai of Nepal: some successful cases. Proceeding of Crop-livestock integration workshop, 1991. Asian Rice Farming Systems Network, IRRI, Los Banos, Philippines.

USDA. 2010. Slaughter Availability to Small Livestock and Poultry Producers- Maps. United States Department of Agriculture Food Safety and Inspection Service, USA

Vaidyanathan, A., 1988. Bovine Economy in India. Oxford and IBH Publishing Co. Pvt. Ltd, New Delhi, India.

Van Keulen, H. and Schiere. H. 2004. Crop-livestock systems: Old wine in new bottles? In: 4th International Crop Science Congress: 'New directions for a diverse planet'. 26 Sep - 1 Oct 2004, Brisbane, Australia

Walsh, M. and Feinstein, M. H. 2003. Grazing preferences of hill sheep in relation to physiography, soils and semi-natural vegetation using field observations and satellite tracking. *Agricultural Research Forum*, p26.

Integrated Farming System Practices: Challenges & Opportunities, pp 183-195

New India Publishing Agency, New Delhi, India

7

Poultry Rearing: A profitable Enterprise in Integrated Farming Systems

N. Panda and *S. Nanda

Introduction

Poultry farming is an emerging activity for enhancing nutrition and providing livelihood support to the farmers of this country. Since two decades, this component emerge as the fastest growing segment of agriculture where the broiler segment having a growth rate of more than 10% and the egg production 5-6% per year. As per the report by USDA's Agricultural Marketing Service, India has one of the world's largest and fastest growing poultry, ranked third in egg production (FAOSTAT) and sixth in broiler meat production. Poultry farming is lucrative because there is minimum investment to start, rapid return, high feed conversion ratio and provides continuous source of income for the farm.

Review of literature

Integration of chicken with fish farming might be an economically viable and productive system for both rural farmers and commercial entrepreneurs. Poultry manure is a complete fertilizer, with the characteristics of both organic as well as inorganic fertilizers (Banerjee *et al.,* 1979) and fresh chicken manure contains

College of Veterinary Science & Animal Husbandry, Orissa University of Agriculture & Technology, Bhubaneswar - 751003

*National Diary Development Board, State Office-Bhubaneswar, Odisha

1.6% nitrogen, 1.5% phosphorous and 0.9% potassium (Woynarovich, 1979). This excreta is an excellent feed for fish due to its high content of soluble organic salts, more nitrogen and phosphorous as compared to other livestock manure. It has been reported that one kg of fish can be produced by using about 17 kg of chicken manure (Fang *et al.,* 1986). Hoq *et.al.* (1999) conducted an experiment in the Bangladesh Fisheries Research Institute comparing the growth of fish getting direct droppings from poultry sheds over the pond. After 12 months of culture, a significant higher production of 4,290 kg/ha was observed from ponds receiving direct poultry dropping, whereas, 3,365 kg/ha production was resulted from manure treated ponds. However, freshwater prawn survival was better in the ponds of manure application than the direct dropping.

In another experiment in Bangladesh Alam *et al.* (2009), reported that integrated farming with poultry, fish and crops can play a significant role in increasing manifold production, income, nutrition and employment opportunities of rural populations. The research work was carried out at nine farmers' ponds covering an area of 0.12 ha each in Farming Systems Research and Development (FSRD) site, taking three different components like fish, poultry and vegetables production. The results indicated that in integrated pond management, two additional enterprises viz., poultry and year round vegetables exhibited encouraging production over traditional management. The average fish production obtained from integrated pond management was more than 3 times higher over the traditional management. The total economic return in terms of gross return and gross margin achieved from integrated pond management was 1297.94 and 1496.53% higher over traditional pond management respectively. The utilization of family labour round the year in pond based integrated production system contributed to improve the production as well as to create employment opportunity for income generation.

Utilization of family labour in pond based production system was found encouraging during the study period. The utilization of male labour was 100, 90, 90, 85, 75 and 70% in case of fish harvesting and marketing, fish feeding, pond preparation and releasing of fingerlings, trellis preparation, vegetables harvesting and marketing and marketing of poultry produce respectively (Table 2). It was observed that among different activities, the participation of female members were encouraging. Female members were participated in almost all activities except poultry shade preparation and fish harvesting and marketing.

Table 1: Utilization of family labour (%) under integrated farm activities

Activities	Labour Utilization (%)		
	Male	Female	Hired
Poultry shed preparation	60	0	40
Poultry feeding care	10	90	0
Poultry product marketing	70	30	0
Pond preparation and realising of fingerlings	90	10	0
Fish feeding	90	10	0
Fish harvesting and marketing	100	0	0

Gangaware *et. al.*(2013) has made IFS approach with livestock-fish integration in Bihar to utilize the livestock farm wastes and conversion of waste into valuable fish protein. The spilled over feed or feed derived from livestock manure may be utilized as direct feed or the manure from livestock helps in production of planktons which form the feed for fishes in the pond. The livestock-fish farming may be extensive, intensive or semi-intensive system depending upon the availability of resources and capital. The objective of integrated livestock cum fish farming is to produce maximum plankton in water through manuring which is rich in protein and a natural feed for fishes. The species of fishes which are consumed by the people are efficient utilizer of phyto and zooplankton. The macrophytic feeding nature of fishes is excellent for integrated livestock fish culture.

Duck and fish farming

At present combination of duck with fish farming is considered as a means of reducing the cost of feed for ducks and a convenient and inexpensive way of fertilizing ponds for the production of fish (Pillay, 1990). In this integrated system, ponds provide living and foraging areas for the ducks and fish.

Ducks are reared in shelters built on the banks of the ponds or constructed over the ponds on silts, or sometimes built on floating platforms. They are known to eliminate almost all the snails in ponds in depths up to 30–40 cm, thus controlling the immediate host of bilharziasis. The duck strains mostly Khaki Campbell, Indian Runner Muscovy and White Pekins are the preferred breeds in India for Integrated farming (Fig. 1.7.). Ducks are reared at different densities, depending on the climatic conditions, the method of raising (extensive or intensive), water quality, and other factors. Around 750-1000 ducks are stocked per ha. Demonstration trials conducted in India in polyculture of Indian and common carps (at a stocking density of 6,340 fingerlings/ha) raised with ducks (100/ha) have yielded 4,323 kg of fish/ha/year, 250 kg of ducks and 1,835 eggs (Jhingran and Sharma, 1980). Raising 1000–2000 ducks/ha on ponds in Vietnam, increased the average fish yield to 5 t/ha/year compared to 1 t/ha/year without ducks (Delmendo, 1980).

Either broiler or layer birds can be raised in the integrated system (Fig. 3, 4, 6). For one hectare fish pond 500 to 600 birds and on an average 60 kg poultry manure is required per day. Duck-fish farming system is mostly practiced in Assam, West Bengal, Bihar, Odisha, Andhra Pradesh and Kerala. Duck droppings directly fall in water or collected and used for fertilization in pond. Fish gather duck droppings as direct food or consume spilled feed. Ducks consume mosquito larvae, tadpoles, dragon fly larvae and snails which also serve as vector for certain parasites. The dabbling habit of ducks increases the available oxygen in pond water. For commercial farming or for maximum profit high egg producing ducks like Khaki Campbell or Indian Runner is preferred instead of local ducks. An average of 250 ducks/ ha is recommended for duck cum fish farming.

Rama Rao *et al.*(2006) conducted an integrated research suitable for a tribal region in Chhattisgarh taking 6 combinations with crop (grains, fodder) and livestock (cow, buffalo, bullock, goat poultry and duck). A model having 2 bullocks + 1 cow + 1 buffalo + 10 goats + 10 poultry + 10 ducks along with crop cultivation was the best with a net income of Rs 33,076 per year against crop farming alone (7,843 per year). The cost returns of 1: 2.24 and employment generation of 316 days was observed in the mixed farming model.

Gangwar *et.al.* (2013) analyzed socio-economic impact of poultry based farming system on the farmers of the hills of Kumaon region for their livelihood security and women empowerment. The analysis is based on the data collected from 95 poultry farmers selected from three hill districts of Kumaon region for two production years, 2011-12 and 2012-13. The economics of prevailing poultry production systems has been worked out and it has been found that cost on rearing of chicks up to 3-5 weeks is nearly Rs. 46 per chick and a small unit of 10-15 birds in backyard poultry gives a net income of Rs. 11470/ annum. The chicken broilers could be reared successfully by farmers in the remote hills villages. The feed cost has a lion's share (72%), followed by cost of chicks in the total cost. The study has revealed that poultry could be successfully reared in backyard as well as intensive broiler farming. The adoption of integrated poultry-fish farming provides fetch additional income of Rs. 4,000-5,000 and employment opportunities for 45-50 human days. Additionally, the consumption of eggs/fish and meat adds to food quality and livelihood security of the resource-poor family. High cost of feed and chicks have been identified as the major constraints of integrated poultry farming.

In the NAIP component- 3 livelihood project operated under OUAT an integrated farming model was tested for three years (2011-13) with crop, vegetables, mushroom, poultry and pisciculture taking 20 farmers in three districts of Odisha like Dhenkanal, Phulbani and Kalahandi (Panda *et al.,*2013) Fig. 2, 5, 8). The integrated model was divided into two categories taking 0.8 ha and 1.6 ha of land (Table 2 and Table 3). The net return in the 0.8 ha model was Rs. 1,37,907 compared to Rs.12, 739 in the conventional method and total 7man days created was 555 vs. 204 (Behera *et al.,*2014). Similarly in the IFS model 1.6 ha of land holding the net return was 1,98, 968 vs. 17,052 and the employment generated in man days was 899 vs.400 in the conventional method.

Table 2: Performance of IFS model of 0.8 ha of land

Enterprise	Component yield (REY in t)	Gross return (Rs)	Net return (Rs)	Employment generation (man days)
0.8 ha IFS model				
Crop (Rice-onion) (7110 m^2)	11.96	1,33,964	70,013	320
Mushroom(45 m^2)	10.31	1,14,441	50,755	141
Poultry (45 m^2)	5.20	57,753	12,222	58
Pisciculture (800 m^2)	1.68	18,680	4,917	36
Total system yield	29.15	3,24,838	1,37,907	555
Conventional cropping (rice- green gram) (8000 m^2)	4.05	44,999	12,739	204

Table 3: Performance of IFS model of 1.6 ha of land

Enterprise	Component yield (REY in t)	Gross return (Rs)	Net return (Rs)	Employment generation (man days)
1.6 ha IFS model				
Crop (Rice-onion) (14310m^2)	22.87	2,56,061	1,29,006	633
Mushroom(45 m^2)	9.95	1,10,386	46,700	141
Poultry (45 m^2)	5.13	56,897	10,834	58
Pisciculture (1600 m^2)	3.50	38,855	11,628	67
Total system yield	41.45	4,62,199	1,98,168	899
Conventional cropping (rice-green gram) (16000 m^2)	7.21	79,932	17,052	400

Channabasevanna and Biradar (2007) have tested different rice-fish-poultry models in Karnataka during 2004-05 and 2005-06 and found that integrated farming system recorded higher system productivity (15,555 kg/ha/year) and net returns (Rs. 48,603/ha/year), over conventional rice-rice system (6667 kg/ ha/year and Rs. 21,599/ha/year, respectively). The productivity per day was

Fig. 1: Pond based Duck rearing

Fig. 2: Mushroom in IFS model

Fig. 3: Broiler bid rearing in IFS

Fig. 4: Coloured bird rearing

Fig. 5: Pond based crop/Vegetables

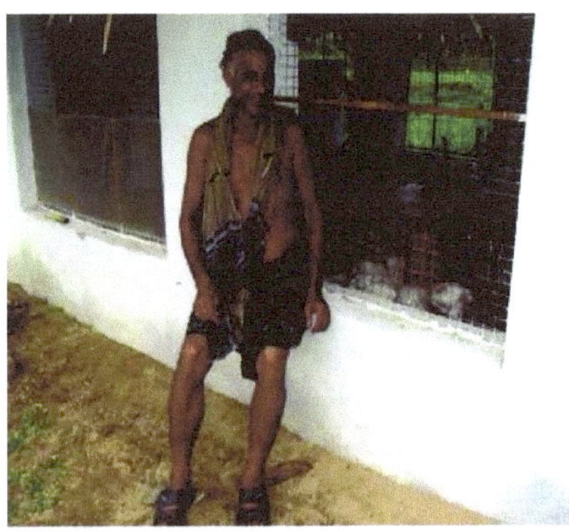

Fig. 6: Broiler rearing near the pond

Fig. 7: Duck farming in IFS

Fig. 8: Fishing from the IFS pond

2.3 folds higher (42.6 kg/ha/day) in IFS over conventional system (18.2 kg/ha/day). The IFS also recorded the highest water use productivity (43.2 kg/ha.cm) and labour use efficiency (25.17 kg/ha/labour). Among different models, Rice-fish (pit at the center of the field) – poultry (reared separately) recorded maximum total productivity in terms of rice grain equivalent yield (17502 kg/ha/year) and net returns (Rs. 62977 /ha/year) and B:C ratio (1.91).

Methodology

Breeds suitable for integrated farming

In an integrated farming system we can go both for meat and for egg producing chicken. Specially developed strains are now available with the traits of quick growth and high feed conversion efficiency. For broiler production commercial strains like Vencobb, Hubbard, Hypeco, Lohmann, Ross etc. and for layer farming the popular strains like BV 300, Bovan, Hyline, Key stone etc. The Coloured chicken varieties like Gramapriya, Vanaraja, OUAT synthetic strain etc have been developed by poultry institutes of ICAR and some agricultural universities are more suitable for integrated farming for promoting family poultry production in rural and tribal areas (Mandal *et al.,* 2004, Arya *et al.,*2009). Gramapriya variety is used for production of eggs where as Vanaraja and Giriraja are used both for production of egg and meat. OUAT synthetic breed is suitable for meat purpose and can be reared in both intensive and semi intensive condition. Gramapriya layers lay about 150 eggs annually where as Vanaraja birds lay about 120-130 eggs annually. Some other varieties like Girirani, Krishnaj, Gramalakhami and Cari Gold have been evolved in different SAUs and ICAR institutes.

Housing and management

The commercial birds are reared intensively with proper housing and feeding. But the coloured birds can be either reared intensively or semi intensively (Doley *et al*., 2009). In rural poultry system housing is provided only during night and adverse weather conditions. Locally available materials are used for construction of houses. Housing with cement concrete roof and concrete floor can be done. Many farmers use country tiles as roof materials. In the fish- poultry integrated system the house can be prepared on the bond or over the water. In the pond ducks can be reared and they feed on the plankton of the pond. The droppings of the birds fall on the pond directly. Poultry dropping is a good manure as they contain more nitrogen, phosphorus and potash as compared with other manure. The production of vegetables can be taken in the bond which can be fertilized with this manure.

Feeding strategies

Feed is the major input that determines the cost of production of eggs and chicken meat which accounts for 65-75% in broiler and 55-65% in layer. Maize, the principal cereal used in poultry diet as energy source. Otherwise broken rice, wheat or barley, deoiled bran or rice polish can be given to the birds for energy source (Table-4). As far as the protein source is concerned the birds can be fed either from vegetable source of protein like soyabean meal, ground nut cake, til oil cake , sun flower oil cake etc. Most of the vegetable source of proteins are deficient of critical amino acids like lysine and methionine. Among the vegetable source of protein soyabean meal is rich in lysine and til oil cake is comparatively rich in methionine. Animal source of proteins like fish meal and meat meal are well balanced with essential amino acids. Most of the commercial feed manufacturers are producing feed without any animal protein supplement but adding synthetic lysine and methionine in the diet. Minerals and vitamins also play a major role in the production and growth of the chicks. Number of growth promoters and anti stressors are also being used to optimize growth and egg production and reduce the thermal stress. (Swain *et al*., 2007, Beura *et al*., 2012). Number of deficiency diseases can be occurred without addition of minerals and vitamins in the ration of the birds. Especially for layers the requirement of calcium is more which is 3- 4% of the total diet. To meet the calcium requirement of the layer birds oyster shell meal or lime stone is used in the ration. There is a great scope for preparation of low cost feed by utilizing the locally available ingredients not only for the broiler but also in the back yard poultry production. A farmer can prepare a low cost feed by taking maize/ broken rice, oil cakes like ground nut/ til/ sunflower and animal protein supplements like fish meal. The feed should be offered at the evening after returning from scavenging in a rural poultry model. Supplemental feed helps to grow better and produce eggs in much higher rate than non supplemented one. Layer birds should be fed another 5% of oyster shell meal to meet the calcium demand. Because egg shell constitutes 11-12% of the total egg mass and 94% of the egg shell contains calcium. If there is deficiency of calcium it will lead to decrease egg production, thinning of egg shell, breaking of the eggs while handling and even eggs with no shell have been produced.

Table 4: Composition of a backyard Poultry Feed

Sl. No.	Ingredients	Quantity (%)
1.	Maize/Wheat/Barley	40-50
2.	Broken rice	12-20
3.	Oil cakes like Sunflower/Sesame/ GN/ Soyabean	20-30
4.	Fish meal/ Dried fish	8-10
5.	Deoiled Rice ban/Rice polish	7-10
6.	Mineral Mixture	2.5
7.	Common salt	0.5
8.	Oyster shell meal (Layer)	5

Backyard poultry production in IFS

Back yard poultry production which involves keeping poultry in backyard is considered as a primitive system of producing eggs and meat, since input is very much less compared to modern intensive system of production. The name backyard poultry production implies rearing of poultry in small numbers in the backyard under free range or semi intensive system. The system is designed to utilize the natural food base (fallen grains, kitchen waste, insects and grass) available in free range condition for nutritional needs, gainful employment and income generation by population living in rural and especially tribal areas.

It is however surprising to note that family poultry production continues even today after 4 and ½ decades of industrial poultry production since production goals of intensive/ industrial system of production is not compatible with the livelihood and socio cultural practices of the rural poor and tribals. Further intensive system of poultry production requires high inputs and sophisticated management, nutrition and health care practices which poor people can not afford. Inputs require for intensive system of production is also not available in local markets. Knowledge base of rural and tribal people centers around local poultry breeds which has been developed by the community /tribe over the years. Family poultry production is used by weaker sections of the society as an insurance against crop failure, walking bank and to ensure basic economic returns besides empowerment to village women and children. Coloured plumage domestic fowls are preferred by rural and tribal people for socio cultural purposes and ceremonial offer.

Birds raised in family poultry production system scavenge their feed requirements. Some families do provide surplus home grown grains, by products and kitchen wastes. They drink water during scavenging if available outside. Water is also provided at home. Usually earthenware pots are used for providing feed and water. Broody hens are used for hatching of chicks. Broodiness increases duration of pauses hence desi hens lay few number of eggs varying from 40 to 110 eggs during the year. Poultry raised in backyard are lighter. It helps them to fly and to escape predation. The maximum weight of hens does not exceed 1.5 kg and those of cocks 2.0 kg (Panda *et al.,*2004 and Ramarao *et al.,* 2005). The average age at first egg is about 5-6 months. They lay tinted or brown coloured eggs. Most of the eggs produced are used for hatching. Only few eggs are eaten at home or sold. In certain tribal belts of the country like Odisha, Jharkhand, Chhatisgarh and MP cock fighting is a popular sports. Fighting cocks are raised with special care and sold at high price. Those cocks with history of winning comparatively fetch more money than others. Optimum size for family units varies from 2 to 10 hens. Not more than 2 cocks are maintained for breeding purpose.

The birds raised in family poultry system are sold on their visual appearance and not by weight. The eggs and meat from deshi birds fetch a premium price compared to improved birds raised in farms. Niche market exists for sale of eggs and meat of desi fowls in rural and tribal areas. The family poultry production is auto generating rural base.

Health management

Tough poultry farming is very popular and remunerative but proper health care should be undertaken both small and back yard farms. Regular vaccination should be practiced to prevent from fatal diseases. The most common and frequent occurring is the Ranikhet Disease (RD) which occur in endemic form and one of the killer disease in poultry. Proper vaccination schedule should be practiced for prevention of the diseases. If not vaccinated mortality is very high and sometimes entire family poultry units or even in villages are wiped out. Parasitic infections are also common those are reared in semi intensive condition. Morbidity becomes very high when birds are not treated against worms. At least once in three months the birds should be dewormed against broad spectrum anthelmentics.

Vaccination schedule for broiler birds

Disease	Vaccine	Age	Quantity	Route
Marek's disease	Marek's	1st day	0.2 ml	Under neck skin
Ranikhet Disease	F-1 or Lassota strain	5-7 day	one drop	Eyes/nostrils
Gumboro Disease	Gumboro	15-17 day	one drop	Eyes
Ranikhet Disease	F-1 or Lassota strain	21-23 day	one drop	Eyes/nostrils
Gumboro Disease	Gumboro	30-35 day	one drop	Eyes

Vaccination schedule for layer birds

Disease	Vaccine	Age	Quantity	Route
Marek's disease	Marek's	1st day	0.2 ml	Under neck skin
Ranikhet Disease	F-1 or Lassota strain	5-7 day	one drop	Eyes/nostrils
Gumboro Disease	Gumboro	15-17 day	one drop	Eyes
Ranikhet Disease	F-1 or Lassota strain	21-23 day	one drop	Eyes/nostrils
Gumboro Disease	Gumboro	30-35 day	one drop	Eyes
Fowl Pox	Fowl pox vaccine	6 week	0.2-0.5 ml	Foot pad/leg
Ranikhet Disease	R_2B strain	8 weeks	0.5 ml	Wing
Ranikhet Disease	R_2B	every 6 month	0.5 ml	wing

The estimate and return for a unit of 100 broiler birds

Non recurring Expenses

- Construction of poultry shed (Pucca house with asbestos roof) @ 300/sqft for 100 sqft = 30,000.00
- Feeder, drinker, brooder, Electric appliances etc. @ 50/bird = 5,000.00

Total = 35,000.00

Recurring Expenses

- Cost of day old chicks @ 30.00 for 100 chicks = 3,000.00
- Feed consumed @ 3kg/bird (100x3=300kg) @ 22.00 = 6,600.00
- Litter materials, Medicine and vaccine @10/bird = 1,000.00
- Electric bulb and energy consumption @5/bird = 500.00

Total Income = 11,100

- Sale of birds (mortality @3%) weight of the birds 1.65 kg @ 80/kg
  97 birds xRs.132.00/bird=13, 604 say= Rs. 12,804
- Poultry manure Rs. 200.00
- Sale of gunny bags Rs. 150.00

  Total = Rs. 13,154

Net Return 13,150-11,100=2,150/ batch

In a year 8 batches of 6 weeks duration and return will be Rs.2,150 x 8 =17,200. The broiler can be raised in a single house or two poultry sheds can be prepared and in a cycle wise new batches can be reared at a gap of 15 days (Panda, *et al.,* 2013). In a year 16 batches can be run and the farmer can get more benefit with little labour. No separate labour is required either women or men can spare 1-1½ hrs for this type of integrated farming. In the integrated farming this litter can be very well used by the fish or for crop/vegetable production.

Conclusion

Poultry farming is an important activity in the integrated farming system. Both the broiler and layer can be reared. Poultry farming not only gives boost to the economy of the farmer but also provides good manure for the crop and horticulture plants. In the fish- poultry integrated system the house can be prepared on the bond of the pond or over the water. The droppings of the birds fall on the pond directly which help to regenerate planktons for feeding of fish and ducks. Thus poultry farming is an integral part of the farming system prevailing in India which can be guided properly for employment and rural livelihood support.

References

Alam, M. Robiul, M. Akkas Ali, M. Akhtar Hossain M. S. H. Molla and F. Islam 2009. Integrated approach of pond based farming systems for sustainable production and income generation. *Bangladesh Journal of Agricultural Research*. 34(4):577-584.

Arya, R., Kumar, A., Prasad, S., Soren, S.K.R. and Singh, D. K. 2009. Studies on production performance of desi x exotic crosses of chicken under backyard farming. Proceedings of XXVI IPSACON, 22-24 Oct. 2009, Mumbai, 39:140.

Banerjee, R.K., P. Roy, C.S. Singit and B.R.Dutta 1979. Poultry dropping- its manurial potentiality in aquaculture. *Journal of the Inland Fishery Society of India.,* 2(1) : 94-108.

Behera, B., M. Nedunchezhiyan, S. Mohanty and N.Panda. 2014. Final Report of the NAIP-3 on Sustainable Rural Livelihood and Food Security to rainfed farmers of Odisha.pp. 29-30.

Beura, T.K., Panda, N., Mishra, P.K., Panigrahi, B., Panda, H.K. and Pati, P.K. 2012. Effect of vitamin E and C on the growth and immune competence of coloured birds during summer. *Animal Nutrition and Feed Technology*, 12:13-24.

Channabasavanna, A.S. and D. P. Biradar. 2013. Relative Performance of Different Rice-Fish-Poultry Integrated Farming System Models with Respect to System Productivity and Economics. *Karnatak Journal of Agricultural Science.* 20(4). 706-709.

Delmendo, M.N. 1980. A review of integrated livestock-fowl-fish farming systems. In: Integrated Agriculture Aquaculture Farming Systems, R.S.V. Pullin and Z.H. Shehadeh (eds), ICLARM Conf. Proc. 4, pp. 59–71

Doley, S., Barua, N., Kalita, N. and Gupta, J.J. 2009. Performance of indigenous chicken of North-Eastern region of India under different rearing systems. *Indian Journal of Poultry Science,*44: 249-52.

Fang, Y.X., X.Z. Guo, J.K. Wang, X.Z. Fang and Z.Y. Liu 1986. Effects of different animal manures on fish farming. *In* : J.L. Maclean, L.B. Dizon and L.V. Hosillos (Eds.), *The First Asian Fisheries Forum.* Asian Fisheries Society, Manila, p. 117- 120.

FAO (1996) Towards Universal Food Security . FAO Document s for world food summit,1996. Food and Agriculture Organization, Rome, Italy.

Gangwara , L.S., Sandeep Saran and Sarvesh Kumar 2013. Integrated Poultry-Fish Farming Systems for Sustainable Rural Livelihood Security in Kumaon Hills of Uttarakhand. *Agricultural Economics Research Review* 26. 181-188.

Hoq ,M. Enamul, G.B. Das and M.S. Uddin.1999. Integration of fish farming with poultry: effects of chicken manure in polyculture of carps and freshwater prawn. *Indian Journal of Fish.* 46(3):237-243.

Jhingran V.G. and B.K. Sharma, 1980. Integrated livestock-fish farming in India. In: Integrated Agriculture Aquaculture Farming Systems, R.S.V. Pullin and Z.H. Shehadeh (eds), ICLARM Conf. Proc. 4, pp. 225–238.

Mandal, A., Patel, M., Kumar, A., Singh, B., Ghosh, A.K. and Bhardwaj, R.K.2004. Performance of different cross bred chicken in intensive system. *Indian Journal of Poultry Science,* 39:211-14.

Panda, B.K., Padhi, M.K. and Sahoo, S.K. 2004.Comparative growth and performance studies of exotic birds in the intensive and extensive system of rearing at Orissa state. *Indian Journal of Poultry Science* , 39:285-88.

Panda, N., Panigrahi, B.P.,Nayak,S.,Behera, B. and Dash, S.N., 2013. Small scale broiler farming: An alternate livelihood for the landless in Rural Odisha. Book chapter in the book Natural Resource Conservation. Emerging Issues and Future Challenges. SSPH publisher.pp793-796.

Pillay, T.V.R., 1980. Aquaculture. Principles and Practices. Fishing News Books, London

Rama Rao, S.V., Panda, A.K., Raju, M.V.L.N., Shyam Sunder, G., Bhanja, S.K. and Sharma, R.P 2005.Performance of Vanaraja chicken on diet containing different concentration of metabolizable energy at constant ratio with other essential nutrient during juvenile phase. *Indian Journal of Poultry Science,* 40 :245-48.

Ramrao, W. Y., S P Tiwari and P Singh. 2006. Crop-livestock integrated farming system for the Marginal farmers in rainfed regions of Chhattisgarh in Central India. *Livestock Research for Rural Development.* 18.(7). Article #102.

Singh, J.P., Gangwar, B., Pandey, D.K. and Kochewad, S.A. 2011. Integrated Farming System Model for Small Farm Holders of Western Plain Zone of Uttar Pradesh. PDFSR *Bulletin No. 05, pp. 58.* Project Directorate for Farming Systems Research, Modipuram, Meerut, India

Swain, A.K., Sahu, B.K., Das, S.K. and Mishra, S.K. 2007.Responses of growth promoters to the performance of broilers. *Indian Journal of Poultry Science,*42 :323-25.

Woynarovich, E. 1979. The feasibility of combining animal husbandry with fish farming with special reference to duck and pig production. *In* : T.V.R. Pillay and WA. Dill (Eds.), *Advances in Aquaculture.* Fishing News Books Ltd., Farnham, Surrey, England, pp. 203-208.

Woynarovich, E., 1980. Raising ducks on fish ponds. In: R.S.V. Pullin and Z.H. Shehadeh (eds.). Integrated agriculture aquaculture farming systems. ICLARM Conf. 4, pp. 129–134.

Integrated Farming System Practices: Challenges & Opportunities, pp 197-213

New India Publishing Agency, New Delhi, India

8

Prospects of Agroforestry in Integrated Farming Systems

P. J. Mishra

Agroforestry is a collective name for land-use systems and technologies in which woody perennials including trees, shrubs, bamboos etc. are deliberately combined on the same land-management unit with herbaceous crops or animals either in some form of spatial arrangement or temporal sequence. In agroforestry system there are both ecological and economic interactions amongst the different components (Lundgren and Raintree, 1982). The origin of agroforestry practices, i.e. growing trees with food crops and grasses, is believed to have been during Vedic era (Ancient period, 1000 B.C.), the agro-forestry as a science is introduced only recently. Agro-forestry is the science of designing and developing integrated, self-sustainable, land-management systems that involves the introduction and retention of woody components such as trees, shrubs, bamboos, canes and palms along with agricultural crops including pastures or animals, simultaneously or sequentially on the same unit of land and time, to satisfy the ecological as well as socio-economic needs of people. This practice of growing trees and arable crops together on the same piece of land is an age-old practice with the dry land farmers. Agroforestry results in multiple products, increases income, reduces runoff and soil loss, utilizes off-season rainfall and radiation and gives stability to dry land agriculture. Further it makes the future generation farmers to inherit not only the land of their predecessors, but also the wealth of standing biomass of trees.

AICRP on Agroforestry, Orissa University of Agriculture & Technology, Bhubaneswar-751003, Odisha, India

National Forest Policy (1988) has set a goal that 33% of the country's geographical area should be under forest and tree cover. Under Tenth Five Year Plan, the Government of India has set a target to increase the country's forest cover up to 25%. According to the latest report of Forest Survey of India (2013), the forest cover in the country is 697,898 km^2, constituting 21.23% of its total geographical area. Out of this, dense forest constitutes 12.24 % and open forest 8.99 %. The only solution to achieve these targets is through the practice of agroforestry. Therefore, in the quest of optimising productivity, the multi-tier system came into existence.

Suitable tree species for agroforestry

The selection of tree species should be such that the cultivation of arable crops traditionally grown on that site can be continued under the new system. While selecting the species suitability for growing under the prevalent agro-climatic conditions, utility of trees for meeting the needs of farmers for timber, fodder, fuel, fruit and fibre etc. and preference may be given to indigenous and fast-growing species having no interference with arable crops, easy in establishment, fast growth and short gestation period, non-allelopathic effects on arable crops, ability to fix atmospheric nitrogen, easy decomposition of litter, ability to withstand lopping, multiple use and high return and ability to generate employment. It is extremely difficult to select species having the ability to fulfill all these criteria. Therefore, the farmers who are directly taking part in adopting agroforestry on their fields should select the species as per his choice considering all these characters.

Tree species based on specific utilization purpose

Fodder-cum-fuel wood species

Albizia amara, Albizia procera, Albizia lebbeck, Erythrina indica, Gliricidia sepium, Gmelina arborea, Hardwickia binata, Leaucanea leucocephala, Pithecellobium dulce, Prosopis cineraria, Sesbania grandiflora, Sesbania sesban etc.

Fuel-wood and timber species

Acacia nilotica, Acacia mangium, Albizia lebbeck, Albizia procera, Azadirachta india, Cassia siamea, Casuarina equisitifolia, Dalbergia sissoo, Dendrocalamus strictus, Pongamia pinnata, Melia azadirach, Parkinsonia aculeate, Tectona grandis, Thespepsia populnea etc.

Softwood and pulpwood species

Ailanthus excelsa, Ailanthus tryphysa, Bombax ceiba, Paraserianthes falcataria, Populus deltoids, Bamboo species etc.

Fruit and vegetable species

Fruits: Annona reticulata, Annona squamosa, Artocarpus heterophyllus, Emblica officinalis, Mangifera indica, Psidium guajava, Ziziphus jujuba, Carica papaya etc.

Vegetables: *Moringa officinalis,* greens, Okra, Brinjal, Aroids, *Muraya koeinigi, etc.*

Priority for selection of species

After the screening of the species to suit the agro-climatic conditions, the next aspect is profitability. Ideally the agricultural production of agricultural crop should not be affected. But in reality, farmers want to earn higher total returns. Rich farmers are able to take bold decisions but not small farmers, who like to earn more money but not at the cost of reduced food production. This is because home-grown food is the main source of subsistence and failure to harvest an adequate quantity of grains severely threatens their subsistence. Uncertainties about the marketing of wood can be greatly reduced by establishing a suitable marketing network to handle the agroforestry produce.

Selection of crops

To obtain full utilization of environmental resources and to increase economic benefits, an appropriate crop should be inter-planted with or after the establishment of tree plantation in accordance with site and climatic conditions. During initial years of establishment the crops that require sufficient amount of sunlight are inter-planted between the tree species. As the tree grows and their crown enlarges, causing considerable shade on the floor, especially during the growing season, some shade-tolerant crops or medicinal plants such as ginger, turmeric, elephant foot-yam etc. are selected for under planting.

Agroforestry practices in integrated farming system

Agroforestry is a form of natural resource management, since the resource management issues interact with other production systems as well. Technologies related to integrated production systems involving crops, livestock, perennials and fisheries needs to be addressed further with the changing economy. The value of commerce in agriculture would be appreciated and farmer's decisions on land use would be increasingly based on comparative advantage rather than on subsistence needs of the farm household. Therefore a major shift might be expected in agricultural production through diversification. These shifts in production system will get further boost because of integration of world markets, urbanization and rising personal incomes. Further, to design appropriate farming systems suited to diverse farming situations, farmer's participation beginning from planning till execution of research programme will be inevitable and they would actually participate in research efforts . This bottom approach will help the scientist in having a better perception of farm problems and enable them suggest eco-friendly resource use efficient and economically profitable farming-system strategies.

Resource conservation

Water is a critical element for reclamation of degraded lands for sustainable biomass production, ultimately leading to a better quality of life and enabling conditions. Watershed management is an approach for area planning of natural resources, especially land, water and plants, to observe socio-economic needs of human society. The watershed management plan for treating for different parts should be decided by simple measures for protection of the resource base and created assets, improvement of productive system, generation of employment opportunities and ensuring higher income on a recurring basis. Degraded land because of many limitations can only be improved through agroforestry, which control erosion, reduce run-off, improve *in-situ* soil-moisture conservation , increase water-table but also improve productivity as well as profitability. Researches involving agroforestry are still required on many areas like watershed hydrology, control of sedimentation, situation- specific cost-effective technologies, production of fodder, forages, industrial grasses and medicinal plants for quick-return and livestock improvement programme for harnessing maximum benefits from limited biomass.

Improvement in soil fertility and structure

Plantation of compatible and desirable species of woody perennials on farmland results in an improvement in soil fertility. There are several possible mechanisms of this, which include: (a) Increase in organic matter content of the soil through the addition of leaf litter and other plant parts, (b) More efficient nutrient cycling within the system and consequently more efficient utilization of nutrient that are either inherently present in the soil or externally applied, (c) Biological nitrogen fixation and solubilization of relatively unavailable nutrients, e.g. phosphate through the activity of mycorrhiza and phosphate-solubililising bacteria, (d) Increase in the plant-cycling fraction of nutrients, with their resultant reduction beyond the nutrient-absorbing zone of the soil, (e) Complementary interaction between the component species of the system, resulting in a more efficient sharing of nutrient resources among the components, (f) Enhanced nutrient economy, because of different nutrient-absorbing zones of the root system of the component species and (g) Moderating effect of additional soil-organic matter on extreme soil reactions and consequently improved patterns of nutrient- release ability.

Improvement in the organic matter status of the soil can result in an increased activity of the favourable micro organisms in the root zone. In addition to the nutrient relations, such micro-organisms may also produce growth-promoting substances through desirable interaction and cause commensalism effects on the growth of plant species. Inclusion of trees and woody perennials on farm lands can, in the long run, result in marked improvements in the physical conditions of the soil, e.g. its permeability, water-holding capacity, aggregate stability and soil-temperature regimes. Although these improvements may be slow, their net effect is a better soil medium for plant growth.

The role of trees in soil conservation and erosion control is one of the most widely acclaimed and compelling reasons for including trees on farm lands prone to erosion hazards. The beneficial effects of trees in this regards extends beyond protecting the immediate farm lands under consideration, to impart stability to the ecosystem and reduce the rate of siltation of downstream aquatic ecosystems, dams and reservoirs.

The influence of trees on hydrological characteristics can extend from the microsite to the farm and regional levels. Although the effect of water use by a tree component on water availability to crop plants in different climatic conditions is not yet fully understood, there is evidence that the hydrological characteristics of catchment areas are favourably influenced by the presence of trees .

Management options in agroforestry systems

To achieve higher growth of components in agroforestry systems management options includes : (i) Suitable species, (ii) proper land preparation, (iii) fertilization, (iv) application of mulch or manure, (v) irrigation, and for reduced growth are: (i) shoot and root pruning, (ii) pollarding, (iii) excessive shading, (iv) herbicide and (vii) browsing. In a trial conducted at CAZRI, Jodhpur, application of N and P fertilizers positively influenced the production of grasses below trees. The effect was more pronounced for *Chrysopogon fulvus* compared to that for other grasses, viz. *Cenchrus ciliaris, Cenchrus setigerus* and *Sehima nervosum*. The tree species were *Prosopis cineraria, Acacia tortilis* and *Albizia lebbeck.*

Types of agroforestry systems

The following are common agroforestry systems prevailing in different agro-ecological regions of India:

1. Agri-silviculture (trees+crops)
2. Agri-horticulture (fruit trees+crops)
3. Agri-silvi-horticulture (trees+fruit trees+crops)
4. Agri-silvi-pasture (trees+crops+pasture or animals)
5. Silvi-olericulture (tree + vegetables)
6. Horti-pasture (fruit trees+pasture or animals)
7. Horti-olericulture (fruit tree + vegetables)
8. Silvi-pasture (trees+pasture/animals)
9. Silvi or Horti-sericulture (trees or fruit trees+sericulture)
10. Horti-apiculture (fruit trees + honeybee)
11. Aqua-forestry (trees + fishes)
12. Boundary plantation (tree on boundary + crops)
13. Block plantation (block of tree+ block of crops)
14. Energy plantation (trees+crops during initial years)
15. Alley cropping (hedges+crops)
16. Forage forestry (forage trees+pasture)
17. Shelter-belts (trees+crops)
18. Wind-breaks (trees+crops)
19. Live fence (shrubs and under- trees on boundary)
20. Homestead (multiple combinations of trees, fruit trees, vegetable etc).

Besides these common agroforestry systems, there are many more types followed in different agroecological regions of India.

Traditional agroforestry practices in India

Shifting cultivation

In this system the natural vegetation, usually forests, is cleared by slash-and-burn method, cropped with common arable crops for a few years, and then left unattended till natural vegetation regenerates. Due to the increasing demographic pressure, the fallow period became greatly reduced to even 4-5 years though previously the fallow period was 10-20 years. As a result there is heavy soil erosion and decline in fertility and productivity of the soils. Shrinking land area per family and increase in population has forced them to look for land management systems by which they can get something from the land even during the fallow phase. Thus intercropping under or between fast-growing trees in a fallow phase is one of the approaches while finding alternative to shifting cultivation. A farming-system approach based on watershed management has been advocated as an alternative to shifting cultivation.

Home gardens

Home gardens preserve the bio-diversity in tropical regions. They consist of plants, which may include trees, shrubs, vines and herbaceous plants growing in or adjacent to a homestead or home compound. These are intended primarily for household consumption and there is intimate association of woody perennials with annual and perennial crops and invariably livestock, within the compounds of individual houses, with the whole crop-tree-animal unit being managed by family labour. Plantation crops such as coconut, cacao, coffee, arecanut and black pepper (vine) often are the dominant components of many home gardens of the humid tropics. Banana, papaya, mango, guava, custard apple and jackfruit are the common fruit and vegetable plants. In spite of very small average size of the management units, home gardens are characterized by high species diversity and usually 3-4 vertical canopy strata. The important aspect of these gardens is that the production for home consumption occurs throughout the year. In the recent years in these home gardens farming systems approach has been incorporated and their various profitable components have been adopted by the farmers to improve the land productivity.

Plantation based cropping systems

Commercial plantation crops like rubber, coffee, coconut and oilpalm represent a well- managed and profitable models in the tropics. During the early phases of plantation when some intercropping is feasible, the commercial production of these crops is aimed at having an additional income. Coconut-based integrated/ mixed farming is still popular in small holdings. Important cereals grown with coconut include rice, finger millet and maize, pulses such as pigeon pea, greengram and blackgram and cowpea and oilseed crop groundnut; root crops such as sweet potato, yams, elephant foot-yam and taro; spices and condiments such as ginger; turmeric, cinnamon, clove, chilies, and black pepper; fruits like pineapple, mango, banana, and papaya; and cash crops such as sugarcane, sesame; among improved pasture grasses species of *Brachiaria, Dichanthium, Panicum, Setaria, Paspalum* and *Pennisetum* improved forage legumes such as *Stylosanthes, Desmodium, Glycine, Leucaena* and *Macroptilium*.

Scattered trees on farm lands

The practice of growing agricultural crops under scattered trees on farm lands is quite old. In this system trees are grown scattered in agricultural fields for many uses such as shade, fodder, fuel wood, fruit, vegetables and medicinal uses. Some of the practices are very extensive and highly developed in India. Farmers retain trees of *Acacia nilotica, A. catechu, Bambusa, Dendrocalamus, Butea monosperma., Dalbergia sissoo, Mangifera indica, Azadirachta indica, Moringa oleifera, Tamarindus indica, Ceiba pentendra, Anacardium occidentale, Psidium guajava, Cocos nucifera Syzyzium cuminii, Zizyphus mauritiana* and *Gmelina arborea.* In coastal areas of peninsular India *Borassus flabellifer* is found scattered in the fields of groundnut, rice and greengram. Every part of the palm is used by common man: the leaves for thatching, trunk as pillar or timber, fruit is roasted and consumed the radicle of germinating seeds is roasted, a beverage (alcohol) is extracted from the spadix, which is also used to prepare jaggery and vinegar.

Suitable agroforestry systems for Odisha

Agri-silviculture system

In agrisilvicultural systems where agricultural crops are grown in association with trees have been proved to be more remunerative and can avoid the menace of drought to a great extent. The agrisilviculture system is recommended for land capability class IV with annual rainfall of 750 mm. A large number of tree-crop combinations, particularly of N-fixing trees with sorghum, groundnut, castor and pulses were evaluated in Alfisols and Vertisols. Short duration dryland crops such as pearl millet, black gram and green gram, combined with widely spaced tree rows of *Faiderbia albida* and *Hardwickia binata*, have been found compatible in semi-arid tropical areas (Korwar, 1992). For rainfed upland conditions of Odisha, *Acacia mangium, Gmelina arborea, Dalbergia sissoo* and *Tectona grandis* are some of the suitable tree species for this system. *Acacia mangium* is most suitable as it is a fast growing, N-fixing tree which could attain a height of 18 m with dbh 25 cm after 11 year of planting when grown at 8 m X 2 m spacing and intercropped with different agricultural crops. The leaf litter accumulation of 3-4 t/ha/year and pruning material of about 4-5 t/ha/year under *Acacia mangium* is very common. In this system 2-7 % higher soil moisture retention was observed at different depths compared to open field condition during post rainy season. Similarly, the available nutrient content was also higher in tree + crop combination plots than sole pineapple grown in open field condition. Pineapple was the best suitable crop to be included from 9th to 12th year of tree plantation in this agrisilvicultural system with mean net returns of about Rs. 1,00,000/ha/year with B:C ratios 2.5 as against a net return of Rs. 56000 ha/yr. with B:C ratio 1.8 when grown as a sole crop.

Agri-horticulture and agri-horti-silviculture system

The Fruit based agroforestry system is a planting system comprising combinations of plants with various morpho-phenological features to maximize the natural resource use efficiency and enhanced total factor productivity. The system comprises of a combination of perennial and annual plant species as different components in the same piece of land arranged in a geometry that facilitates maximum utilization of space in four dimensions (length, width, height and depth) leading to maximum economic productivity of the system (Fig. 1&2). Fruit based multitier cropping systems were found to be effective alternatives to the traditional rice based monocropping system for increasing the profitability under rainfed upland conditions (Fig. 3).

Fig. 1: *Dalbergia sissoo*+Pineapple

Fig. 2: *Acacia mangium* +*Aloe vera*

Fig. 3: Coconut + Rice

Mango based Agri-horti-silvicultural system

In mango based agrihortisilvicultural system, mango grafts are planted in tree rows between two *Acacia mangium* trees leaving 3.0 m on either side of the tree at 6 m x 6 m spacing and annual crops like ragi, sesamum and cowpea are grown in alleys of tree up to four years., shade tolerant crops such as colocasia (*Colocasia esculanta*), arrowroot (*Maranta arundinacea*) and turmeric (*Curcuma longa*) are intercropped thereafter with about 80% yield recovery from 5th to 8th year of tree-crop growth period. Short duration fruit crop like pineapple (*Ananas comosus*) is the best suitable crop to be included in the agrihortisilvicultural system (Fig. 4) from 9th to 12th year with mean net returns of Rs. 87348, 77110 and 75954 /ha/year with B:C ratios 2.12, 1.98 and 1.97 when intercropped with *Acacia mangium*, *Dalbergia sissoo* and *Gmelina arborea* in between mango trees, respectively, as against a net return of Rs. 56000 ha/yr. with B:C ratio 1.82 when grown as a sole crop.

Fig. 4: *Acacia mangium* + mango + Pineapple

Guava based Agri-horti-silvicultural system

Guava grafts can be planted in-between two *Acacia mangium* trees within a row, planted at 6 m x 6 m spacing. Intercrops like ragi and sesame are grown in the system up to four years and thereafter shade tolerant tuber crops like colocassia, arrowroot (Fig. 5) and turmeric are grown suitably up to seven years with 80-85% yield recovery and benefit: cost ratio of 2.48. Among the tuber crops, arrowroot in association with *Dalbergia sissoo* gave the highest net return (Rs.51640/ha) as well as the benefit: cost ratio (2.48) followed by turmeric in association with *Dalbergia sissoo* (Rs 47964/ha, 2.37) and arrowroot in association with *Gmelina arborea* (Rs 47458/ha, 2.36). Similarly in Ber based agri-horti system Pearl millet + pigeon pea (Solapur, Pigeon pea + black gram (Rewa), Castor (Dantiwada) and Cluster bean (Hyderabad) showed promising

Fig. 5: *Acacia mangium* + Guava + Arrowroot

results in rainfed environment. Ber on an average gave 40 kg fruits/tree along with the 100 kg of horse gram and 450 kg of cowpea cultivated in interspaces (Osman *et al.*, 1989).

Silvi-pasture system

Fig. 6: *Acacia mangium* + Guinea grass

Tree pattern is an important factor for any successful silvipastoral system. Trees can be evenly distributed over the area; either in rows or clusters, and pruning or thinning is to be managed for higher efficiency. Shelter for livestock is planned for severe weather events. Stress to livestock is reduced through moderation of pasture microclimate which a well-planned silvipastoral system provides. There is indeed a climate stabilizing effect of trees that reduce wind and provide shade to reduce heat stress and wind chill on livestock. Perennial grass components, besides imparting stability to crop production in arid areas, also act as vegetative filter strips for prevention of wind and water erosion. Moreover, the grass component improves the soil organic matter and starts giving production from the establishment year onwards.

The silvipasture systems are recommended for land capability class V and VI. *Acacia mangium* is the most suitable as it is fast growing N-fixing tree which can be planted at 6 m X 2 m spacing and guinea (Fig. 6) and thin napier grasses can be grown in between up to 12th year of tree plantation in this silvicultural system which could produce green fodder of about 20 t/ha/year with a mean net returns of about Rs. 15,000/ha/year with B: C ratios 2.2. Both the grasses recorded yield recovery above 91 % as compared to sole crops. Higher soil moisture and nutrient content was also observed under this silvipasture system.

Horti-silvi-pastoral system

In a guava based hortisilvipastoral system of rainfed uplands of Odisha, guava grafts are planted in between two *Acacia mangium* or *Dalbergia sissoo* trees planted at 6 m x 6 m spacing and perennial grass like guinea (*Panicum maximum*) or annual legume fodder i.e. Stylo (*Stylosnthes hamata*) is intercropped in alleys of trees with 80-85% yield recovery and benefit:cost ratio of 1.88. Fodder grass guinea is found to be more compatible than stylo in guava + *Acacia mangium* based hortisilvipastoral system. Similarly at CRIDA, the hortipastoral system with *Cenchrus / Stylos* in rainfed guava and custard apple, *Cenchrus* yielded dry forage 7 t/ha with 17.5% of crude protein during the first year while *stylos* recorded 5.6 tons of dry fodder during the second year of plantation (Fig. 7&8).

Farming system with multiple components

Conventional farming in rainfed areas is becoming risky and unable to meet the minimum requirement of farm families. Adoption of developed technology, diversification with different high value crops e.g. vegetables, fruits, medicinal plants, spices and inclusion of backyard poultry and goatery will increase the income, meet the household requirements and minimize the risk of farmers. It is observed in on-farm integrated farming system research units in rainfed areas of Odisha that integrated farming system gives higher overall production, net return and return per rupee spent and generates more employment than conventional farming. Integrated Farming System experiment at Kantabania village of Nayagarh district from 1.2 ha land holding produced a total income, net return and benefit cost ratio (BCR), of Rs. 163200, Rs. 91800 and 2.29 as against Rs 70600, Rs 32600 and 1.86 with conventional farming systems respectively. Inclusion of high yielding varieties of paddy, vegetables, and high value vegetables e.g. pointed gourd, arrowroot and backyard poultry with dual purpose *Vanaraja* breed not only increased the overall income but also generated additional employment for rural farm families and minimized the risk and uncertainty of conventional farming (Mishra and Nanda, 2013). The Agri-horti-silvi-duckery-fishery system (Fig. 9) studies North-East India showed that multi enterprise model comprising cereal crops, pulses, oil seeds, horticultural crops such as mango and pineapple, vegetables crops and livestock components of duckery, piggery, and fishery with harvesting structures in 1 ha land gave 5 times more profit than traditional mono crop rice cultivation which gave maximum production of 1t/ha of rice (Rs.5000-6000/ha/year).

A farming system involving crop, animal and silviculture has been the right combination of enterprises that existed even in early periods of agricultural development. An animal like the goat, which feeds on all types of vegetation, is a hardy species that can withstand the adverse climatic conditions of drylands.

Fig. 7: Guava + Guinea grass

Fig. 8: *Acacia mangium*+Guava+Guinea grass

Fig. 9: Farming system model for coastal Odisha

Besides it can provide additional returns to farmer. The manure can aid in improving the soil fertility. Radhamani (2001) studied integrated farming systems involving three cropping systems with *Ailanthus excelsa* + crop + goat, *Ceiba pentandra* + crop + goat and *Embica officinalis* + crop + goat to identify the most suitable component linkage. They also reported the additional employment gains (314 man-days / year) through integrated farming system with crop + goat under rainfed vertisols. Integration of sorghum + cowpea (grain or fodder) and *Cenchrus glaucus* intercropped in *Emblica officinalis* with goat had highest productivity and economic returns under Coimbatore conditions. The highest B: C ratio and employment generation was also found in this system.

Homestead agroforestry system or home garden

This agroforestry system supplements a variety of basic needs such as timber, fruits, vegetables, drugs, milk, meat, eggs, fodder, live fence etc. These home gardens are highly productive and sustainable (Fig. 10). The species diversity and varying production cycles of different components ensure continuous production throughout the year from the home garden system. These are characterized by high species diversity and usually 3-4 vertical canopy strata. The lower most being at less than 1.0 m in height dominated by different vegetables like brinjal, okra, chili, tomato, cabbage, dioscorea, etc. and the second layer is 1.0 to 3.0 m height comprising small fruit plants

Fig. 10: Homestead agroforestry

such as banana, papaya, and lemon. The third layer is at 5.0 to10.0m height occupied by fruit trees like Guava, drumstick, Custard apple, etc. The upper most layers is the tree layer which can be divided into two, consisting of the emergent full grown timber and fruit trees having height more than 20.0 m and medium size trees of 10-20 m. In the upper layer, species like *Samanea saman, Bambusa vulgaris, Bambosa tulda,* jack fruit, bael, tamarind, neem, subabul, etc. are grown. These home gardens act as total insurance against crop failures under aberrant weather condition.

Apiculture with trees

In the rainfed uplands of Odisha, apiary with combination of tree species such as *Glyricidia sepile, Grieviella robusta, Gmelina arborea, Eucalyptus spp* are practiced for production of honey where the trees spp act as bee forage. With minimum cost involvement, farmers are getting good return from the system.

Trees on farm boundaries

Trees grown in agricultural fields or on field bunds are usually grown on farm boundaries. *Eucalyptus* and *Populus* in north India*, Acacia nilotica, Dalbergia sissoo* and *Prosopis juliflora* in paddy field bunds, bamboo in Sikim, *Borassus flabelliferin* and *Casuarina equisetifolia* in coastal areas of Andhra Pradesh and Odisha are commonly grown along field boundaries or bunds. Some farmers also grow *Gliricidia sepium*, *Jatropha* spp. *Vitex negundo* and *Erythrina indica* as live hedges (Fig. 11).

Fig. 11: *Acacia mangium* on farm boundaries

Aqua-forestry

This system is extensively seen throughout the coastal regions along Andhra and southern Odisha coast. Farmers are cultivating fish and prawn in saline water and growing coconut and other trees on pond dykes (Fig. 12). These trees help in producing flower, fruit and litter feed to fishery and generate extra income to the farmer. Now prawn culture in the mangroves is also practised, which from a rich source of nutrition to the aquatic life and breeding ground for juvenile fish, prawn and mussels. An integrated system of animal husbandry including goatery, poultry, duck farming and fishes in the small ponds in home gardens make a balanced system with higher nutrient-use efficiency per unit area. The leaves of many leguminous trees, viz. *Gliricidia sepium, Leucaena leucocephala, Moringa oliefera* etc. have been found to serve as good fish feed when offered as pellets and improved its productivity.

Fig. 12: Aquaforestry in coastal area

Medicinal and aromatic plants in agroforestry

Medicinal and aromatic plants (MAPs) play an important role in the healthcare of people around the world, especially in developing countries. Until the invention of modern medicine, man depended on medicinal plants for treating human and livestock diseases. Human societies throughout the world have accumulated a vast body of indigenous knowledge over centuries on medicinal uses of plants and for controlling pests and diseases of crops and livestock. About 80% of the population of most developing countries still use traditional medicines derived from plants for treating human diseases (de Silva, 1997). The increasing demands for medicinal plants by people in developing countries have been met by indiscriminate harvesting of spontaneous flora including those in forests. Consequently, many species have become extinct and some are endangered. It is therefore, imperative that systematic cultivation of medicinal plants be developed and steps introduced to conserve biodiversity and protect threatened species.

Many plants in traditional agricultural systems in the tropics have medicinal value. These can be found (either planted or carefully tended natural regenerations) in homegardens, as scattered trees in croplands and grazing lands, and on field bunds. Similarly, 'arjun' (*Terminalia arjuna*) in India, chinaberry (*Melia azederach*) in Asia and Erythrina species in Latin and Central America combine many uses including medicinal. Holy Basil or 'tulsi' (*Ocimum sanctum*), drumstick (*Moringa oleifera*) and curry leaf (*Murraya koenigii*) are backyard plants in many Indian households and they are routinely used for common ailments or in food preparations. Three categories of medicinal plants could be noted in home gardens: species used exclusively for medicine, horticultural or timber species with complementary medicinal value, and 'weedy' medicinal species. While the first two categories are deliberately planted the latter group is part of spontaneous growth. The species composition, plant density and level of management vary considerably depending on the soil, climate and market opportunity and cultural background of the people.

Two types of intercropping systems can be distinguished involving MAPs: as upperstory trees and as intercrops in other tree crops. Medicinal tree species that grow tall and develop open crown at the top can also be used for this purpose. Tall and perennial medicinal trees that need to be planted at wider spacing such as *Prunus africana*, *Eucalyptus globules* (for oil), sandalwood (*Santalum album*), ashok (*Saraca indica*), bael (*Aegle marmelos*), custard apple (*Anona squamosa*), amla (*Emblica officinalis*) and drumstick or moringa (*Moringa oleifera*) can be intercropped with annual crops in the early years until the tree canopy covers the ground. Some of the medicinal trees may allow intercropping for many years or on a permanent basis depending on the spacing and nature of the trees. The intercrops give some income to farmers during the period when the main trees have not started production. Many tropical MAPs are well adapted to partial shading, moist soil, high relative humidity and mild temperatures (Vyas and Nein, 1999), allowing them to be intercropped with timber and fuel wood plantations, fruit trees and plantation crops. Some well known medicinal plants that have been successfully intercropped with fuel wood trees (e.g., *Acacia auriculiformis*, *Albizia lebbeck*, *Eucalyptus tereticornis, Gmelina arborea,* and *Leucaeana leucocephala*) in India, include safed musli (*Chlorophytum borivilianum*), rauwolfia (*Rauwolfia serpentina*), turmeric (*Curcuma longa*), wild turmeric (*C. aromatica*), *Curculigo orchioides*, and ginger (*Zingiber officinale*) (Chadhar and Sharma 1996; Mishra and Pandey 1998; Prajapati et al. 2003). The trees may benefit from the inputs and management given to the intercrops. Short stature and short cycle MAPs and culinary herbs are particularly suited for short-term intercropping during the juvenile phase of trees.

Aromatic plants as potential intercrops

Aromatic Mentha spp. (*M. arvensis, M. piperita, M. citrata* var. citrata, *M. spicata, M. cardiac* and *M. gracilis*) and *Cymbopogon* spp. [lemongrass (*C. flexuosus*); Java citronella (*C. winterianus*); and palmarosa (*C. martinii*)] can be grown intercropped with *Populus deltoides* or *Eucalyptus* spp. for three to five years after planting the trees. Tree growth was generally little affected or improved probably due to the inputs given to the intercrops. Herb growth and oil yields of these aromatic plants were not affected during the first two years of intercropping, but they decreased slightly from the third year compared with sole crops. Aromatic plants can be profitably grown in association with fuel and timber trees such as *L. leucocephala, Casuarina* spp., and *Grevillea robusta*.

References

Agricultural Research for Development :Potential and challenges in Asia, ISNAR, The Hague, The Netharlands, pp 37 – 49.

Bene, S.G., Beal, N. W. and Cote, A. 1977. Tree, Food and people. IDRC., Ottawa, Canada

Chadhar S.K. and Sharma M.C. 1996. Survival and yield of four medicinal plant species grown under tree plantations of Bhataland. *Vaniki Sandesh* 20(4):3–5.

De Silva T. 1997. Industrial utilization of medicinal plants in developing countries. pp. 38–48. In: Bodeker G., Bhat K.K.S., Burley J. and Vantomme P. (eds), Medicinal Plants for Forest Conservation and Healthcare. Non-Wood Forest Products No. 11, FAO, Rome, Italy.

King, K.F.S. and Chandler, M.T. 1978. The wastelands. ICRAF, Nairobi, Kenya.

Korwar, GR (1992). Alternate land use systems. In LL Somani, et al., (Ed) Dryland Agriculture in India State of Agricultural Research in India. Scientific Publishers, Jodhpur, India p. 143-168.

Lundgren, B.O. and Raintree, J.B. 1982. Sustained Agroforestry. In : Nestel, B. (Ed.).

Mishra, P.J. and Nanda, S.S. 2013. Integrated Farming System for Livelihood security in rain fed areas of Odisha, Book chapter 89 in Natural Resources Conservation: Emerging issues & Challenges. Published by Satish Serial Publishing House, New Delhi, pp. 981-984.

Mishra R.K. and Pandey V.K. 1998. Intercropping of turmeric under different tree species and their planting pattern in agroforestry systems. *Range Manag Agroforest* 19: 199–202.

Nair, P.K.R. 1984. Soil Productivity aspects of Agroforestry. ICRAF, Nairobi, Kenya.

Osman, M., Venkateswarlu, B. and Singh, RP. 1989. Agroforestry research in the tropics of India a Review. In National Symposium on Agroforestry Systems in India. CRIDA. Hyderabad.

Prajapati N.D., Purohit S.S., Sharma A.K. and Kumar T. 2003. A Handbook of Medicinal Plants. Agribios (India), 553 pp.

Vyas S. and Nein S. 1999. Effect of shade on the growth of Cassia ungustifolia. *Indian Forester* 125: 407–410.

Integrated Farming System Practices: Challenges & Opportunities, pp 215-230

New India Publishing Agency, New Delhi, India

9

Organic Farming in Integrated Farming Systems

S. C. Sahoo

Organic farming primarily aimed at cultivating the land and raising crops in such a way so as to keep the soil alive and in good health by use of organic wastes and beneficial microbes. Organic farming is performed with an intension for maintaining diversity of crops, keeping soil healthy for future years, producing high quality products and recycling nutrients. Unlike modern agriculture, organic farming is not depended on synthetic chemicals. Instead, it depends on natural products to boost soil fertility through microbial activity. Organic farming uses a multiple cropping system, which benefits from growing several crops. This crop diversity supports a range of plants, insects, soil micro-organisms and animals that add to a healthy farming system. Organic farmings are of different types such as bio-farming, nature farming, eco-farming and biodynamic farming. Organic agriculture is developed with concented efforts by various people to create the symbiotic relationship between the nature and man. In the present day, organic agriculture has shown potentiality in commercial, social and environmental sector.

Organic farming is a production management system, which promotes agro-ecosystem and agrobiodiversity. The objective of organic farming is to generate all the required plant nutrients inside the farm and adopt crop protection using local resources. Organic farming emphasizes on the use of farm yard manure,

Directorate of Extension Education, Orissa University of Agriculture & Technology Bhubaneswar-751003, Odisha, India crop residues, composts, vermi-compost and bio-fertilizers to enrich soil fertility and use of bio-control measures to manage insect and diseases. In particular, organic agriculture is intended to produce high quality, nutritious food which helps in preventive health care. Organic farming increases biodiversity at every level i.e. from micro-organism to mammals. It strengthens biodiversity by using bio-pesticides and inorganic fertilizers.

During the past few years, organic agriculture has become popular in many parts of the world. Despite this growth only a small share of total agricultural land is under organic agriculture. The small market shares in developed countries may be attributed to the premium price at which organic food is marketed. About the future for organic agriculture, some argue that it could become the conventional production system of the future, while others do not agree with. In future, the role of organic agriculture will be decided by the seasonal issues.

Organic agriculture's role can be determined when it can become economically competitive with conventional agriculture. This depends on productivity of organic agriculture, demand for its products and the extent of consumer prices. This factor therefore also has a strong policy component. Feeding the world with organic agriculture may require more land than with conventional agriculture. It will reduce the area of natural ecosystems and the quality of biodiversity on and around agricultural land will be improved. As global food security has become a primary concern, the productivity of organic agriculture for feeding the world is an important factor. Important factors in this respect are that some practices in conventional agriculture cause environmental damage and that the scarcity of natural resources will also become a limitation in conventional systems.

Indian farmers were basically organic farmers. Over the years, use of synthetic inputs has come to the level of causing a concern to the environment and human health. To make organic farming successful, it is essential that eco-friendly technologies, which can maintain agricultural productivity, have to be developed and made available to the farmers.

Certified organic products including all varieties of food products namely basmati rice, pulses, honey, spices, oil seeds, fruits, processed food, herbal medicines etc. are produced in India. Apart from edible sector, organic cotton, cosmetics etc are also produced. Banana, pomegranates, pineapple, grapes, amaranth, ginger, onion, sugar/ jaggery are some organic commodities produced in India.

Organic agriculture is as old as Indian agriculture. No standards were existing except the traditions set by our ancestors. Increase in population and per capita income are obviously pushing up the food demand, which needs to be met through enhanced productivity per unit area by application of modern tools of technology. In this context, suitable technology needs to be devised to maintain higher production level by use of organic inputs.

Advantages of organic farming

- Farmers can reduce their production costs as it is not necessary to buy expensive chemicals and fertilizers.
- In the long term, organic farms save energy and protect the environment.
- Organic food has no harmful chemicals such as insecticides, fungicides and herbicides.
- Organic plants are more drought tolerant.
- Protecting the environment from pesticides leaking into the soil and then the water table. So, pollution of ground water is checked.

In the present day context, organic farming is gaining importance in the era of conservation agriculture for sustainable crop production. Huge sources of organic matters e.g. agricultural waste, debris of forest, animal body parts, rural & urban residues go waste. With consistent effort, these can be appropriately utilized to provide plant nutrient by maintaining soil health. In the developed nations, there is great demand for organic food. Market for organic food is growing day by day, both inside the country and abroad.

Due to abundant availability of biomass, agricultural wastes and natural resources; there is vast scope for expanding organic agriculture in Odisha. Several plants present in the forest eco-system, can be effectively used for production of bio-pesticides to be utilized for management of insects, diseases etc. Various methods have been developed to manage pest and diseases without use of harmful chemicals. Use of botanicals, pheromone trap, light trap, clean cultivation, line sowing, proper water management techniques, summer ploughing, advocating tolerant crop varieties and use of biological agents are prominent among non-chemical plant protection measures.

Disadvantages of organic farming

- Working on an organic farm takes great amounts of time and energy. In order to meet organic requirements detailed methods and techniques are used.
- Organic farming requires tremendous amounts of skill. They are not allowed to use chemicals actions that conventional farmers are allowed to use. This requires immense amounts of knowledge and experience.

- An organic farmer cannot produce as much food as a conventional farmer over a short period of time.
- The organic products are quite expensive for the common people to afford and hence most of the organic products remain out of reach to the common people.
- The farmers have to control weeds, diseases and insects by manual methods/organic pesticides.
- It is difficult for a traditional farmer to adopt and learn the technology and practices of organic farming. In addition to that, a new organic farmer will require proper guidance from a trained organic farmer from time to time.
- Organic food is more expensive because farmers do not get as much out put of their land as conventional farmers do. Organic products may cost up to 40% more.
- Marketing and distribution is not efficient because organic food is produced in smaller amounts.
- Organic farming cannot produce enough food that the world's population needs to survive. This could lead to starvation in countries that produce enough food today.

Integrated organic farming

Integrated organic farming is a commonly and broadly used word to explain a more integrated approach to farming as compared to existing monoculture approaches. It refers to agricultural systems that integrate livestock and crop production. It denotes a holistic system of farming, which optimizes productivity in a sustainable manner through creation of interdependent agri-eco systems where annual crop plants, perennial trees, fish and animals are integrated on a given field (Panda, 2013). Integrated organic farming is the system of ongoing crops in a way, to keep the soil in good health. Use of organic wastes (crop, animal and aquatic wastes) and other biological materials, mostly produced in situ along with beneficial microbes helps organic farming. Concept of organic farming is based on following principles:

- Nature is the best role model for farming, since it does not use any external input.
- The entire system is based on understanding of nature's ways of replenishment. The system does not degrade the soil.

- The soil in this system is considered as a living entity
- The living population of microbes and other organisms in soil are significant contributors to its fertility on a sustained basis. It should be protected and nurtured.
- The total environment of the soil is more important and should be preserved.

Resource recycling in integrated farming system

Integrated farming system deals with several components including crop and household components. If planned appropriately, the bi-product of one component can be effectively used as raw material for production of another component. For example, the bi-product of paddy crop is used for mushroom cultivation; the bi-products of mushroom cultivation can be used as substrate for vermi-composting. The excreta of poultry farm can be used as feed for pisciculture (Rana, 2014).

Harnessing the nutrient from farm renewable resources include crop residues and farm wastes. They are valuable sources of plant nutrients. In tropical and subtropical soils, there is a general deficiency of organic carbon and plant nutrients due to rapid loss of these components by bio-degradation. To make up for these losses, extensive utilization of organic residues in agriculture is essential. In addition, they also protect the soil from erosion. The manurial value and quality of these wastes could be improved by composting and enriching these organic sources along with inexpensive materials such as rock phosphate. In India, there is great potential for utilization of crop residues / straw of some of the major cereals and pulses. Approximate availability of straw is to the tune of 141.2 m t, which contributes about 0.7, 0.84 and 2.1 m t of N, P_2O_5 and K_2O respectively, after deducting 50% quantity utilized as animal feed (Sharma, 2009). Farm manures should be returned to the fields. As a rule, organic matter fit for soil application should not be burned. In India, the estimated production of dung and urine from bovine population works out to 1002 and 658 m t respectively. They contribute about 5.71 m t of N, P & K. Cow dung is an important input for biogas plants having dual advantage of providing both fuel (gas) and fertilizer (slurry) (Sharma, 2009). Agro – forestry systems can lead to more nearer to the nutrient cycling than agriculture. The following renewable resources in farm can be utilized to develop business set-up in rural environment.

Recycling of farm wastes and crop residues has been found to economize farm production as well as safe environment. Korikanthimath and Manjunath (2004) working in Goa conditions found that adoption of intensified cropping systems helped to recycle the crop residue more efficiently than the rice crop alone.

The organic manure, FYM, poultry manure, mushroom residue smeared fully intact for reuse but the crop residue available under different system is about 62% and rest is used for dry or other purpose. In IFS model at PDFSR, Modipuram, all the farm wastes and crop residues were recycled either in situ incorporation in to the soil (green manure crops, cowpea intercropped in sugarcane, cane trash, leaves of potato and redgram, roots of berseem and other leguminous crops and green biomass added after picking of pods etc.) or by composting (Vermicompost, FYM) of cow dung & urine mixed with farm wastes (Singh *et al.*, 2011). A detail account of recyclable farm resources and nutrients availability is given in the table 1.

This nutrient budgeting indicates that through recycling of all the available farm resources, plant nutrients equivalent to 121.7 kg N, 226.8 kg P and 411.9 kg K could be added in to the soil. Considering a realizable amount of 30% of the total nutrient incorporated in to soil through recycling, a saving of 228 kg of NPK (44.6% of 511 kg of NPK – annually required for field and plantation crops) was observed. The average annual requirement of NPK however, was 285.3kg, 116.3kg and 109.9kg, respectively). In addition to this, nutrient rich pond silt and pond water recycled for crop production also add a total amount of 18.56 kg N, 6.21 kg P and 74.24 kg K with a market value of rupees nine hundred fifty or more. The OC% of pond silt was as high as 1.20 with an average value of 0.95. Addition of pond silt and water was found to increase

Model of resource recycling

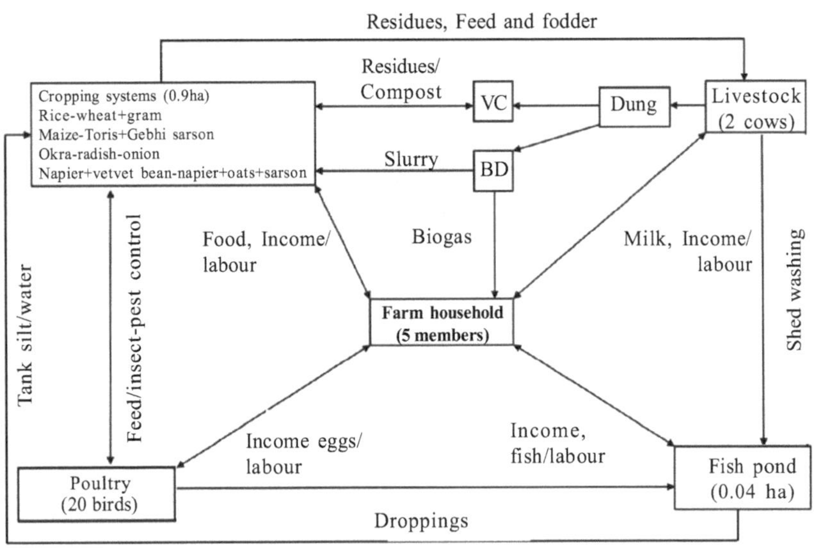

Resource flow model of integrated farming system – wet land (1.0 ha) (Crops + Livestock + Fish + poultry)

Table 1: Nutrient budgeting under integrated farming system at PDCSR, Modipuram

Source of nutrients and per cent nutrient content (N:P:K) on dry wt. basis	Available quantity at farm (kg)	Approx. released nutrient N (kg)	Approx. released nutrient P (kg)	Approx. released nutrient K (kg)	Total released/ required in the IFS (kg)
Green manure crops					
Sesbania spp. (1.29:0.36:1.64)	8800	18.9	5.3	24.0	48.2
Cowpea (1.29:0.36:1.64)	8500	18.3	5.1	23.2	46.6
Crop residues (dry wt.)					
Sugarcane leaves (0.4:0.18:1.28)	900	3.6	1.6	11.5	16.7
Arhar leaves (1.29:0.36:1.64)	232	3.0	0.8	3.8	7.6
Potato leaves (0.52:0.21:1.06)	1450	7.5	3.0	15.4	25.9
Cow dung (dry wt.) (0.4:1.2:1.9)	17600	70.4	211.0	334.0	615.4
Total nutrients added in to soil	-	121.7	226.8	411.9	760.4
% of total requirements	-	42.6%	>100%	>100%	
Nutrient requirement/year (field + plantation crops)	-	285.3	116.3	109.9	511.5

(*Source:* Singh *et al*, 2011)

the yield of rice and wheat by 3.48 q/ha and 2.41 q/ha, respectively. Organic source of nutrients are rather cheap than chemical fertilizers and also help in maintaining soil health and keep environment safe.

Residue recycling – Odisha experiences

The dairy based integrated farming system model with dairy pisciculture, goatery, mushroom, poultry, vegetables and paddy cultivation in 1.15 ha area at Nimapara in Puri district generated net return of Rs 4, 53, 595/year as compared to Rs 38, 480 in the conventional method involving only rice cultivation and pisciculture (Table 2). The B:C ratio was the maximum (4.5) from banana cultivation followed by pisciculture (4.1) (Nanda 2015). The residue obtained from rice was utilized in mushroom cultivation and the residues of live stock and horticultural crops were utilized in vermi composting. The poultry litters were partially used in pisciculture.

A small farmer of Baragarh district in the west central table land zone with 1.6 ha land area, with investment of Rs 56800 has earned net profit of Rs 67680 per year from crop-horticulture-animal based integrated farming system as against Rs 39200 per year from the conventional cropping practices (Table 3). Maximum B:C ratio (2.7) was obtained from papaya cultivation followed by banana and vegetable crops (2.5) (Nanda and Sahu 2012). The residue obtained from crops was utilized as feed in pisciculture.

Table. 2: Performance of vegetable-pisciculture-poultry-dairy system at Nimapara, Puri

Component	Area(ha/No*.)	Expenditure(Rs)	Net returns		B.C.ratio	Employment generation (man days)	
			(Rs)	% contribution		Male	Female
Crop component							
Paddy	0.4	22,000	23,080	5.1	2.0	90	60
Horticulture component							
Vegetables	0.2	27,000	60,500	13.3	3.2	105	45
Banana	0.1	5,090	17,810	3.9	4.5	25	18
Papaya	0.05	775	425	0.1	1.5	3	4
Animal component							
Dairy	5*	64,800	1,51,200	33.3	3.3	151	30
Goatery	5*	7,000	15,500	3.4	3.2	25	10
Poultry	100*	15,520	23,980	5.3	2.5	20	18
Other enterprises							
Fishery	0.4	25,000	77,000	17.0	4.1	30	61
Vermicompost	4 units	6400	18,100	4.0	3.8	2	4
Mushroom	1,800*	54,000	66,000	14.6	2.2	40	60
Apiary	3 units	5600	10,600(3 yrs)	2.3	2.9	2	1
Total	1.15	2,33,185	4,53,595	100	2.9	493	311
Conventionna method							
Paddy	0.6	12,700	8,480	22.0	1.6	45	25
Fishery	0.4	18,000	30,000	78.0	2.6	52	12
Total	1.0	30,700	38,480	100	2.2	97	37

Table 3: Rice based integrated farming system from Baragarh

Component	Area/ No.	Expenditure(Rs)	Net Profit(Rs)	B.C. Ratio
Paddy	0.6 ha	16800	16680	1.99
Fish	0.2 ha	12000	8000	1.67
Banana	80 nos.	3200	4800	2.5
Papaya	80 nos.	4800	8200	2.7
Vegetables	0.8 ha	20000	30000	2.5
Total	1.6 ha	56800	67680	
Conventional method	1.6 ha	37600	39200	

Organic products in integrated farming system

Now-a-days, there has been focus on production of organic products in agriculture and allied areas. In integrated farming system, efforts can be made to produce various products like food grain, oilseeds, vegetable, fruits, mushroom, fish, meat, milk etc. organically to fetch good price in the market. Several components of organic farming can be utilized in integrated farming system models to reduce the cost of production and enhance the quality of produce. Similarly, this will help to recycle the resources properly for enhancing production and profit of integrated farming system models.

Vermicompost

Vermicompost is the end-product of the breakdown of organic matter by earthworm. These products have been shown to contain reduced levels of contaminants and a higher concentration of nutrients than other organic materials before vermi-composting. Practice of vermi compost production for selling or for own use will substantially increase the family income.

Farmyard manure (FYM)

In many villages, cow dung cakes are used as alternate source of fuel. Cow dung has great potentiality to be used for production of farm yard manure. To prevent use of cow dung directly for burning, fuel wood can be used as an alternate in place of cow dung cakes in household usage. FYM also contains plant material to which is used as mat for animals to absorb the faeces and urine. Farm yard manure in liquid form, known as slurry, is produced by more intensive livestock rearing systems where concrete is used, instead of straw bedding. Manure from different animals has different qualities and requires different application rates. For example; cattle, pigs, sheep, chickens, goats etc. have different properties. For instance, sheep manure is high in nitrogen and potash, while pig manure is relatively low in both. Chicken litter, coming from a bird, is very concentrated in nitrogen.

Processing of farm yard manure is also not up to mark by the rural people. Farmers should be trained on scientific method of preparing farm yard manure. This will help in propagating the right usage of FYM.

Green manure

Green manuring crop such as dhaincha, glyricidia, cowpeas should find place in crop rotation. These green manuring crops enrich the fertility of soil.

Azolla

Azolla is a fern grown in water for use as fertilizer and cattle feed. It is mainly used in rice farming. Azolla is very rich in proteins, essential amino acids, vitamins (vitamin A, vitamin B_{12}, Beta Carotene), growth promoter intermediaries and minerals like calcium, phosphorous, potassium, ferrous, copper, magnesium etc. Azolla, on a dry weight basis, constitut 25-35% protein, 10-15% mineral and 7-10%, a combination of amino acids, bio-active substances and biopolymers. Azolla can be easily digested by livestock due to high protein and low lignin content. It doubles its biomass within 2-3 days. Generally, Azolla multiplies vegetatively. Azolla covering water surface reduce light penetration to soil surface, resulting in reduction of germination of weeds. Thus, growth of azolla reduces occurrence aquatic weeds in flooded rice fields. Azolla has been used as feed for pig, duck and fish.

NADEP compost

The NADEP Compost is unique compost that can be prepared in certain time frame and with less human effort. The process of making the compost involves layering of several combustible materials in a mud-sealed structure with bricks and water. Due to its simplicity and ease of applicability in fields, the method became popular among the farmers in Western India. Rural women can easily use this method to generate good quality of compost.

Biogas plant

Biogas is a methane rich flammable gas that results from the decomposition of organic waste material. It is produced by the anaerobic digestion or fermentation of biodegradable materials such as biomass, manure and plant materials. Owing to simplicity in implementation and use of cheap raw materials in villages, it is one of the most environmentally sound energy source for rural areas.

Women can use Bio-gas for cooking, lighting and regular income by utilizing biogas slurry. This slurry is good manure for kitchen gardens. Bio-gas also reduces drudgery of collecting fuel wood.

Water hyacinth compost

The water hyacinth (*Eichhornia crassipes*) is a floating "obligate" (requiring a wet habitat) plant. This alien species grows in all types of freshwater ecosystems. Water hyacinth varies in size from a few inches to over three feet tall with showy lavender flowers. Its leaves are rounded and leathery, attached to spongy and sometimes inflated stalks. The plant has dark feathery roots. It is very efficient in utilizing aquatic nutrients and solar energy for profuse biomass production. Depending on the time of the year and location, the plants double in number and biomass every 6 to 15 days. It makes sound economic sense to utilize this species as an organic input to soils. Water hyacinth waste provides mulch that assists in both water retention and weed suppression.

Organic poultry production

It is a positive step to rear poultry organically. From a large commercial enterprise to a child's training ground for animal care and business, poultry flocks provide several opportunities.

Organic feed for any animal suitable of high quality and nutritious. Good rearing practices; feed care, housing, lighting and cleanliness, all contribute to the health and productivity of the birds.

Organic feed should be 'certified organic' and so labelled by a recognized certification organization. Then the producing farmer will meet expectations set out in organic standards. Many conventional birds are treated with antibiotics to improve feed conversion and prevent disease. One should avoid drug residues in the eggs and meat by raising the birds organically, but it may increase the cost of production. The increasing demand for organic food shows that this choice can pay in long run. Chickens need high protein feed which should be supplemented.

Feeding tips

- Chickens don't like dusty or powdery mash. Moistening the mash can stop the valuable protein powder from filtering down and being left in the feeder.
- Sprouted grains help better egg colour in winter and are a good source of vitamins. They can be used to replace synthetic amino acids.
- Supplement the flocks' feed with kitchen scraps; bread, vegetable peelings, apples etc. Excess feeding may cause nutritional imbalance and affect the health and production of the birds. Don't feed raw potato peels.

- Moistening their food may stimulate their appetite.
- Foraging for plants, seeds, grubs and insects allow a flock to balance its own diet and can reduce feed costs by about 30% in summer.

Organic method of dairy management

Organic farming is a system of production, a set of goal-based regulations that allow farmers to manage their own situations. To get good quality organic dairy products, the following points are to be kept in mind.

- Cows and calves are fed 100% organic feed.
- Organic crops, hay and pasture are grown without the use of synthetic fertilizers and pesticides.
- Genetically modified organisms (GMOs) are not to be used.
- Land used to grow organic crops should be free of all prohibited materials for at least three years prior to the first organic harvest.
- Synthetic milk replacers are prohibited. Calves may be fed organic milk.
- Animals over six months of age should have access to pasture during the growing season.
- Only approved health care products can be used. Many of these are restricted in how and when they can be used. Antibiotics are not allowed.
- An organic farmer should keep sufficient records to verify the compliance with the standards.

Organic fish production

Organic aquaculture aims to have natural fish production, improving traditional production and high efficiency production with low stocking density. Non-use of chemical products and hormones are encouraged. Because of these reasons, organic aquaculture takes much attention all over the world. However, lack of certification organisations and standards for organic aquaculture in many countries cause lower production for organic aquaculture.

The fish that are produced under natural conditions without being exposed to any protective additives or genetic modification are called "organic fish". Organic aquaculture is the production process every stage of which is controlled and certificated by the certification agency. Fish can be grown with organic method in, domestic waters, pools, net cages, barrages, lakes, ponds, fish traps in farms. Organic products have richer and more beneficial attributes and term of food value. Organic fish production is a model of production without using any chemicals, pesticides or the products modified genetically. Although this alternative production model used in many countries in the world constitutes very meager portion of world's aquaculture production.

The organically development status of fishery products which is one of the world's fastest growing sectors is highly similar to that of organic agriculture. Besides, organic aquaculture has fallen behind the agriculture in terms of certificated product diversity and quality.

Production environment

The facility which will be used in organic fishery production should be constructed in a region that will not be affected negatively from conventional production units. The natural landscape in the field should be given importance. It should not be given any harm to the plants that are dying out. Besides, a part of the field should be kept for natural vegetation. Water source which will be used is needed to be of high quality. This water source (stream, river etc.) should continue its ecological functions in its natural condition.

For the protection of the field, it is suitable if measures that will not harm other living creatures in the system. Domestic animal species should not be disturbed. For instance, protective materials such as nets can be used for this purpose. If there are any constructions such as barrages in the farm field, some ways suitable for fish movement can be constructed.

In organic fishery production, farm wastes should not be released to water that will affect biological diversity negatively or cause pollution.

Fish health and welfare

Organic management practices provide a high level of disease resistance and prevention of infection. Principally, some precautions for the fish should be taken. All management techniques, especially when influencing production levels and rate of growth maintain the good health of the living organisms. To maintain this, the stock density should be low, regular health checks should be made; dead fish should be immediately removed from the pool and stressing factors should be lowered to the minimum level. The choice of the species resistant to diseases and subspecies should be considered.

In case of disease, it is proper to use natural treatment methods. Synthetic chemical medicines, antibiotics are not allowed. If any medicine is given in 3 months before the sale of the product the products obtained from them cannot be sold as organic. The preparations obtained from some plants like garlic (*Allum satium*), euphorbia (*Euphorbia sp.*), horse chestnut (*Aesculus hippocastanum*) & neem (*Azadirachta indica*) and the preparations of *Bacillus thuringensis* can be used. In the fight against diseases and the disinfection of pool equipment, the usage of some organic components (hydrogen peroxide, rock salt, quicklime, sodium hypochlorite) and the components that are not toxic in nature (formic acid, citric acid, alcohol) areallowed.

Conclusion

No process is perfect and has its own pros and cons. Organic farming has its own benefits that are bigger than the related disadvantages. Thus, it is essential to create awareness among the people about the benefits of this process so that it can be adopted on a large scale..

Certification of organic produce plays a vital role in production and marketing of organic food. Certification is a step to make organic product distinguishable from the conventional product. This will help to fetch good price in domestic and international market for the producers. Certification of organic produce is a positive distinction for organic products thus avoiding confusion. It also helps in market planning and gives organic agriculture more credible position to the outside world.

Organic way of integrated farming system will not sustain the farm productivity over the years, it will maintain the soil and environment and provide higher in come to the farmer.

References

Bio Suisse (Association of the Swiss Organic Agriculture Organisations), 2001. Standards for the production, processing and marketing of produce from organic farming. http://www.homesteadorganics.ca/poultry-production. aspx

Jim Pierce, Oregon Tilth. 2014. Introduction to Organic Dairy Farming. http://www.extension.org/pages/18325/introduction-to-organic-dairy-farming#.VKOpPNKUfy0

Korikanthimath,V.S and Manjunath, B.L. 2004. Resource use efficiency in Integrated farming Systems. In Proceeding of the Symposium on Alternative Farming Systems; Enhanced income and employment generation options for small and marginal farmers, PDCSR, Modipuram, pp.109-118.

Nanda, S.S. and Sahoo, S.C. 2012. Integrated farming system for livelihood security of farmers in Odisha. National Symposium on rice based farming system for livelihood security under changing climate J. Censes. 27-29. Feb. 2012 Chiptima, Odisha. Pp 21-27.

Nanda S.S. 2015. Sustainable livelihood options for poverty eradication. Presentation made at Global Social Science Conference, OUAT, Bhubaneswar, 14-17, Feb-2015.

Panda, H. 2013. Integrated organic farming handbook. Asia Pacific Business Press Inc. Pp 472. (Source : http://www.niir.org/books/book/integrated-organic-farming-handbook/isbn-9788178331522/zb,18b58,a,3,0,3e8 /index. html

Rana, S.S. 2010. Presentation on resource recycling and flow of energy in different farming systems. (Source :http://hillagric.ac.in/edu/coa/agronomy /lect/agron-4711 /Lecture% 2010%20Resource%20 and%20 energy% 20 flow%20 in%20 different%20 IFS.pdf of energy in different farming systems

Sharma, A.K. 2009. Bio – fertilizers for sustainable Agriculture. Published by Agro bios (India), Agro House, Jodhpur. Pp 407.

Singh, J.P., Gangwar, B., Pandey, D.K. and Kochewad, S.A. 2011. Integrated Farming System Model for Small Farm Holders of Western Plain Zone of Uttar Pradesh. PDFSR Bulletin No. 05, pp. 58. Project Directorate for Farming Systems Research, Modipuram, Meerut, India

Singh, S.K. 2013. Recycling of within farm renewable resources: An entrepreneurial opportunity for woman folk. Lecture delivered during model training course on Gender Perspective in Integrated Farming System w.e.f 17-24 January 2013 at ICAR Research Complex for Eastern Region, Patna, Bihar.

Sridhara, S., Nagachaitanya, B., Chakravarthy, A.K., Shetty, T.K.P. 2009. Women in Agriculture & Rural Development. Published by New India Publishing Agency, New Delhi. Pp. 358.

Yesim Ötles, Osman Ozden, Semih Ötles. 2010. Organic fish production and the standards. Acta Scientiarum Polonorum, Technologia Alimentaria 9(2) 2010. http://www.food.actapol.net/pub/1_2_2010.pdf.

Integrated Farming System Practices: Challenges & Opportunities, pp 231-268

New India Publishing Agency, New Delhi, India

10

Integrated Farming Systems: An Initiative for Climate Resilient Agriculture

S.C. Mohapatra and *S.K. Satapathy

The vulnerability of agriculture to climate change has become an important issue resulting in reduced productivity due to adverse climatic changes. The overall purpose is to assess farming systems change to enable adaptation to climate change. It deals with the predicted climate change scenarios and the adaptations necessary to reduce climate change impacts and examines improvement of indigenous breeds for adaptation to climate change. It also explores the use of crop modeling as avenues to address adaptations of crop production systems and draws the recommendations to outlines of future research needs.

Although the adverse effects of climate change are likely to be borne disproportionately by the small farmers of the developing world, it is the integrated farming systems with their high degree of biodiversity which are best equipped to withstand the shocks of climatic extremes. Apart from the landless and urban poor, small farmers are among the most disadvantaged and vulnerable groups in the developing world. It is obvious that climate-related environmental stresses are likely to affect individual households differently compared to more market-

Directorate of Extension Education, Orissa University of Agriculture & Technology, Bhubaneswar-751003, Odisha, India
*Krishi Vigyan Kendra, Ganjam, Berhampur, Odisha

oriented farmers. The climate change reduces crop yields. The effects on the welfare of subsistence farming families may be quite severe, especially if the subsistence component of productivity is reduced. Changes in quality and quantity of production may affect the labour productivity of the farmer and negatively influence his/her family health(Rosenzweig and Hillel 1998).

Global warming is predicted to result in a variety of physical effects including thermal expansion of sea water, along with partial melting of land-based glaciers and sea-ice, resulting in a sea level rise which may range from 0.1 to 0.5 metres by the middle of the next century, according to present estimates by the Inter-governmental Panel on Climate Change (IPCC). The IPCC has projected potential impacts of climate change could adversely affect agricultural production and food security. A sea level rise could pose a threat to agriculture in low-lying coastal areas, where impeded drainage of surface water and of groundwater, as well as intrusion of sea water into estuaries and aquifers, might take place. In parts of Egypt, Banglasdesh, Indonesia, China, and other low-lying coastal areas already suffering from poor drainage, agriculture is likely to become increasingly difficult to sustain. Some island states are and coastal areas particularly at risk (Rosenzweig and Hillel 2008).

A climate change impact potentially significant to small farm production is loss of soil organic matter due to soil warming. Higher air temperatures are likely to speed the natural decomposition of organic matter and to increase the rates of other soil processes that affect fertility. Under drier soil conditions, root growth and decomposition of organic matter are significantly suppressed, and as soil cover diminishes, vulnerability to wind erosion increases, especially if winds intensify. In some areas, an expected increase in convective rainfall - caused by stronger gradients of temperature and pressure and more atmospheric moisture - may result in heavier rainfall, which can cause severe soil erosion.

Conditions are usually more favourable for the proliferation of insect pests in warmer climates. Longer growing seasons may enable a number of insect pest species to complete a greater number of reproductive cycles during the spring, summer and autumn. Warmer winter thus causing greater infestation temperatures may also allow larvae to winter-over in areas where they are now limited by cold, during the following crop season A gradual continuing rise in atmospheric carbon dioxide (CO_2) will affect pest species directly and indirectly. Models on plant diseases indicate that climate change could alter stages and rates of development of certain pathogens, modify host resistance, and result in changes in the physiology of host-pathogen interactions. Climate change could have positive, negative or no impact on individual plant diseases, but with increased temperatures and humidity many pathogens are predicted to increase in severity.

The possible increases in pest and disease infestations may bring about greater use of chemical pesticides to control them, a situation that may enhance production costs and also increase environmental problems associated with agrochemical use. Of course, this may not be the case with farmers who use integrated farming systems or other forms of diversified cropping systems that prevent insect pest buildup either because one crop may be planted as a diversionary host, protecting other, more susceptible or more economically valuable crops from serious damage or because crops grown simultaneously enhance the abundance of predators and parasites which provide biological suppression of pest densities (Altieri and Nicholls 2004).

We should address climate change related issues in systematic manner through constructive and developmental programme to increase in productivity and profitability of small and marginal farmers. Climate change effects will be more in Indian conditions as there are different agro-climatic zones. The most affected will be marginal and small farmer which constitute 80% of Indian farming community. Therefore, a comprehensive and multidimensional strategy is needed to mitigate climate change effects.

Climate change is a reality. No longer it can be considered a mere possibility. This is evident from the increase in the occurrence of odd and extreme weather events. Naturally, the agriculture sector has started to feel the pinch. It is now clear that though climate change may be a global phenomenon, its consequences are felt locally. Thus, we need to take situation-specific actions to mitigate the impact of climate change and adapt the farming system to the changed weather conditions. Fortunately, the fast-developing field of information and communication technology (ICT) is coming in handy for the agriculture sector. The farm sector is finding it harder to meet the growing and diversifying demand because of climate change and depleting natural resources. Besides, farming has lost its remunerative edge.

Effect of climate on farming systems

The agricultural sector has, broadly, a dual system comprising a well developed, capital intensive and export oriented *commercial sector* and a subsistence-based *community farming* sector. It is estimated that, later sector has low in technology and external inputs and highly dependent on labour near about 90% households are directly involved in crop, cattle and small live stock production respectively. Two major farming systems exist and these are: small scale cereals & livestock and Intensive agriculture. Owing to climate, soil types and evapo-transpiration rates it is better adapted to crop-based agriculture. At present, livestock contribute 40% to the agricultural output. Crop production activities are limited, mainly due to general dry conditions and poor soils. Low and sometimes poorly distributed rainfall have limited rain-fed crop production to only those areas receiving 1000 mm and above annually it is about 64% of the Indian states. Rain-fed crops include upland rice, wheat, maize, sorghum and ragi. Irrigated agriculture is concentrated in fertile areas with high annual rainfall or abundant water resources from rivers, streams, dams or borewells. Horticultural production occurs across a wide range of environmental conditions, with distribution restricted primarily by access water and soil quality and topography. Horticulture supports more than 50 crop types, categorized into three commodity groups. Horticultural production systems have good potential for expansion and are seen as a source of adaptation to climate change. The development of farming systems is supported by an agricultural policy which is pro-poor and driven by a need to reduce poverty.

Although rice and wheat cultivation on an average contributed around 70% to total agricultural economic output, it is important in terms of its contributions to food security. This sector is sensitive to climate variability and change. The periodic fall in cereal production due to drought or floods, create cereal deficits, which increases cereal imports and prices. With climate change, these occurrences are expected to be frequent thus necessitating huge budget appropriations for the national calamity (drought & flood) relief fund.

The effect of climate change on the livestock sector is experienced through changes in quality and quantity of vegetation, availability of fodder and the occurrence of climate related animal diseases. Indirectly, these factors reflect themselves on livestock productivity parameters such as conception and calving rates, mortality rates and meat quality. During periods of drought, livestock production volumes decrease, while the number of livestock marketed increases. The value of cattle and small ruminants may drop due to increased supply and the deteriorating conditions of the animals. Farmers often opt to sell their animals before the value reaches unacceptably low levels (Cosbey *et al.* 2005).

The impact of climate change on the dairy sector is mainly through input shortages and prices. Input prices of dairy products increase in times of drought. Farmers have to purchase additional fodder to feed their milchy cows (Henson, 1992).

Climate change issues can be handled in four impressive issues through Natural research management (NRM), crop production management strategy, Management of livestock and fisheries and other intervention of public and private. The details of these activities can be depicted as follows. The important findings related to natural resources, crop production livestock and fisheries and institutional interventions are highlighted from Table 1 to 10 and Fig. 1 to 10.

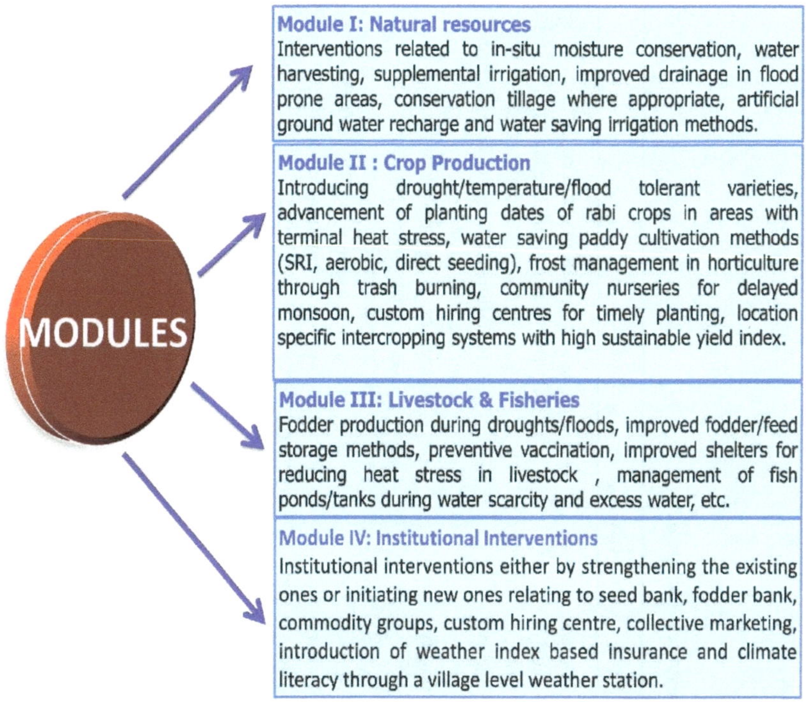

Fig. 1: Different natural resource management activities at village level

Table 1: Thematic areawise intervention on crop and enterprises for natural resource management

Thematic area	Area/ No	Intervention	Crop / Enterprise	Remarks
In-situ moisture conservation	500 no	LLDP mulching	Mango	Cost of irrigation reduced 20%
	20 ha	Summer deep ploughing	Paddy	Conserve soil moisture about 8%
	8 ha	Ridge & furrow method	Cowpea ,Radish	Water requirement is 18% less
	10 ha	Line sowing with seed cum fertilizer drill	Paddy	Germination 12% higher & moisture content 16% higher
	35 ha	Green mannuring	Paddy	Increased in soil pH from 5.5 to 5.8
Recycling of organic matter	-	Vermicomposting	-	38 q vermicompost produced
Soil reclamation	650	Soil testing	Cereal, PulsesVegetable	Balanced fertilization
	16 ha	Acid soil management	Maize	14% increase in yield(45.4 q/ha)
			Groundnut	12% increase in yield (24q/ha)
Resource conservation	4 no	Construction of check dam	Water storage capacity increased by 2250 cum, Irrigation area increased by 9.0 ha.	
	25 No	Percolation ponds	14% increase in soil moisture content	
	3 ha	Renovation of defunct farm pond	Water storage capacity increased by 2400 cubic Meter, Irrigation area increased by 8.0 ha. & cropping intensity increased by 2.6 %.	
	3 ha	Shelter belt plantation	Cashew+casuarina in river embankment	

Table .2. Effect of different varieties on productivity and economics under crop production management

INTERVENTION	Crop	Area (ha.)	Yield (q/ha)		Net return(Rs)		B:C ratio	
			FP	IP	FP	IP	FP	IP
Varietal replacement (Characteristics-Short duration, high yielding/hybrid, drought resistant)								
Khandagiri(Short duration)	Paddy	12.0	22.6	26.4	111.00	14950	1.96	2.24
Sahabhagi dhan(Drought resistant)	Paddy	20.0	21.4	28.8	10200	16300	1.91	2.32
MTU-1001 (High yielding)	Paddy	18.0	34.0	39.4	16700	21400	1.88	2.16
Pratikshya (High yielding)	Paddy	10.0	36.4	43.5	17500	22600	1.85	2.2
Swarna sub-1 (Flood resistant)	Paddy	8.0	36.8	43.9	18400	23100	2.0	2.3
Super-36 (High yielding)	Maize	15.0	-	47.5		26400		2.4
MH- 9468 (High yielding)	Maize	8.0	-	41.6		20300		2.1
3OR77 (High yielding)	Maize	6.0	33.5	42.3	13500	21000	1.8	2.2
TARM-1 (Short duration)	Greengram	12.0	5.8	8.2	11200	17400	1.8	2.1
OBGG 52 (Short duration)	Greengram	8.0	5.3	7.5	10100	15600	1.7	1.94
Prasad (Short duration)	Blackgram	5.0	4.9	7.7	7200	16800	1.7	2.2
PU-31 (Short duration)	Blackgram	7.0	4.5	7.2	7100	14300	1.6	2.1
Utkal kumari (High yielding)	Tomato	4.0	187	214	51100	60800	2.20	2.32
Chiranjivi (hybrid)	Tomato	4.0	219	307	63200	81300	2.3	2.71

FP : Farmers' practice, IP: Improved practice

Table 3: Effect of crop diversification on productivity and economics under crop production management

Thrust Area	Crop	Area (ha.)	Yield (q/ha)		Net return(Rs)		B:C ratio	
			FP	IP	FP	IP	FP	IP
Crop diversification / Crop alternation								
Green star	Brinjal	4	188	243	67900	91100	2.08	2.45
Sufala	Brinjal	2	184	202	64000	72000	2.1	2.3
Pusa jwala	Chilli	2	68.2	81.4	72500	94500	2.13	2.38
Utkal manika	Cowpea	2	49.2	65.2	27280	38580	2.60	2.92
Damini	Cauliflower	4	135	152	60700	71600	2.1	2.31
Krishna-1	Cauliflower	4	142	164	64400	78800	2.2	2.47
Konark	Cabbage	2	196	234	60300	74000	2.54	2.68
Cannon ball	Watermelon	4	220	325	60000	92000	2.3	2.8
Ajaya	Okra	2	87	120	46600	54500	2.1	2.32
Guamal local	Pumpkin	2	260	350	55000	83000	2.2	2.65
Nakhara	Bittergourd	1	85	110	33600	49000	2.0	2.4

Table 4: Effect of integrated nutrient management on productivity and economics under crop production management

Thrust Area	Crop	Area (ha.)	Yield (q/ha)		Net return(Rs)		B:C ratio	
			FP	IP	FP	IP	FP	IP
Integrated nutrient management (Micronutrient, biofertiliser along with RDF)								
Micronutrient application (B &Zn)	Paddy	22.0	38.1	42.2	17700	24000	2.01	2.11
Application of Boron	Cauliflower	8.0	170.0	192	55600	76400	2.63	2.8
Application of Rhizobium with RDF	G nut	10.0	12.4	16.8	21200	33900	2.3	2.7

Fig. 2: Different crop production practices

Fig. 3: Paddy straw and oyster mushroom production for income generation

Table 5: Intervention on different Livestock management activities

Intervention	Area (ha)	Remark
Animal health camp	487 animals	Increased resistant to different diseases
Feeding management in livestock	272 animals	Milk yield/lactation (Lt.)- 1125(FP) 1400 RP)
Fodder production	0.04	242q/ha
Introduction of Rainbow Rooster Poultry birds	500 birds	Avg. body wt. at 6 months- 2.4 kg with a net return of Rs 240/bird & B:C ratio of 3.0
Backyard Poultry Variety- Rooster Poultry birds	975 birds	Avg. body wt at 6 months 2.5 kg with a net return of Rs 190/bird & B:C ratio 2.6
Breed upgradation of buck var. Black Bengal	-	Black Bengal buck even it in extreme climate have good growth
Eco-friendly Shelter house for ruminants	-	Biodegradable wire net • Less heat generation inside the shelter Improvement of ventilation

Fig. 4: Livestock management activities

Table 5: Intervention on different fish production activities

Interventions	Area(ha)	Results
Plankton management in fish culture (Fermented mixture of mustard oil cake 50kg/ha, Rice bran @50kg/ha, Sugar jaggery @25kg/ha and yeast @20-25gm/ha)	02	T_1 16.5q/ha, T_2 – 24.7q/ha; T_1 -Plankton population (ml/54.5lt water) 0.75 T_2 - Plankton population (ml/54.5lt water) 2.4

Fig. 5: Fish production activities

Fig. 6: Capacity building and extension activities for dissemination of farm technologies to the rural mass through institutional intervention

Table 7: Impact assessment of custom hiring centre

Sl. No	Implements hired	Used in crop/ Any other	Working Efficiency (Ha/Hr)		% Inc. in	saving of Mandays /ha
			Before Hiring	After Hiring	Work use efficiency	
1.	M.B. plough	paddy	0.06	0.2	220	3
2.	Power sprayer	Insecticide spray	0.07	0.33	367	2
3.	Diesel Pumpset	Irrigation	0.08	0.20	140	4
4.	Power weeder	Tomato, chilli	0.01	0.18	1700	15
5.	Sprinkler	Irrigation	0.07	0.25	250	6
6.	Power tiller	Land preparation	0.06	0.20	220	6
7.	Thresher	Threshing of paddy	1q/hr	11q/hr	1100	30

Fig. 7: Farm implements of custom hiring centre used by farmers

Table 8: Contingent measures in post *phailin* period

Crops	Stage of damage in crop as applicable	Damage %	Contingent measures
Paddy	Panicle	>50 %	Excess water drained out to save the crop. Monitoring & management of BLB (Copper Oxychloride + Streptocycline) & BPH (Imidacloprid) was done. Contingent crops like Greengram ,Black gram were grown in damaged field.
Vegetable	Flowering & fruiting	30-35%	Excess water drained out to save the crop. Management of wilt (Carbandazim + Plantomycin) was done.
Mango	Vegetative	60-65 %	Renovation of orchards with intercropping of vegetables
Cashew nut	Vegetative	55-60 %	Planting of seedling to be done
Papaya	Fruiting	100 %	New plantation to be done

Few examples on impact of climate change and climate resilient farming

Seasonal variability in rainfall and long term warming trends are posing serious threats for sustainable crop and livestock production. (Annual report, ICAR 2012-13)

Rice varieties to mitigate the climate change effect

The climate change and its effect on agriculture are interrelated processes draws utmost importance to enhance the resilience of agricultural production systems to various climatic variability. Rice (*Oryza Sativa L*) is the major crop in Odisha covering about 69% of the cultivated area and 63% of the total food grain production. On the basis of thirty years weather data, three drought prone districts viz. Ganjam, Jharsuguda & Subarnapur and one flood & cyclone prone district Kendrapara were selected for intervention, in farmers' field through respective Krishi Vigyan Kendras. The improved high yielding rice varieties Khandagiri, Sahabhagi dhan, Konark, Pratikshya, Lalat and hybrid variety JKRH-401 were allotted randomly to compare with local genotypes with respect to their yield and other effects under changing climate. The influence of abnormal climatic situation and location specificity on the grain, straw, above ground biomass yield of rice and carbon content atmospheric CO_2 fixation in its aerial portion and economic feasibility were studied in the NICRA Project during 2012 and 2013. The mean data of the two consecutive years showed that under drought prone situation HYV rice Pratikshya recorded highest grain yield (38.34 q/ha), biomass (80.44 q/ha), carbon content (25.74 q/ha), CO_2 fixation (2.15 X

Table 9: Pre and post evaluation of different crops in NICRA adopted villages

Change in Area/Nos. in crops		Change in Yield in crops	
Before NICRA	After NICRA	Before NICRA	After NICRA
Drought tolerant paddy varieties = 0 ha.	var. Sahabhagi dhan = 23 ha.	Paddy(Upland) = 21 q./ha (medium land) = 34 q./ha. (Low land) = 28 q./ha	Paddy (Upland) = 32 q./ha. (Medium land)= 55 q./ha. (lowland) = 45 q./ha.
Maize var. Navjot= 3 ha.	Hybrid Maize (Bisco 740, 30-R-77) = 16 ha.	Maize (Navjot) = 23 q./ha.	Maize (Hybrid) = 38 q./ha. seeds
Radish= 1.5 ha. Cowpea = 1.2 ha. Cauliflower = 3 ha. Cabbage = 2.3 ha. Brinjal = 1.2 ha.	Radish = 1.3 ha. Cowpea = 3 ha. Cauliflower = 5 ha. Cabbage = 5.4 ha. Brinjal = 2.5 ha.	Radish= 130 q./ha. Cowpea = 48 q./ ha. Cauliflower = 150 q./ha. Cabbage = 132 q./ ha. Brinjal = 140 q./ha.	Radish = 173 q./ ha. Cowpea = 88 q./ha. Cauliflower =182 q./ha. Cabbage = 192 q./ ha. Brinjal = 214 q./ ha.

Table 10: Pre and post evaluation of different income generating enterprises and use of farm machineries (no.) in NICRA adopted villages

Enterprises	Before NICRA	After NICRA	Before NICRA	After NICRA
Mushroom production	Individual -4	Individual – 8 SHG - 3	Power tiller – 1	Power tiller – 06
Floriculture	Individual – 5	Individual – 5 SHG – 3	Tractor – 1	Tractor – 3
Soil Testing	5 %	30 %	Paddy Winnower -0	Paddy Winnower-6
Micro-nutrient application in vegetable	25 %	85 %	Power Sprayer – 0	Power Sprayer -2
			Power Weeder – 0	Power Weeder - 2

10^5 moles/ha) and net return (Rs. 26,200/ha), whereas, Hybrid rice JKRH-401 significantly recorded maximum yield of grain (52.34 q/ha), straw (57.52 q/ha), biomass (109.83 q/ha), carbon content (35.14 q/ha), CO_2 fixation (2.93 X 10^5 moles/ha), net return (Rs. 26,800/ha) and B:C (2.94) in flood and cyclone prone district Kendrapara. The local genotypes of rice crop were found less suitable in the changing climatic situation (Mohapatra *et al.*2015).

Rising minimum temperature trends over India

The rising minimum temperature has become a concern for *rabi* crops. In order to have a clear assessment of the minimum temperature trends over the entire country.. The magnitude of change on annual basis over the entire country is 0.24°C for a 10 year period. The extent of area with a strong increasing trend forms about 81.8% of the total geographical area. Temperature rise during the *kharif* season is 0.19°C over a10 year period with profound regional variations.

Table 11: Magnitude of changes in minimum temperature over different seasons, regions and time periods

Season	Districts cluster based on temperature rise	No. of districts	1971-2009 (°C/10 yr)	1990-1989 (°C/10 yr)	1990-1999 (°C/10 yr)	2000-2009 (°C/10 yr)
Kharif	Entire country		0.19	-0.18	0.50	0.09
	Slightly warm	42	0.12	0.00	0.35	0.22
	Moderately warm	90	0.16	-0.17	0.47	0.10*
	Strongly warm	366	0.24	-0.16	0.59	0.03*
Rabi	Entire country		0.28	-0.06	0.36	0.25
	Slightly warm	56	0.17	-0.62	-0.1	0.51*
	Moderately warm	112	0.21	-0.49	0.12	0.37*
	Strongly warm	359	0.34	-0.15	0.39	0.41*
Annual	Entire country		0.24	-0.05	0.36	0.25
	Slightly warm	13	0.12	-0.66	0.20	0.36*
	Moderately warm	50	0.15	-0.31	0.00	0.21*
	Strongly warm	508	0.26	0.00	0.41	0.25*

(* t test significant at 5%)

Minimum temperatures during the *kharif* showed strong warming trend in southern states, Indo-Gangetic Plains (IGP), northeastern parts, most parts of Jammu & Kashmir, Gujarat and entire Himachal Pradesh. A strong warming trend was noticed over 52.7% of geographical area with a warming of 0.24°C/ 10 yr (Table 11).

Though the magnitude of change was more during 1999, the trend line for the sub-period 2000–2009 for moderately and strongly warm regions was statistically significant. Minimum *rabi* temperatures during 1971–2009 showed strong warming over IGP, West Bengal, northeastern states, Chhattisgarh, Rajasthan, Gujarat and eastern parts of Madhya Pradesh. The rise during *rabi* over the country as a whole was greater in magnitude compared to *kharif*. The temperature rose @ 0.28°C/10 yr. The warming during 2000–2009 decade (0.25°C/10 yr) for *rabi* was about three fold more than *kharif*. A strong warming tendency was noticed over 54.9% of the geographical area, which was 2.2% more than the area during *kharif* season under the same category. The magnitude of change in area classified as slightly warm covering 7.7% of the geographical area was relatively high and proceeded @ 0.51°C/10 yr during 2000–2009 period, which was the highest across regions and sub-periods.

Monsoon rainfall pattern in north eastern region

Long term rainfall trends in the NEH region was assessed to know its impact on agriculture and horticulture. Average amount of monsoon rainfall decreased from 900–3000 mm (1951–90) to 850–2350 mm (1991–2007), indicating an average reduction of 18% rainfall in the recent period. The study indicates significant ($P<0.01$) decrease in rainfall in Ukhrul and Senapati districts of Manipur and Phek, Zunheboto and Wokha districts of Nagaland. Similarly, the number of rainy days reduced from 65 to 91 days (1951–90) to 57 to 85 days (1991–2007) indicating an average reduction of 9% rainy days over the region. The reduction was significant ($P<0.01$) for all the districts of Nagaland; upper Assam districts of Tinsukia, Dibrugarh; and Tirap, Changlang, Lower Dibang valley districts of Arunachal Pradesh. The traditionally wet north eastern region has shown a tendency of moving towards a drier monsoon regime in recent times as evidenced through standardized precipitation index (SPI). More locations in the region moved to the negative side of SPI indicating water stress of varying degrees during the *kharif* crop season.

Sea water intrusion in the coast of Karnataka

To evaluate the extent of salt water intrusion in the coastal Karnataka, soil and water samples were collected from 24 benchmark sites interspersed at every 10 km interval in the 320 km stretch along the coast in January 2012. At each location, five samples representing approximately 0km, 0.5 km, 1 km, 3 km and 5 km were collected from shore towards inland to assess salt water incursion in land. To study the depth to which salt water has leached into the soil samples at 10, 50, 100 cm depth were collected. In total, 120 water samples and 360 soil samples were analyzed. Chloride, sulphate, bromide, fluoride, sodium, potassium, calcium and magnesium concentrations in ground water were estimated during pre- and post monsoon to assess the extent of coastal salinity. Fluoride concentration was below 1 mg/L except in a few locations, but it was more in winter. Bromide was present in 50% of the locations and very high concentrations of bromine (0.50–1.80 mg/L) was observed at Mukka and Kumta indicating that the region has experienced intrusion of sea water from the ocean.

Snowmelt water harvesting saves orchards during dry spells in Kashmir

About 3.3% of total geographical area of Kashmir is under cultivation, out of which 60% is rainfed. Horticulture based farming systems are predominant. However, frequent dry spells during kharif affect cultivation of apples and pears, which are major crops of Pulwama region. In Drubgam village of the district where NICRA is being implemented, most of the cultivated area is rainfed and the apple and pear orchards generally suffer due to erratic rainfall. Further, the productivity of these crops is low due to lack of irrigation and low input use In particular, irrigation is essential to ensure higher fruit retention if dry spells occur. To cope with this climate related problem. KVK, Pulwama introduced rainwater/snowmelt water harvesting interventions and micro irrigations systems in the village Rainwater/ snowmelt water harvesting structures have been demonstrated at selected farmers' fields as effective means of collection and storing runoff and snow melt water. The stored water is used for providing irrigation at critical stages of growth of fruit crops and vegetables. This water is also useful for spraying orchards, vegetables in the absence of water source in nearby areas. During the past two years, it was observed that not only the growth of apple trees recover but also there was improvement in the yield and quality of the produce due to efficient utilization of harvested rainwater and snow. Impressed with the benefits, many farmers are coming forward to adopt this technology

Productivity variation of potato in Punjab

Climate change impact on potato productivity in Punjab was carried out using WOFOST crop growth model. It was estimated that productivity of potato cultivars in Punjab will not be affected in 2020 over the baseline scenario but will decline in 2055 (–2.62%–5.3%). However, if the present distribution of potato acreage in Punjab remains unaltered in future, there will be benefits from climate change as the potential productivity of potato will increase (+3.1 to +3.6%) in 2020 but it will again decline to baseline values in 2050.

Aeration and feeding management to minimize the effect of cyclone on aquaculture in Sundarban Islands

Flooding the freshwater ponds with saline water during cyclones is a major problem affecting the fish/ prawn culture in Sundarbans. Studies over a 2 year period revealed that two hours of continuous aeration (0.2 kg/m3 pressure with 4 L / m3 / minute volume) at 4 hr interval is an effective adaptation strategy for salinity up to 5 ppt for freshwater fishes (*Cyprinus carpio* and *Puntius sarana)*. Inclusion of additives like immune-stimulant (Immutron)/ probiotic (Gut Act)/ prebiotic @ 10 ml/kg high energy floating feed (having 30% protein, 4% fat and 8% fiber) increased the growth rate under salinity stress. High energy feed fortified with immune stimulant showed best growth ($P< 0.05$) followed by probiotics and then prebiotics in *Cyprinus carpio, Labeo rohita* and *Oreochromis mosambicus*.

Shift in fish maturity and breeding period

A shift in fish maturity and breeding period of Indian Major carps (IMC) rainbow trout, golden mahseer and snow trout, was recorded recently due to climatic variations along Himalayan and sub Himalayan regions. The maturity of IMCs in North-eastern region of country has advanced by one month (from April to March) and the period extended one month (from July to August).The full matured rainbow trout was observed during December end to January, golden mahseer during April to June and snow-trout in August in Uttrakhand. Increase in water temperature due to the global warming affects habitat, metabolism, growth and reproduction of fishes. Hatchery survey data comprising various aspects of breeding and spawning of fishes was compiled in the form of E-Atlas for West Bengal and Assam.

Water footprints for milk production

Consumptive water use (CWU) in milk production comprises direct (drinking, bathing and servicing) and indirect (feed and fodder) components. For stall-fed animals, the entire direct water requirement is met from groundwater and/or surface water sources (blue component of water footprint). For animals partly or fully under grazing system, however, rainwater may partly account for the water use (e.g. drinking and bathing). Direct water use by animals was estimated by assessing seasonal drinking water requirement of different breeds of dairy animals. Based on the diameter of water pipe, time of water flow and number of animals in the enclosure, the average water use of servicing and bathing worked out to be 50 L /day at NDRI farm and marginally lower (40 L/day) on the farmer's field. Based on the monthly feed intake data of crossbred cattle, buffaloes and local cows at NDRI farm and the seasonal (three seasons) data from the farmer's field, total water footprints (m3/tonne) of milk production were estimated. Results indicated that total consumptive water use (m3/tonne) at organized farm was less than that of unorganized farms. Water use was more for buffalo and crossbred cattle than *desi* cattle breeds. Unorganized farm used 1950 to 1980 m3/tonne water (cwu) for crossbreds and buffaloes as against 1540 m3/tonne for local cows. Total water footprints (m3/tonne) of milk production in organized and unorganized farms

Table 12: Total water footprints (M^3/tones) of milk production in organized and unorganized farms

Breed	Average milk yield (l/day)	CWU (m^3/t		Total
		Blue (Direct+Indirect)	Green(indirect)	
Organized farm				
Karan fries	9.0	996	216	1212
Miurrah	7.4	1031	238	1269
Sahiwal and Tharparkar	7.2	1279	304	1583
Unorganized farms				
Cross bred	8.3	1166	812	1977
Buffalo	5.2	1201	746	1947
local cow	4.5	981	563	1544

Knowledge gap between farmers having exposure to climate change activities

Two hundred forty farmers of Jagannath Prasad block of Ganjam district, Odisha were selected on random sample basis to test their knowledge level through NICRA (National Initiative Climate Resilient Agriculture) Project . The study revealed that the overall knowledge on crop production practices related to Integrated Nutrient Management, seed management, water management, Integrated Pest Management practices covering cultural, mechanical, biological and chemical control measures benefited the farmers in NICRA project adopted village to enhance the adoption of improved practices which has shown the importance of Integrated Crop Management methods as the important tool of extension to enhance the farmers knowledge.

Table 13: Knowledge gap between farmers having exposure to climate change issues

Practices	Knowledge of farmers on climate change issues (%)		Knowledge gap (%)
	Having exposure	Have not any exposure	
Integrated nutrient management	62.16	38.83	19.44
Seed Management	77.50	60.22	14.40
Water management	94.50	89.17	04.45
Cultural pest control measures	74.72	64.44	08.57
Mechanical pest control measures	68.06	44.44	19.68
Biological pest control measures	77.92	45.00	27.43
Chemical pest control measures	48.00	15.83	26.81

Climate change trends and adaptation actions

There is sufficient evidence suggesting climate change trends and adaptation actions in relation to the changes in global climate over the past century, and this phenomenon will continue throughout the 21st century due to anthropogenic activities as well as natural cycles. Climate change predictions by the 2050s decade include:

- Expected increase in potential evapo-transpiration of 4 - 8% for central and western part of the Indian sub-continent.
- Shortened length of the rainy season. Predicted changes in climate have implications for agricultural productivity. With changes in precipitation and hydrology, temperature, length of growing season and frequency of extreme weather events, considerable efforts would be required to prepare the deal with climate change related impacts in agriculture.

There are number of mitigation and adaptation strategies employed within the various farming systems:

Mitigation strategies

Enhancing soil organic matter (SOM) concentration is necessary to reduce agricultural emissions of greenhouse gases (GHG). Agricultural soils are a significant sink for carbon through formation of SOM. Conservation of natural to agricultural ecosystems depleted the global soil organic carbon (SOC) pool by 50 to 100 Pg (billion tones) of C and this trend is continuing by conversion of forests and savannas to agriculture in the tropics.

Conservation practices are recommended to minimize losses of SOC and plant nutrients. Both SOC and plant nutrients are concentrated near the soil surface and therefore are prone to soil erosion. The SOC is greatly associated with the finer and more reactive clay and silt fractions.

Effectiveness of soil conservation practices can be evaluated on the basis of the amount of SOC concentration associated with eroded sediments showed that conservation tillage decreases SOC associated with eroded sediments by about half and additional decrease is obtained by conversion to no-till.

Adaptation strategies

Adaptation is defined by Inter-governmental Panel on Climate Change (IPCC) as an adjustment in natural or human systems in response to actual or expected climatic stimuli or their effects, which moderates harm or exploits beneficial opportunities. Historically, people whose livelihoods depend on agriculture have developed ways to cope with climate variability autonomously. Today, the current speed of climate change will modify known variability patterns to the extent that people will be confronted with situations they are not equipped to handle. Thus, anticipatory and planned adaptation is an immediate concern. However, vulnerabilities are mostly local and, thus, adaptation should be highly location specific. Some simple adaptations such as change in planting dates and adapting resource conservation technologies(RCTs) could help in reducing impacts of climate change to a large extent, Adopting RCTs will provide increases in water use efficiency and it is estimated that climate change will induce water scarcity. Techniques like laser assisted land leveling, system of rice intensification (SRI), direct sowing rice(DSR), permanent raised bed planting system would ensure crops production using less water (Reilly, 1996).

Resource conservation techniques

Resource conservation techniques are the practices, when followed results in saving of energy, cost and also reduce the environmental pollution over the conventional practices,. The resource conservation technologies (RCTs) primarily focus on saving of resources through minimal tillage, ensuring soil nutrients and moisture conservation through crop residues and cover crops and adoption of spatial and temporal crop sequencing. These sustainable technologies and the practices therein have long been practiced by the farmers in India and world. Adoption of RCTs which allow alterations in water use, tillage and surface residue management practices can have a direct effect on emissions of greenhouse gases (GHGs) and increase the carbon stocks in the soil. Soil submergence and puddling in rice cultivation lead to unique processes that influence ecosystem sustainability and environmental services such as carbon storage, nutrient cycling and water quality. For example the submergence of soils promotes the production of methane through anaerobic decomposition of organic matter (Abrol and Sanger, 2006).

Conservation agriculture through RCTs can assist in the adaptation to climate change, by improving the resilience of agricultural cropping systems and making them less vulnerable to abnormal climatic situations. Better soil structure and higher water infiltration rates reduce the danger of flooding and erosion following high intensity rainstorms increased soil organic matter levels improve the water-holding capacity and hence the ability to cope with extended drought period (Bazzaz and Somnrock., 1996).

But RCTs also help to mitigate the effects of climate change, at least with regard to the emission of greenhouse gases. With the increasing soil organic matter, soils under RCTs can retain carbon from carbon dioxide and store it safely for long periods of time. This carbon sequestration continues for 25 to 50 years before reaching a new plateau of saturation. The consumption of fossil fuel for agricultural production is significantly reduced under RCTs and burning of crops residues is totally eliminated, which also contribute to a reduction in greenhouse gas release. Soils under zero tillage-depending on the type of management – might also emit less nitrous oxide. With paddy rice in particular, the change to zero tillage systems combined with adequate water management can positively influence the release of other greenhouse gases, such as methane and nitrous oxides (Modak and Behera, 2015).

Advantages of RCTs

- Increased soil organic carbon
- Saving in irrigation water
- Increase in energy efficiency
- Promotes timely sowing
- Saving of resources
- Yield advantage
- Lower production cost
- Improved soil health
- Increase in profitability

Zero tillage

No-tillage or zero tillage is a farming system in which the seeds are directly deposited into untilled soil which has retained the previous crop residues. It is also referred to as no-till. Special no-till seeding equipment with discs (low disturbance) or narrow tine coulters (higher disturbance) open a narrow slot into the residue covered soil which is only wide enough to put the seeds into the ground and cover them with soil. The aim is to move as little soil as possible in order not to bring weed seeds to the surface and not stimulating them to germinate. No other soil tillage operation is done. The surface residue of such a system are of critical importance for soil and water conservation. Weed control is generally achieved with herbicides or in some cases with crop rotation. No till agriculture is widely promoted as a climate friendly farming system and the IPCC Fourth Assessment Report attributes the greenhouse gas mitigation potential to no-till system. It reduces greenhouse gas effects through the following ways.

1. Soil carbon sequestration
2. Reduction of GHG emissions from soils

3. Reduction of fossil fuel use
4. Reduction of synthetic nitrogen fertilizer use

Data of Table.4 showed that net global warming potential was least in no tillage system compared to conventional tillage system, low input with legume cover system and organic with legume cover system. However, net primary production was lower in no tillage system than conventional tillage system but it is higher than the other two systems(Solomon *et al.* 2007).

Laser land leveling

Precision land leveling using laser assisted land leveler equipped with drag scrapper is a process of smoothening the land surface within ± 2 cm of its average micro-elevation. It is contemplated that laser levelers may play a significant role in improving resource use efficiency under surface irrigated systems.

This crop establishment technique also facilitates crop diversification and intercropping of wheat, chickpea and Indian-mustard with sugarcane, maize with potato, mint with wheat, rice with soybean, and pigeon pea with sorghum or greengram.

The benefits of laser land leveling over other land leveling methods include the following:

- Precise level and smoother soil surface
- Reduction in time and water required to irrigate the field
- Uniform distribution of water in the field
- Uniform moisture environment for crops
- Less seed rate, fertilizer, chemicals and fuel requirements

Direct seeded rice

Direct-seeded rice is a feasible alternative to conventional puddle transplanted rice having good potential to mitigate and adapt to climate change. Climate change is expected to increase the variability of monsoon rainfall and the risks of early or late-season drought. The DSR increases the capacity of poor farmers to cope with climate induced change by offering a choice of rice establishment methods and by reducing the amount of water required for crop establishment and subsequent crop growth. Further, farmers faced with early drought, can direct seed with minimal soil moisture, rather than wait for sufficient rainfall for transplanting. Earlier crop establishment through DSR also reduces the risk of yield loss from late-season drought, and the cost of additional irrigation. In DSR

Table 14: Above ground net primary production and relative global warming Potential

Practice	Net primary production (kg/ha/yr)	C sequestration (kg/CO_2eq/ha/yr)	Net Global warming potential (kg/CO_2eq/ha/yr)		Per NPP Net Global warming potential (kg/ CO_2eq/ton)	
Conventional tillage	9294	0	1140	100%	123.38	100%
No till	9190	100	140	12%	15.23	12%
Low input with legume cover	8840	400	530	55%	71.27	58%
Organic with legume cover	7790	290	410	36%	52.63	43%

technology seeds can be sown directly on residue retained fields by using Happy-Seeder machine. It will reduce burning of rice straw that is environmentally unacceptable as it leads to the release of soot particles and smoke causing human health problems, emission of greenhouse gases such as carbon dioxide, methane and nitrous oxide causing global warming. In addition, the entire amount of C and N, 25% P and 20% of K present in the straw are also retained in soil, which may be lost due to burning. Table.15 data shows that number of tractor operation, time and fuel consumption was less in case of zero tillage system compared to conventional drill tillage that result in 83.42% saving of time and 90.76% saving of fuel.

Table 15: Time and fuel consumption as influenced by tillage practices in wheat at farmers' field

Tillage Practice	Tractor operation	Time (hr/ha)	Fuel (ltrs/ha)	Time saving (%)	Fuel saving(%)
Zero tillage	1	1.56	6.00	83.42	90.76
Conventional tillage (Drill)	10	9.41	65.00	-	-

It is evident from the data of Table.16. that rice grain yield and water productivity is higher in laser land leveling both under direct seeded rice and traditional puddle rice compared to traditional landed leveling. On the other hand, total water use was lower in case of laser land leveling than the traditional leveling.

Table 16: Grain yield and irrigation water productivity of rice under different crops establishment technique and land leveling practices

Crop establishment technique	Rice grain yield (t/ha)		Total water use (m3/ha)		Water productivity (kg/grain/m3 water)	
	Laser leveling	Traditional leveling	Laser leveling	Traditional leveling	Laser leveling	Traditional leveling
DSR (Drill sown)	5.25	5.10	11200	12471	0.50	0.41
TPR (puddle)	5.41	4.98	13718	15056	0.39	0.33
Mean	5.33	5.04	12459	13763	0.45	0.37

System of rice intensification (SRI)

The system of rice Intensification (SRI) is a technique of rice cultivation which modified the management of soil, water and nutrients that support optimal environment for growth of rice. The SRI showed that keeping the paddy soil moist gives better results, both agronomically and economically, than continuous flooding the soil throughout the crop cycle. The beneficial effects are enhanced by complementary agronomic practices that greatly increase the growth of roots and of soil micro organisms which make it possible to grow more productive phenotypes from any rice genotype. Thus SRI is currently attracting the greatest attention to address the issue of producing more rice with less water. The SRI can also reduce GHG emission and improves soil health. A study at Indian Agricultural Research Institute, New Delhi showed that the global warming potential (GWP) in the SRI was only 28.9% over the conventional method. It increased the water productivity by 44% compared to conventional planting method.

Permanent raised bed

In the context of conservation agriculture, where soil is essentially biologically tilled, bed planting has significant role in enhancing the eco-friendly cultivation with higher productivity and profitability of various cropping systems. The important crop rotations, which virtually can directly go for permanent bed planting are soybean-wheat, maize-wheat, pigeonpea-wheat, maize-vegetable-wheat, maize-toria, mustard-wheat, pigeonpea+mungbean/urbean-wheat etc. In this system 40 cm wide raised beds and 30 cm wide furrows are made with the help of bed planter. These would have the potential to conserve soil moisture, reduce production costs, improve input-use efficiency, reduce lodging of crops, produce higher yield, permit more rapid movement of air,water and light between crops and provide more sustainable soil management while still allowing the use of the existing, widespread gravity irrigation system.

Mulching and green manure

The supply of organic matter to the soil through mulching and green manure is important for maintaining and enhancing soil fertility. Mulching materials can come from crop residues or green manure crops; it provides feed for the soil life and mineral nutrients for the plants. If legume crops are used as green manure they can supply up to 20 kg/ha of nitrogen to the soil. In the case of rice, this can result in mineral fertilizer savings of 50 to 75 %. The spreading of mulch on the soil surface reduces evaporation, saves water, protects from wind and water erosion, and suppresses weed growth.

Nitrification inhibitors

Enhanced efficiency fertilizers (Slow and controlled-release fertilizers and stabilized N fertilizers) have been defined as products that minimize the potential of nutrient losses to the environment. Urease and nitrification inhibitors have shown good potential in increasing soil retention and plant recovery of applied fertilizer N, but less is known about their impacts on reductions in total N_2O emissions. Slow release, controlled release and stabilized fertilizers have been shown to enhance crop recovery and reduce losses of N via drainage or atmospheric emission.Their benefits in reducing N_2O emissions have been explored to a lesser degree.

Irrigation scheduling

Good irrigation scheduling means applying the right amount of water at the right time. In other words, making sure water is available when the crop needs it. Scheduling maximize irrigation efficiency by minimizing runoff and percolation losses. This often results in lower energy and water use and optimum crop yields, but can result in increased energy and water use in situations where water is not being properly managed.

Pressurized irrigation system

It is one of the latest methods of irrigation which is becoming increasingly popular in areas with water scarcity. Water is directly applied into the root zone of plants and it permits to limit the watering closely to the consumptive use of the plants. Thus, this system minimizes conventional losses such as deep percolation, runoff and evaporation. These systems include sprinkler system, drip irrigation.

Crop rotation and diversification

Crop rotation refers to a planned sequence of crops grown in a regularly recurring succession on the same area, in contrast to continuous monoculture or growing a variable sequence of crops. Crop rotation affects soil quality and growth and yield of subsequent crops in many ways, including changes in SOC concentration, soil structure and aggregation, nutrients cycling and incidence of pests. In combination with minimum or no-till management, crop rotations can reduce soil erosion, enhance SOM and sequester SOC(Lal,2004).Thus, crop or cropping system selection has a major effect on C inputs to the soil. Crop rotation is more effective in retention of C and N in soil than monoculture. Rotation effects on SOC sequestration are altered by tillage and soil properties. (Table.17.)

Crop modeling

There are two basic approaches for evaluating crop's and farmer's responses to changing climate:-

1. Structural modeling of the agronomic response of plants and the economic/ management decisions of farmers based on theoretical specifications and controlled experimental evidence;
2. Reliance on the observed response of crops and farmers to varying climate.

Table 17: Diversification in agricultural system and potential benefits under climate change.

Type of diversification	Nature of diversification	Benefit	Examples
Increased structural diversity	Makes crops within the field more structurally diverse	Pest suppression	Strip-cutting alfalfa during harvest allows natural enemies to emigrate from harvested strips to adjacent non-harvested ones
Genetic diversity in monoculture	Growing mixed varieties of a species in a monoculture	Disease suppression	Genetic diversity of rice varieties reduces fungal blast occurrence
Crop rotations	Temporal diversity through crop rotations	Increased production	Manipulating diversity through crop rotations of greater cover crops and nitrogen-fixing crops increased the yield of the primary crop

For the first approach, sufficient structure and detail are needed to represent specific crops and crop varieties for which responses to different conditions are known through detailed experiments. Similar detail on farm management allows direct modeling of the timing of field operations, crop choices, and how these decisions affect costs and revenues. The ability of simulation models to predict growth and development as affected by soil and weather conditions, agronomic practices and cultivar traits makes such models attractive tools for crop improvement.

The availability of data with the necessary geographic detail, coupled with human resource capacity in modeling, is currently the major limitation rather than the computational capability or basic understanding of crop responses to climate. In response, it facilitated a capacity building initiative in the use of the Decision Support System for Agro-technology Transfer (DSSAT) model to the situation (Rewick *et al.* 1999).

Climate change impacts in social and economic aspects of farming system

The capacity of a farming system to adapt to changing weather and climate conditions is chiefly based on its natural resource endowment and associated economic, social, cultural and political conditions. The socio-economic environment is currently characterized by the following trends which potentially have a great influence on its ability to adapt to climate change:

- The semi humid condition prevailed in Indian sub-continent. Rainfall is medium and variable and droughts occur frequently.
- Internal migration, mostly from rural areas to urban and mining centres is very common.
- The income distribution is extremely skewed with a large part of the population living in poverty.
- Levels of vulnerability to natural and personal disasters are high among some sections of the population.
- Traditional management of the land with land consolidation system is losing its effectiveness

Climate change induced drought and floods has serious social and economic implications. The attainment of noble human development efforts as contained in National Development Plan, Vision 2030, and the Millennium Development Goals (MDG) are undermined by climate change. While many other factors affect the social and economic circumstances either individually or collectively, climate change is said to be an emerging important parameter for the 21st century

Conceptual and analytical frameworks

The impact of climate change on farming can be conceptualized within the framework of its impact on household food security and poverty. Understanding the interactions between climate change and poverty is critical to understanding local and global trends of climate change impacts Such an understanding is, in turn, vital to the development and implementation of effective strategies to mitigate the effects of climate change. Fig. 8 illustrates a conceptual framework on how climate change affects the farming households While the impact of climate change is felt across all farming communities, being rich or poor, communal or commercial, its impact transcend it more on the poor people living in marginalized areas.

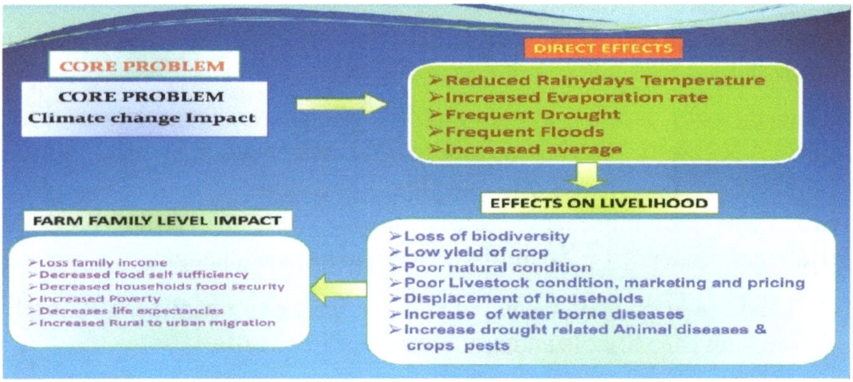

Fig. 8: Effects of Climate Change on Farmers Involved in Integrated Farming System

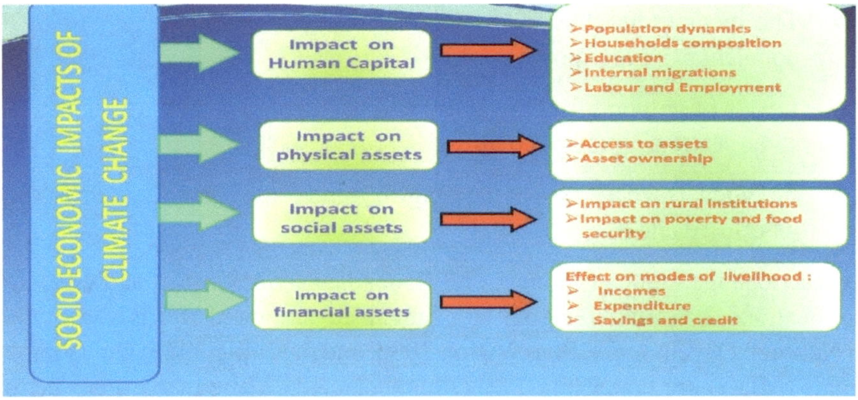

Fig. 9: Socio-Economic Impacts of Climate Change

Climate social stressors

Socio-economic impacts of climate change are four fold; they affect the physical, human ,financial and social assets of humans (Fig. 9). This study adopted the above analytical framework to discuss the socio-economic impacts of climate change

Impacts on human capital

The status of a population, growth and the underlying factors influencing the dynamics of a population, such as fertility and mortality are important players in the human development equation. A country's population influences access to resources, resource allocation and the general pressure on resources. The impacts of climate change, which in Nation takes the form of frequent drought, rainfall variability, occurrence of floods, increasing temperatures, and water low availability, on population is determined by its size relative to resource endowment, access and distribution. A lot more of the population are self-employed in agriculture and its related industries i.e. over 60% of the population practice some form of agriculture for a livelihood. This situation makes the effects of climate change on agriculture and subsequently on labour obvious. The vast majority of farm workers both on communal and commercial farms is unskilled and lives under precarious situations (UNDP, 2007). On average, farm workers are spending 70% percent of their income on food that confirms the high levels of poverty among farm workers. It makes farm workers highly vulnerable to economic shocks associated with loss of agricultural production due to climate change. For example, the most likely rational action that a livestock farmer will take if faced by drought is to downsize his herd (e.g. sell off male animals) or try to cut on costs by reducing the number of labour on the farm.

Impacts on social assets

1. Impact on rural institutions

The traditional institutions such as kinship systems, traditional political structures, co-operatives, trading groups, and SHG(mutual assistance groups) are involved in many aspects of rural development and poverty alleviation. Climate change and poverty are highly correlated, and impoverished people are simply too busy scraping a living to care about participating in organizational and voluntary life. According to a study on the impacts of poor health on the farming sectors the participation of elderly people in rural institutions is already curtailed by them having to provide support to the young, thereby subverting the role of this critical institution in traditional society

2. Impact food security

Food Security: The key recognition in linking climate change to food security is that there are multiple factors, at all scales, that impact on individual and household's ability to access sufficient and nutritious food: these include household income, human health, government policy, conflict, globalization, market failures, as well as environmental issues Building on this recognition, three principal components of food security may be identified (Brinkman and Sombrock, 1996).

i) The *availability* of food (through the market and through own production);

ii) Adequate purchasing and/or relational power to acquire or *access* food

iii) The acquisition of sufficient nutrients necessary from the available food, which is influenced by the ability to digest and absorb *nutrients* necessary for human health, access to safe drinking water, environmental hygiene and the nutritional content of the food itself .The above mentioned principal components of food security are illustrated in Fig. 10. The realization of the ideal of adequate access to sufficient nutritious food which meets dietary needs and food preferences for an active and healthy life is one of the most daunting challenges two indicators are crucial in assessing progress in the attainment of food security goals, namely; dietary energy supply and food availability (Fischer *et al.* 1996)

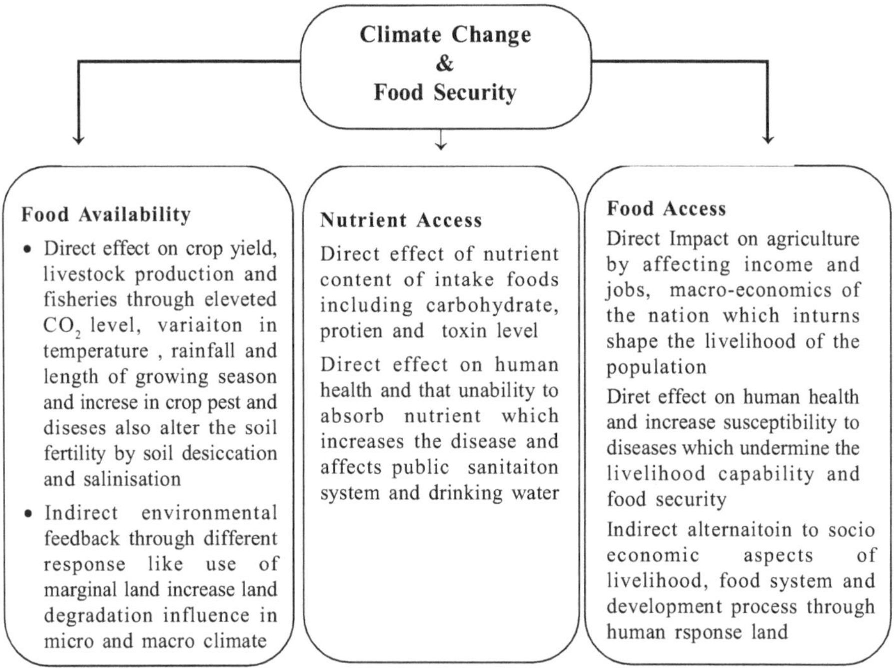

Fig. 10: Climate Change and Food Security

Impact on financial assets

Effect of climate change on income and expenditure:

The major households obtained their main household income from farming activities though with great Regional differences. Farm family income indicates that subsistence farming as a source of income constitutes 31% of total household sources of income, second from wages and salaries which constitute 48%. A comparison of household sources of income between urban and rural households reveals striking differences and reiterates the fact that rural households are especially vulnerable to climate change The majority of urban dwellers (30%), compared to only 7% rural households receive salaries/wages. Salaries and wages do not generally decline in climate induced extreme events such as drought. Rural households relies more on non-farm 'handouts' While the so called 'handouts' may appear insignificant they form the very means of livelihood security for the very poor households dependent on farming

Recommendations for strengthening adaptation to climate change in farming systems

Strategic policy framework

While being cognizant of the effects of climate change, such as reduced rainfall and higher temperature regimes, few policies currently reflect the need to respond to and prepare the economy for the impacts of climate change. There is a need to mainstream climate change issues into national policies and strategies. These can be initiated and enhanced by intensified activities of the different Climate Change Committees and their various working groups at both national and regional levels. Targeted strategies are needed to better ensure that decisions and investments made do not suppress the ability of farming communities and their economic sub-sectors to adapt to future climatic changes.

Research programme formulation and priority setting

Research needs for adaptation to climate change is not sufficiently articulated within the national agricultural research programme and agenda. Programme formulation with clearly set out, widely understood, and multi-stakeholder coordinated priority setting mechanisms will facilitate the implementation, monitoring and evaluation of well-thought government initiated research programmes. A formal multi-disciplinary research programme review, evaluation, and monitoring exercise shall be proposed to ensure that proper guidance is given to all research institutions, professionals and the farming community alike.

Coordinated data management

There is no formal system for farming data (field crops, fruits, vegetables, small-stock, large stock, economic, social, environmental, etc.) management. The current practice is adhoc oriented for incoming, internal, and outgoing information among the diverse farming service providers. There is a need to maintain and enhance the collection of accurate physical, economic and social data that will enable the development of models that predict the impacts of climate change on agriculture

Strengthening communication and awareness through establishing an adaptation information and advice service

Bringing climate change rather than just climate variability into focus as an additional element in normal strategic planning and climate risk management in agriculture requires awareness raising and capacity building. Efficient and planned awareness, incorporating climate change considerations in policy and programme communications, will foster an increased understanding and integration of scientific knowledge into farm management decisions and incorporate issues of climate change adaptation into education and training packages directed at agricultural industries Sources of climate change information and advice on adaptation strategies in the farming communities are not generally available, and it is necessary to propose the formation of climate change information and advise service to satisfy this need.

Investment in breeding, biotechnology and seed technology programmes

Promptly promote the use of indigenous and locally-adapted plants and animals as well as the selection and multiplication of crop varieties adapted or resistant to adverse conditions. The selection of crops and cultivars with tolerance to a biotic stresses (e.g. high temperature, drought, flooding, high salinity content in soil and water, pest and disease resistance) allows harnessing genetic variability in new crop varieties. Biotechnology can contribute to agricultural productivity and food security. The issues in the realm of seed science and technology including seed health and testing, are not adequately addressed and coordinated. There is for instance no seed health laboratory that farmers can resort to for diagnostics, if their crops are beset by seed borne diseases. There is lack of information on seed science issues including seed research and science-based seed testing. There is a need for technical expertise and other resources to utilize existing seed testing facilities in a coordinated fashion with the leading organized seed production and supply organizations in the crop growing Regions.

Empowerment and broadening of early warning systems

There is a need for better cooperation between the Early Warning and Information Systems Unit of IMA with the Emergency Management Unit of the office of the State Agriculture department in order to better utilize the assessment of short- and long-term impact of adverse natural events on agriculture livelihoods, while contributing to disaster preparedness and mitigation of potential risks, through the establishment of a historical climate data archive with monitoring tools using systematic meteorological observations, and information tools on the characteristics of farming system vulnerability and adaptation effectiveness such as resilience, critical thresholds and coping mechanisms.

Researchable Issues for strengthening adaptation to climate change

The identified future research needs to enable adaptation to climate change in the following key areas:

Breeding and evaluation crop varieties

In order to reduce possible adverse consequences to climate change, the agricultural sector should be encouraged to continue develop crop breeding and management programs for heat tolerance and agronomic drought. Collection of germplasm alone is tedious and requires harnessing sufficient human and financial resources. Henceforth, sufficient resources should be made available to adequately support the participation of extension agents in the germplasm collection activity

Climate change monitoring

The clearest objective at present is to prepare for changing climatic hazards by reducing vulnerability, developing monitoring capabilities, and enhancing the responsiveness of the agricultural sector to forecasts of production variations and food crises. A pilot assessment designed to monitor climate variability while enhancing adaptation of farming systems is needed.

DSSAT crop model simulation and validation

Following the capacity-building initiative, there is a need for technical support in validation of the crop models with experimental data from the fields. This will allow for modeling yield changes, and changes in growing season length arising from climate change, and the identification and evaluation of alterations in agricultural practices that would lessen any adverse consequences of climate change (Battisti, 1995).

Ricardian approach-based study

Current initiatives to model climate change impacts are limited to crops. There is a need to model the impact of climate change on farm production units, other than crops. The Ricardian model offers a rigorous alternative model for studying climate change impacts on the agricultural economy. A fully fledged fieldwork-based study, to collect data on the various parameters to be included in the model, is highly recommended.

Conclusion

Agricultural production is limited by its ecologically fragile ecosystems. As a direct consequence, extreme climate change occurrences have a direct impact on the agricultural economy .The. Climate change and variability are among the most important challenges have strong economic reliance on natural resources and rain-fed agriculture. Unfortunately, it is predicted that if global warming is not reduced the country will become hotter and drier. The fact is that the majority of Indians live in rural areas and their socio-economic circumstances makes them especially vulnerable to negative climate change occurrences such as droughts and floods. There is already sufficient evidence that farming systems have changed, land degradation, forest land encroachment and other environmental threats continue to threaten farm families depending on agriculture as a livelihood. Therefore there is a need to fast track policies and strategies that enhance India's resilience to climate change impacts.

- This climatic variability-induced losses in agricultural outputs in recent past, other socio-economic drivers for vulnerability of the sector, traditional knowledge and practices existing within the farming community for coping with climate variability, knowledge and information gaps for better decision making and farming system management. Long-term climate change and variability could increase the frequency and severity of climatic extremes such as droughts and floods. Continuing changes may expose farming systems to conditions not experienced before. For example, there is a risk that changing climate conditions predicted by various studies will shift the areas where agricultural production can occur because of changing rainfall profiles or flood and drought frequency. Such geographical shifts in agricultural land use could threaten the viability of key agricultural sub-sectors, and also disrupt rural communities and the infrastructures that support rural development and livelihoods. There is a need to recognize that droughts and floods are a normal part of the Indian agricultural environment, and that in the future such events may become more common.Moreover, climate change may increase climate variability beyond the range considered normal under past experiences.

- Notably, while such scenarios may pose threats in some regions, they may create opportunities in other regions. Potential impacts of long-term climate change on farming systems could include increased invasion of unwanted plant species, pest and diseases, changes in pasture growth and carrying capacity, and a more limited capacity for diversification that is reduction in the potential for expansion in irrigated agriculture. Changing rainfall patterns, combined with higher temperatures, could reduce water availability and add pressure on water allocation systems. Effective adaptation measures, such as efficient use of resources, low water use crop cultivars, drought-tolerant livestock, and diverse income opportunities for rural communities could reduce the extent of these impacts.
- The climate change impacts are likely to vary across geographical regions and this could impact on the comparative advantage of existing farming systems with possible influences on sectoral commodity trade. Climate change adaptation measures have the ability to enhance opportunities and minimize risks to exposed sectors. Adaptation practices require extensive high quality data and information on climate, and on agricultural, environmental and social systems affected by climate, with a view to carrying out realistic vulnerability assessments and looking towards the near future. Climate change assessment must observe impacts of variability and changes in mean climate (inter-annual and intra-seasonal variability) on agricultural systems. However, agricultural production systems have their own dynamics and adaptation has a particular emphasis on future agriculture.
- The capacity of a farming system to adapt to changing weather and climate conditions is chiefly based on its natural resource endowment and associated economic, social, cultural and political conditions. In responding to climate change, government and development partners have a significant and ongoing role in supporting the efficient allocation of resources, managing distribution of costs and benefits proportionally amongst those potentially affected, and facilitating efficient decision-making by providing information, institutional support and policy frameworks. Developing and implementing adaptation strategies to climate change oblige the involvement and concerted efforts of many stakeholders in ensuring an efficient delivery system. Implementing adaptation measures at farm level requires that research and development, coupled with extension, is well coordinated.

References

Abrol, I.P. and Sangar, S. 2006 Sustaining Indian Agriculture-conservation agriculture the way forward. *Current science*. 91(8): 1020-1025.

Anonymous 2013. Annual report of Indian Council of Agricultural Research 2012-13, New Delhi.

Altieri, M.A. and C.I. Nicholls. 2004. Biodiversity and Pest Management in Agroecosystems. 2nd edition. Haworth Press, New York.

Battisti, D.S. 1995. Decade-to Century Time-scale Variability in the Coupled Atmosphere-ocean System: Modeling Issues. Natural Climate Variability on Decade- to-century Time Scales. *National Research Council, National Academy Press,*419-431.

Bazzaz, F. and Sombroek, W. 1996. Global Climate Change and Agricultural Production: Direct and Indirect Effects of Changing Hydrological, Pedological and Plant Physiological Processes. FAO: Rome, Italy

Brinkman, R. and Sombroek, W.G. 1996. The Effects of Global Change on Soil Conditions in relation to Plant Growth and Food Production. Land and Water Development Division, FAO: Rome, Italy.

Cosbey, A., Bell, W., Murhpy, D., Parry, J., Drexhage, J., Hammill, A., and van Ham, J.2005. Which way forward? Issues in developing an effective climate regime after 2012. International Institute for Sustainable Development. Canada.

Fischer, G., Frohberg, K., Parry, M.L. and Rosenzweig, C. 1996. The Potential Effects of Climate Change on World Food Production and Security. FAO: Rome, Italy.

Harris, D.L. 1998. Livestock improvement: art, science, or industry? *J. Anim. Sci.* 76:2294-2302.

Henson, E.L. 1992. *In situ* conservation of livestock and poultry. Fao Animal Production and Health Paper 99.

Hulme, M. and Viner, D. 1998. A Climate Change Scenario for the Tropics. *Climatic Change* 39: 145–176.

Lal, R.2004. Soil carbon sequestration to mitigate climate change. Geoderma. 123:1-32.

Modak, D.P. and Behera, B 2015. Resource conservation techniques for climate change- Mitigation and adoption. Department of Agronomy, College of Agriculture, OUAT, Bhubaneswar Occasional Note no. 4:83-91.

Mohapatra, S.C. Nanda, S.S., and Sarangi, C. 2015 Rice (Oryza Sativa L) variety as an intervention to mitigate the ill effects of climate change. *Environment and ecology* 33(1 A): 392-395.

Rai, M. (2015). Promising Agricultural Provises to Keep. The 3rd Dr Kissean Kanurgo Memorial Lecture, 22 June, OUAT, Bhubaneswar

Reilly, J. 1996. Climate Change, Global Agriculture and Regional Vulnerability. FAO: Rome, Italy.

Renwick, J.A., Katzfey, K.C., McGregor, J.L. and Nguyen, K.C. 1999. On Regional Model Simulations of Climate Change over New Zealand. *Weather and Climate* 19, 3-14.

Rosenzweig, C. and D. Hillel. 1998. Climate Change and the Global Harvest: Potential Impacts of the Greenhouse Effect on Agriculture. Oxford University Press, New York.

Rosenzweig, C. and D. Hillel. 2008. Climate Change and the Global Harvest: Impacts of El Nino and Other Oscillations on Agro ecosystems. Oxford University Press, New York.

Solomon, S., Qin, D., Manning, M., Alley, R.B., Berntsen, T., Bindoff, N.L., Chidthaisong, Z.C.A., Gregory, J.M., Hegerl, G.C., Heimann, M., Hewitson, B., Hoskins,B.J., Joos, F., Jouzel, J., Kattsov, V., Lohmann, U., Matsuno, T., Molina, M., Nicholls,N., Overpeck, J., Raga, G., Ramaswamy, V., Ren, J., Rusticucci, M., Somerville, R.Stocker, T.F., Whetton, P., Wood, R.A., and Wratt, D. 2007. Technical Summary. Climate Change 2007: The Physical Science Basis. *Contribution of Working Group I to the Fourth Assessment Report of the Intergovernmental Panel on Climate Change*. Cambridge, UK. New York, US, Cambridge University Press.

UNDP, 2007. Human Development Report 2007/2008: Fighting Climate Change— Human Solidarity in a Divided World. Palgrave Macmillan: New York.

Integrated Farming System Practices: Challenges & Opportunities, pp 269-295

New India Publishing Agency, New Delhi, India

11

Gender Balancing in Integrated Farming Systems

M. Behera

In recent years, there is gradual realization of the role of women in the field of agriculture & allied areas. Besides domestic work, they are involved in various types of field works relating to production of crops, milk, fish meat, egg, mushrooms & honey bee and other material. Employees in agriculture sector of the country, contributes nearby 80% of all economically active women (Bhatt, 2013). Therefore, rural India is now going through a process of "Feminization of Agriculture".

Women are critical to the well-being of farm households. Besides child bearing and rearing, women are expected to prepare all meals, maintain the household jobs and assist in crop and animal production. Perhaps, ironically, it is because women who have so many responsibilities that they have been over-looked by agriculturalists and policy makers–it has been more convenient to label men as farmers and women as child raisers and cooks. Rural women form the most important productive work force in the economy of majority of the developing nations including India.

In spite of several efforts there still exists gender gap in the field of agriculture and allied areas. Closing the gender gap is not only the right thing to do; it is the smart thing to do for improved agriculture and food security for all. This was a key message for family farming interventions to close the gender gap.

Directorate of Extension Education, Orissa University of Agriculture & Technology, Bhubaneswar-751003, Odisha, India

World wide women play a major role in agriculture; including fisheries, forestry and livestock production. The nature and extent of their involvement is different in various agro production systems. Their role is diversified from management to landless labourer. In overall farm production, on an average women's contribution is estimated at 55 to 66 per cent of the total labour (Singh, 2013).

Gender balance

Gender balancing is the equal and active participation of women and men in all areas of decision-making and in access to and control over resources and services. The United Nations regards gender equality as a human right. It points out that empowering women is also an indispensable tool for advancing development and reducing poverty. Equal pay for equal work is one of the areas where gender equality is rarely seen.

The United Nations consider gender balance as fundamental right to the achievement of equality, development and peace. To accomplish it in agriculture and rural development, action is needed by rural communities, governments and international development agencies. At the local level, gender balance means men and women are actively involved in decision-making, including those managing community facilities and infrastructure. Institutions responsible for rural development need to improve gender balance among technical and managerial staff, especially in extension work.

Gender equality or balance denotes women having the same opportunities in life as men, including the ability to participate in the public sphere. Gender equity denotes the equivalence in life outcomes for women and men, recognising their different needs and interests and requiring a redistribution of power and resources.

Gender equality is the measurable equal representation of women and men. Gender equality does not imply that women and men are the same, but that they have equal value and should be accorded equal treatment.

Participation of women in farm related activities in different states

The role of women in each activity under farming indicated that independent participation of male member was higher than the female members in all the states except Uttarakhand. Among all the states, participation of women in farm related activities was found highest in Uttarakhand (29.41%) and negligible in Punjab, Haryana, Karnataka and Maharashtra (Singh, 2013). Joint participation with female members was higher in Andhra Pradesh, Himachal Pradesh Maharashtra and Rajasthan in seed selection, nursery raising, transplanting, weeding and harvesting, highest was recorded in Assam whereas joint participation of male members with other male member was higher in land preparation, plant protection, irrigation, engagement of labour and procuring and repayment of loan in Assam, Andhra Pradesh, Maharashtra and Karnataka.

Fig. 1: IFS model developed by farm women

State wise comparison reflected a different picture in Himachal Pradesh; nearly 17.7 per cent rural women were completely responsible for different activities, which is much higher than the national average of 6.67 per cent. The reason behind is mainly because the male members remain outside their home for a long period of the year serving outside their hometown. Inter zonal variation was visible with regard to role and responsibility profile in all the states.

Table 1: Gender analysis of crop farming systems

Sl. No.	Activity	Participation in percentage	
		Male	Female
1.	Land preparation	100	0
2.	Seed preparation for sowing	18	82
3.	Raising nursery and transplanting	25	75
4.	Direct sowing	92	8
5.	Irrigation	83	17
6.	Applying FYM	75	25
7.	Fertilizer application	83	17
8.	Weeding	17	83
9.	Plant protection	83	17
10.	Harvesting	42	58
11.	Threshing	58	42
12.	Bagging/storing of grain	17	83
13.	Marketing of produce	92	8
14.	Storing dry fodder	66	34
	Overall	60	40

Agriculture is a dynamic sector, which is affected by climate change, technological development, market preferences, social changes, migrations and effect of globalization. In spite of all these, the production has to be sustained with effort of both men and women. In crop production system, women play several roles such as seedling raising, transplanting, weeding, intercultural operations, harvesting and storage of the produced. Many a times, it is observed that women along with other family members are deprived of good quality food resulting in nutritional in security to provide required quantity of nutritious food, it is desirable to include several components like dairy, poultry, pisciculture, mushroom, bee-keeping with field crops, vegetables and foods in the farming system activities. Nutritional security of a farm family can be ensured with inclusion of above components on integrated farming system models (Fig. 1). Such models were vested various parts of the country with inclusion of above mentioned components.

The steps to reduce hunger and malnutrition and to help small holder farming families cannot succeed without addressing the role of women in agriculture. We have to work to be responsive to women's roles, responsibilities and priorities in all of our agricultural development programmes. When women farmers are meaningfully included in agricultural development opportunities, not only do farms become more productive but adoption of new technologies increases and overall family health improves.

Evidence shows that if women farmers across the developing world had the same access as men do to resources such as land, improved seed varieties, new technologies and better farming practices; yields could increase by as much as 30 per cent per household and countries could see an increase of 2.5 to 4 per cent in agricultural output. Women have also shown to be more likely than men to reinvest income in the health of children and other family members and in a more varied and nutritious family diet. While working in agriculture, the following priorities may be adopted for ensuring gender responsive.

Know her: Programmes should take into account the context and circumstances of women farmers. Investigation may be done on women's needs, constraints, responsibilities and priorities and anticipate how programmes will affect women's labour, time, current practices and resources.

Design for her : The information collected about women farmers may be used to design the programme. The projects are to be designed specifically for benefit of women farmers and they include goals and milestones that account for women's participation.

Be accountable to her: Work may be done to ensure the programme objectives with active involvement of farm women. The progress may be evaluated in terms of women's success as well as household success. Steps may be taken to collect feedback, measure outputs and ensure that women are rationally participating and benefiting.

The integrated farming system (IFS) has revolutionized conventional farming of livestock, aquaculture, horticulture, agro-industry and allied activities in some countries, especially in tropical and subtropical regions that are not arid. Farming all over the world is not very performing unless relatively big inputs are added to sustain yields and very often compromise the economic viability as well as the ecological sustainability. Evidently, the situation can worsen if high duties are paid on imported materials and energy and the polluter-payer policy is also applied, as it should well be. The IFS can remove all these constraints by not only solving most of the existing economic and even ecological problems, but also provide the needed means of production such as fuel, fertilizer and feed, besides increasing productivity many-fold. It can turn all those existing disastrous farming systems, especially in the poorest countries, into economically viable and ecologically balanced systems that will not only alleviate poverty, but can even eradicate this scourge completely.

Experience from eastern region

Characteristics and potentials of agriculture strategy vary across the regions of India likewise; components of the IFS also vary depending upon the available natural resources and socio-economic factors in these regions. In our country, women constitute 48.47 per cent of total populations and about 74 per cent of the entire female work force is engaged in agricultural operations. Depending on the regions and crops, women's contributions vary but they provide pivotal labour from planting to harvesting and post harvest operations (Shivani, 2013). Over the years, there is a gradual realization of the key role of women in agricultural development. Women work harder and for longer hours than man. Farm women of all ages work in agriculture and 75 to 80 per cent of farm work is carried out in family farm while their contribution in hired agricultural labourer is 52 to 66 per cent (Shivani, 2013). Male farm workers are relatively free during off seasons, however farm women work during these periods too. India has a variety of crops grown in irrigated and rainfed areas. In highly diversified Indian contest, no simple gender division of labour exists with regard to crop production. In certain areas, women play a key role as seeds selector and in seedling productions. Their knowledge on seeds and seed storage contribute to the viability of the agricultural diversity and productions.

Males shoulder the responsibility of agricultural activities such as ploughing, sowing, transport, and sale of agricultural produce. The female on the other hand, carry out the work like transplanting, harvesting, threshing, winnowing, dehusking and storage. Some works are carried out by both men and women. Gender differences exist both in works carried out by rural women and wages paid to them. In Utter Pradesh, rural women are mostly engaged in household activities; but a considerable amount of time is also spent in dairy and agricultural activities. Dairy activities include fodder chuffing, washing cattle and dung cake preparation while the agricultural activities undertaken are fodder transportation from fields home, harvesting and irrigation (Kumar and Ful Zele, 1998).

In Meghalaya, women mainly carry out transplanting weeding and harvesting; while men do the ploughing and threshing. In Chhatishgarh, majority of female farm labourers perform weeding, harvesting, winnowing and transplanting in rainfed areas. In the irrigated areas, grass cutting is the major job for women labourers. In contrast, the tribal women such as Santal, Munda and Araons of Waste Bengal participate in subsistence activities such as trading fire wood and other forest produce. They perform lots of work in agriculture and animal husbandry.

Livelihoods consists of the capabilities, assess and activities required for a means of living. Woman plays a vital role in livelihood security as they are the backbone of agricultural work force. With enterprises combination of alternate farming system, farmers of Bangladesh themselves and their household women where involve in rearing poultry and livestock kitchen gardening and sewing cloths for household use and nursery reforestation (Sadika *et.al*, 2012).

Women friendly farm enterprises

- Processed products of rice/wheat/millets
- Off-season vegetable cultivation
- Vermin composting
- Backyard Poultry rearing
- Dairy management
- Marigold cultivation
- Papaya cultivation
- Mushroom cultivation
- Bee keeping
- Raising of vegetable seedlings and fruit plants
- Preservation of local fruits and vegetables
- Value added agro-products
- Coir/jute work
- Leaf plate making

Table 2: Gender participation of selected farm household (%)

Gender participation	Poultry rearing	Cattle rearing	Milch cow rearing	Goat/ sheep rearing	Sewing	Vegetable production	Vegetabl eseedling raising	Forestry seedling raising
Male	2	55	50	30	0	15	85	40
Female	98	45	50	70	100	85	15	60

- Spices powder making
- Rice processing

Women empowerment through crop based farming system

Agriculture is a dynamic sector and rapid changes occur such as those in environment and climate, technologies, development priorities, impact of changes in other sectors and social changes such as family structure, migration and international policies such as globalization and liberation. Agriculture, the single largest production endeavour in India is increasingly becoming a female activity. Gender mainstreaming is incorporating gender perspective into policies, plan, programmes and projects to ensure that these impact on women and men in an equitable way. For empowering women in crop production/crop based farming system following points are very important;

- Technology development
- Control over resources
- Knowledge of post-harvest process
- Institutional support
- Policy support including research, extension and development
- Recognizing the gender based differences in roles and responsibilities and contribution of different socio-economic groups
- Appropriate technologies and policies for sustainable use of resources

Crop based farming system consist two components: Crop production and Horticulture. Under crop production, depending on crops, several works are performed by farm women. In crop production role of women is very high and women carry out different operations including field preparation, transplanting, weeding, harvesting, post-harvest handling etc.

Rice production

So far as rice cultivation is concerned, studies show that women participate in all major operations for example; sowing/transplanting (86%), weeding (84%), storage of grains (78%), land preparation (72%), cleaning seeds for sowing (70%), gap filling (68%), manure and fertilizer application (68%), harvesting (64%), threshing and winnowing (62%), watching birds (41.6%) and rodent control practices (58%). Nataraju and Lovely (1993) reported that the majority of women participate in pre-harvestings like, reaping the crop (98%), bunding and transporting (94%), operation trampling (80%), transporting and spreading of seeds (60%) whereas, in post harvest activities 84% women are involved in threshing and 92% in winnowing. In cultivation, except ploughing, leveling and irrigating the field, all the other works such as sowing, weeding, transplanting, harvesting, stocking of straw, husking, drying and storing are female dominated tasks. The tasks such as preparation of field, irrigating crops and construction/ repair of field channels are mainly male dominated tasks. In inter cultivation activities (Shivani, 2013) farmwomen play a major role in weeding, application of fertilizers, thinning and gap filling, irrigation and dusting (Table- 3).

Table 3: Participation of men and women in different operations in paddy cultivation

Sl. No.	Farm operations/ carried out	Jointly
1.	Land preparation	Jointly
2.	Cleaning of seed for sowing	Women
3.	Nursery sowing	Women
4.	Nursery aftercare	Women
5.	Seedling uprooting	Women
6.	Transplanting	Women
7.	Irrigation	Jointly
8.	Top dressing of fertilizer	Men
9.	Gap filling	Jointly
10.	Weed control(Manual)	Women
11.	Weed control (Chemical)	Men
12.	Plant Protection	Men
13.	Harvesting	Jointly
14.	Threshing	Jointly
15.	Widowing	Jointly
16.	Marketing	Jointly
17.	Storage	Women
18.	Dehushing	Women

Protected cultivation of vegetables

In the present scenario of perpetual demand of vegetables and shrinking land holding drastically, protected cultivation is the best alternative and drudgery- less approach for using land and other resources more efficiently (Fig. 2). In protective environment (green house/glasshouse or poly house), the natural environment is modified to suitable conditions for optimum

Fig. 2: Dean Ext. Edn OUAT visiting chili demonstration KVK Puri

Plant growth which ultimately provides quality vegetables. Nursery for ornamentals, flowers, vegetables, fruits and plantation crops can be successfully developed inside greenhouse. Women can grow the high- priced vegetables such as tomato, cucumber and capsicum round the year especially during winter season (Fig. 3).

Fig. 3: Dean Ext. Edn OUAT visiting baby corn field in Puri

The horticultural activities such food processing, preservation, packaging, marketing and retail sales of fruits, vegetables, flowers, spices, medicinal plant produce, etc offers enough opportunities in development of agri- business for strengthening and financial empowerment of rural mass. Besides fresh consumption, horticultural crops provide raw material for many ancillary industries. Processed horticultural products have also good export potential in our country.

Production of quality planting materials

The demand for high quality planting material is steadily increasing due to interest in fruit tree cultivation, social forestry, agro-forestry and plantation crops. The need of setting up plant nurseries to meet the demands of the people has been felt by small and marginal farmers as well as by gardeners and farm house owners. In order to meet this demand, there is ample scope for introduction of small nurseries which will serve to augment the incomes of needy sections of rural society. Setting up of a fruit nursery is a long term venture and needs lot of planning and expertise.

Quality planting material is the foundation of enhanced production, profitability and income of horticultural crops. However, sector is experiencing inadequacy of quality planting materials, and the degree of unavailability varies with regions and crops. Women play an active role in the production of quality planting materials of horticultural and ornamental plants for entrepreneurship and employment generation.

Need of women participation in IPM

Gender has a major role to play in IPM because farm women worker's percentage is increasing day by day. Particularly in Bihar where mostly men go out to earn livelihood and women are looking after farm business. But the limited information on the role of women in IPM is available.

Fig. 4: Farm women visiting vegetable field

The indigenous knowledge of pest management is often gendered. It has been observed that the women labour and labour share increased during last 25 years in agriculture. Some field work like paddy transplanting, weeding, harvesting etc. are domain of female workers. Now-a-days, women are no longer secondary worker in small scale and marginal agriculture particularly in Bihar state. Now, women are taking active part in pest management activities and applying management strategies for managing pests in their field crops whereas a decade ago, they did not attead such activities. Even women are handling pesticides very carefully and having great exposure levels than men, even where they do not spray. Women are capable to take decision in purchasing pesticides as per their budget. In present scenario, the organic farming is coming inabig way. In this system of farming women are playing major role in implementing the organic methodology (Fig. 4).

Requirement for gender participations

Policy as well attitude against women participation need be changed. Moreover, every farmer man and woman as well as all persons involved in IPM network should agree to eliminate gender equality and discrimination (Table 2). It requires integration of gender into IPM structure and mechanism. Recently, the FAO has provided sufficient resources to make it possible for gender mainstreaming to take place in the IPM network. Secondly, there has been support and enthusiasm among the farmers themselves to conduct gender training, gender data collection and to build gender information system through participatory approach (Nanda, *et al.,* 2009).

Participatory approach in integration of gender in IPM

Integration of gender in IPM is basically long-term and process oriented activity. Following points are essential for integrating gender in development programme:

- Integrating gender into development programme cannot induce from out side
- Integrating gender into a programme requires process led by both men and women farmers.
- Integrating gender perspective in programme requires political will and leadership.

- Capacity building is essential for integration

Processing and value addition of various crops, fruits and vegetables

Women participation is more in horticultural sector than the food grain production. However most of the post harvest activities either related to food grains, horticulture, livestock and fishery is dealt by women labors. Post harvest activities are much related to the women's day to day activities which they performed at their homes (Fig. 5). Keeping the food material either in fresh form or in processed form for family consumption resembles to an extent of commercial post harvest activities. Women can do both primary and secondary processing of cereals, pulses, spices, oilseeds, fruits, vegetables, flowers, meat and poultry, dairy etc. to supplement family income. Post harvest operations are comparatively less labour intensive than the food production operations. Food processing industry either of cottage level or at commercial level is an enterprise that processes the product of plant or animal origin into the form of consumption.

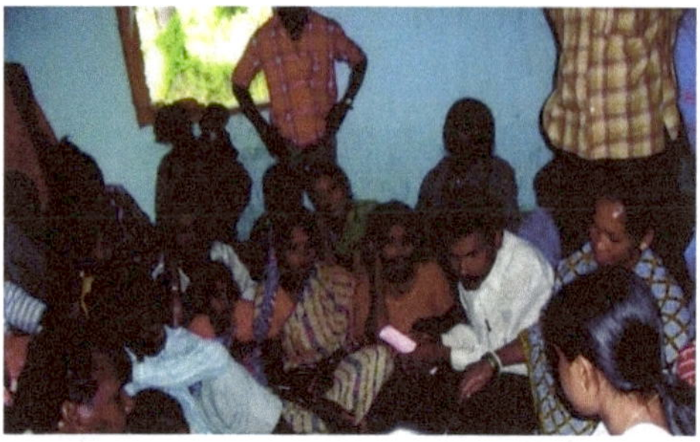

Fig. 5: Demonstration on value addition of vegetables

Vegetables and fruit processing and preservation play a great role to provide employment and industrial base for export of dehydrated and preserved products.

Post harvest management, processing, storage and utilization of vegetables and vegetable products are generally the domain of women at home scale. Cultivation of horticultural crops plays a vital role in prosperity and it's directly linked with health of people. These crops are not only used for domestic consumption but also processed into various products like pickles, preserves, beverages, jam, jelly squash, etc., which offer employment opportunities to the rural women. The focus on the value addition in the horticulture sector is vital for comprehensive development of the rural economy.

Pickles, sauce and soups industries: This is one of the major food processing activities where almost all operations are being done by women worker. Preparing raw materials by peeling, cleaning, cutting and drying, mixing etc. which is required to develop pickles, sauces and soups are done by the women. Therefore working women should have the basic knowledge of technology behind the food preservation through pickling etc.

***Papad* and *Badi* industry:** *Papad* and *Badi* are prepared from cereals and pulses. Now-a-days it emerged as a largest cottage industries in the field of food processing involving most of the women worker (Fig. 6).

Fig. 6: Papad making by women groups

Soybean processing for milk, tofu etc

Soybean processing by women at household level has good scope for income generation for women. Soybean with its high productivity and protein content can make a fourfold impact on malnutrition as compared to pulses. It is the most appropriate options for India as it can be made available at an affordable price to the poorer section of the population. Processed soybean has several health benefits.

Interior plant decoration

Interior plant decoration is also getting momentum due to change in the life styles and particularly women can earn a income by doing interior designing through ornamental and flowering plants not only in various functions but also in various offices, hotels, hospitals etc.

Mushroom cultivation

Now-a-days mushroom is getting much popular in our country, have a good scope for export. About a decade ago, the government promoted the mushroom cultivation for reducing protein-energy malnutrition, generating employment and supplementing the income of the women and earning foreign exchange. As production of mushrooms requires a small area and waste materials utilization and it can be used. It does not require a highly skilled supervisory staff and can be managed by rural women easily. So, rural women can be supported by educating and training them in mushroom production technology (Fig. 7). The Madhya Pradesh Agro-industries Corporation popularized mushroom cultivation in the tribal areas of Chhattisgarh by supplying spawned compost to the prospective growers. Cultivation of paddy straw

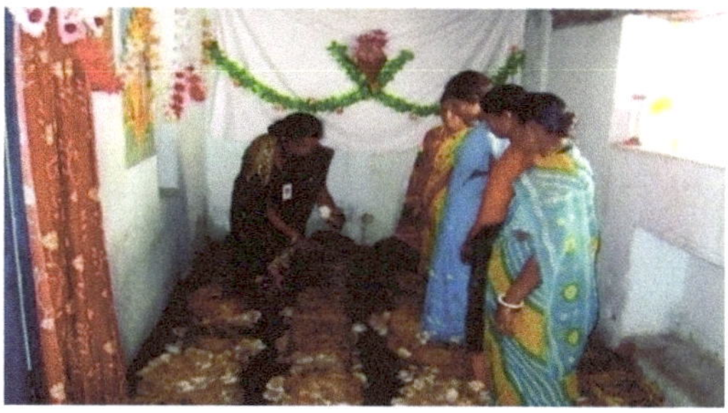

Fig. 7: Mushroom cultivation for additional income

mushroom is popular among farm women in Odisha. Entrepreneurs and growers from Tamil Nadu, Karnataka, Kerala, Andhra Pradesh and Maharashtra have recently taken up large scale mushroom cultivation. Various women shelf help groups in north east region are growing paddy straw mushroom in backyard.

Farming system models help to generate employment for family members including farm women. As observed by Sanjev Kumar (2013), there may be employment generation up to 690 man days in a year from one acre of land.

Women empowerment through animal husbandry practices

In integrated farming system, livestock plays a multi-faceted role in providing draught power for the farm, manure for crops, energy for cooking and food for household consumption as well as the market. In animal husbandry women have a multiple role. With regional difference, women take care of animal production. In eastern states their activities vary widely ranging from care of animals, grazing, fodder collection, cleaning of animal sheds to processing milk and livestock products. In livestock management, indoor jobs like milking, feeding, cleaning, etc. are done by women in 90% of families while management of male animals and fodder production are effected by men. Women accounted for 93% of total employment in dairy production. Depending upon the economic status, women perform the tasks of collecting fodder, collecting and processing dung. Dung composting and carrying to the fields is undertaken by women. Women also prepare cooking fuel by mixing dung with twigs and crop residues. In livestock management, majority of farm women solely took responsibility of maintaining cattle sheds, feeding poultry birds and hatching of eggs/chicks. But when it came to the care of sick animals, type of milk product to be prepared, hatching of eggs, marketing of eggs and milk, the decisions were taken jointly by farm women and men. Though women play a significant role in livestock management and production, women's control over livestock and its products is negligible. The vast majority of the dairy cooperative membership is assumed by men, leaving only 14% to women.

Fig. 8: Involvement of women in poultry rearing

The role and contribution of women in dairying other than usual household responsibilities had been interpreted in social than economical pretext. Contribution of farm women in agriculture is likely to be around 50 to 60 per cent while rural woman contributes a share of more than 75 per cent in animal husbandry operations (Fig. 8). On an average,

Table 4: Employment generation under one acre IFS module

Farming system	Cereals	Vegetable	Poultry	Duckery	Fishery	Goatery	Dairy	Total
Cereals	416	0	0	0	0	0	0	416
Crop + Vegetable	220	310	0	0	0	0	0	530
Crop + Fish + Poultry	376	94	60	0	40	0	0	570
Crop + Fish + Duck	376	94	0	50	40	0	0	560
Crop + Fish +Goat	376	94	0	0	40	110	0	620
Crop + Fish +Cattle	376	94	0	0	40	0	170	680
Crop + Fish + Poultry + Duckery	376	94	60	30	40	0	0	600
Crop+Mushroom + Goat	376	94	0	0	40	110	70	690

a woman contributes about 4-6 hours for animal husbandry operations in rural areas in Bihar.

Women play an important role in animal husbandry activities as manager, decision makers and skilled workers. Caring of animals is considered as an extension of domestic activities in Indian social system and most of the animal husbandry activities like bringing fodder from field, chaffing the fodder, preparing feed for animals, offering water to animals, protection of animals from ticks and lice, cleaning of animals and sheds, preparing of dung cakes, milking, ghee-making and marketing of produce are performed by farm women. Thus, involvement of farm women in farming activities is a common feature in Indian rural setting. They help in farm operations, take their animals for grazing, look after the sale of milk, and in addition, perform the functions related to house management. Most of the work and decision-making by women takes place at the household level, while old men or children take the livestock for grazing and male members participate in public meetings that relate to animal husbandry. Almost all important decisions are taken jointly by both the man and the woman heading the household. These decisions include which animals to sell and at what price, disease diagnosis and treatment of sick animals.

Women's typical role within a livestock production system is different from region to region and the distribution of ownership of livestock between men and women is strongly related to social, cultural and economic factors. Generally, it depends on the type of animals they raise. In many societies, for example, cattle and larger animals are owned by men, while smaller animals such as goats, sheep, pigs and backyard poultry kept near the house are more a woman's domain. When the rearing of small animals becomes a more important source of family income, ownership, management and control are often turned over to the man. Women are crucial in the translation of the products of a vibrant agriculture sector into food and nutritional security for their households. They are often the farmers who cultivate food crops and produce commercial crops along with the men in their households as a source of income. When women have an income, substantial evidence indicates that the income is more likely to be spent on food and children's needs. Women are generally responsible for food selection and preparation and for the care and feeding of children.

Dairy industry

One of the most productive and important aspects of women's farm work in India is dairying. Women performed most of the actual dairy work and were primarily responsible for most dairy production. However, as various aspects of dairying moved from the farm to the factory, women's participation is now shifted from farm to dairy industries.

Role of women in dairy sector

- Feeding of milch animal
- Processing of crop residues
- System of grazing in forest areas and procedure of lopping of trees for fodder
- Care of pregnant animal
- Care of calf
- Clean milk production
- Processing and value addition of milk
- Conservation of fodder

Women and forestry

In rural areas fuel wood contributes 84% of the total household energy consumption. Unfortunately, forests are deteriorating massively due to encroachment of agricultural production, mining, construction of dams, industrial and railway demand. The country has been losing 1.5 million hectares of forest cover annually. In India about 16% of the total geographical area is covered by woodland and forests (Shivani, 2013). Gender roles in using forest resources vary widely depending upon the region as well as socioeconomic class and tribal affiliation. Rural Indian women's interface with the forests is varying - gathering, wage employment, production in farm forestry and management of afforested areas in the community plantation. In India, women are the major gatherers and users of a much more diverse range of forest products than men. Depending upon the socio-cultural variations among different communities, primarily Non-timber Forest Products (NTFP) is collected by women and timber by men. In several parts of India, large proportions of the population depend on NTFP as their main source of livelihood. Apart from fodder and fuel, women collect food, medicinal plants, building materials, material for household items and farm implements. Sal leaves are primarily collected by women. As women are the ones who have traditionally been collecting forest products, they possess the knowledge of properties and potential uses of these products.

Apiculture for livelihood support to women folk

Beekeeping is an important activity for many rural people - both men and women. Increasingly both governments and NGOs are working to encourage women's participation in rural development and beekeeping has been identified in many places as a means of additional income generation that is suitable for women.

Few cultures have any taboos threatening the involvement of women in beekeeping. Apiculture is a very profitable occupation and suitable for small and marginal farmers and even for landless. India has achieved some progress in this industry using its own technologies and developmental Planning. A more than 5000 tonnes of honey is produced annually in our country. Apiculture is an absorbing hobby to some, and to others it is an industry for producing honey and wax. In ancient times, honeybees have been kept in a crude manner in India. Apiculture today is based upon improved methods using the principles of movable frame hive, honey extractor and the smoker.

Beekeeping can be started cheaply and built up as resources allow, there is little need for land ownership and, with some technical know-how, and hives can be located close to home. The demands of time are not great and these can be fitted in with family responsibilities. These are all positive attributes that should encourage women. Nonetheless, beekeeping is frequently perceived to be a male activity and women's participation in beekeeping projects is often lower than might be expected. Therefore, the promotion of beekeeping as an income-generating activity for women folk should be the promoted in especially in rural areas where mostly women's are involved in agricultural activities.

In India, it is very low participation of women in beekeeping. The reasons may be: women were afraid of bees; they could not climb trees; beekeeping was considered a man's occupation. Moreover, traditional ways of living restricted women to carry out domestic activities close to the homestead which hindered participation in beekeeping. However, women commonly use the fruits of beekeeping (wax) to make value added products such as candles. The production of secondary or value added products made from honey, beeswax or other hive products offers a unique space for women's traditional skills. Where work and childcare commitments constrain women to remain within the vicinity of their homes, enabling women to produce value added beekeeping products can be an ideal opportunity for income generation. Male beekeepers are often not interested in this area so it is not challenging to the cultural status quo.

Women and fisheries

About 5 million people in the coastal areas carry out fishing and allied activities for their livelihood. Fish drying/curing, marketing and hand braiding and net-mending are the main areas of women's involvement in Tamil Nadu, Andhra Pradesh and Odisha. Women are also involved in shrimp processing in these states. Among the mangroves of Bhitarkanika on the Odisha coast, both women and men fish estuarine areas. Men cast nets while women and children catch fish with hands. But fishing by boat in the flood tides is exclusively performed by men. In contrast, women's participation in small-scale fisheries is very limited in West Bengal. Even ancillary industry, which in the other Indian east coast states is a women's domain, is dominated by men, as a relatively low number of days in a year is spent on actual fishing. In the fishing villages, fish drying/curing is performed by both women and men who do not belong to the fishing community. In coastal aquaculture, women are involved in prawn and seed collection to a very limited extent (Fig. 9).

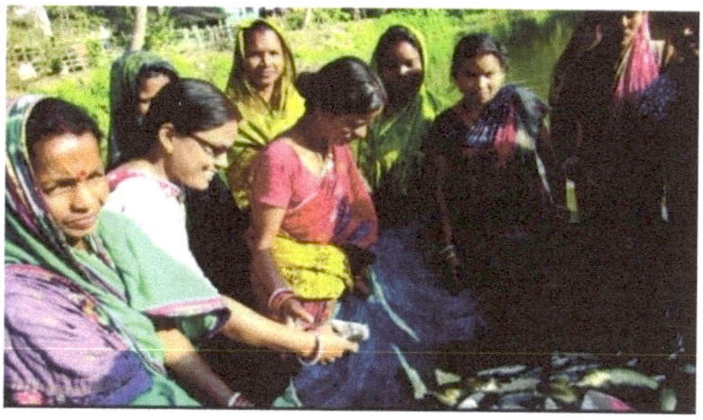

Fig. 9: Involvement of women in pisciculture activities

Fishery sector has witnessed a steady growth from a meager annual fish production of 63.05 lakh tones in 2004-05 to 86.66 lakh tones in 2011-12. In Odisha, the fish production has increased from 2.61 hectares on 1999-2000 to 4.10 lakh tones in 2012-13. The transformation from a purely traditional activity to a full fledge commercial enterprise has generated avenue for employment generation, nutritional security and enhancement in family income. As a general global trend it is estimated that, women accounted for at least 15 per cent of people directly engaged in the fisheries primary sector during 2010. The proportion of woman is considered to be higher (at least 19%) in inland water fishing and as high as 90% in secondary activities such as procession (FAO 2012).

Pisciculture based integrated farming system

Integrated fish farming system is a diversified and coordinated method of farming where one or more farming system (Agriculture, Horticulture, animal husbandry etc.) are integrated with fish farming for efficient utilization of resources and recycling of waste or by-product. This type of farming system is becoming very popular among rural people in different parts of the country.

As it is mostly followed by the marginal small farmers, women are actively engaged in this production system. Jahara(1998) mentioned that integrated aquaculture activities in Malasia are often perceived as an extension of the women's household activities, which makes it easier for women.

Aquaculture plays a multidiscipline role and aims at providing food security, employment generation economic benefit and optimum utilization of water resources. Major aquaculture production comes from the Asian subcontinent. In these countries, women play crucial role in production process. Nandieesha (1994) found that in Cambodia the ponds were women carried out at least of the tasks associated with aquaculture had higher yields than other ponds. In some parts of Thailand and China, because of male migration to cities, women take up the sole responsibilities for farm production including aquaculture (Kusa Kabe, 2003). It was found that women were involved in intensive aquaculture have more resources in terms of land, cash and knowledge than women who are involved in subsistence aquaculture.

Traditional rice fish production system is having importance socio-economic consideration in the life of the farmers in Eastern India. This is mostly practice in the low laying rainfed areas where water level in rainy days. Women's participation in rice-fish is considered to be higher, since women are already involved in rice production (Dehadrai, 1992). Therefore in rice-fish integrated farming system women work force can make significant contribution starting from planting of rice to fish culture. In an manageable vast waterlogged rice environments, naturally occurring fishes and prone enter the field during monsoon season and grow together with the rice crops (Das,2002). In these water bodies, tribal women harvest fish through group fishing by using local devices. In general, small fishes, snails and crabs are the common harvest from most of the rice environment.

Women and rural production

Women in rural India generate income in various ways. Women are highly involved in processing of the NTFP, particularly in small-scale enterprises. This includes basket, broom, rope making, tasar silk cocoon rearing, lac cultivation, oil extraction, and bamboo works, etc. Women constitute 51% of the total employed in forest-based small-scale enterprises. However, this does not mean that men do not have any role in these activities. Among the scheduled-caste weavers in Odisha, men collect grass for basket making while women cure it and make the basket. In the Jeypore tract of Odisha men and women are equally involved in collection, processing and marketing of forest products such as grass, bamboo and resin. The challenge to the sustainability of a production system lies in integrating technology, work, and resources (financial and social) effectively with gender so that both women and men can play an active role in improving the productivity, profitability, stability, and sustainability of major farming systems (Fig. 10).

Fig. 10: Floriculture by farm women

Women friendly tools and equipments

Women are still struggling for activity-specific tools and equipments. Whenever agriculture gets mechanized, women are the first ones to be marginalised. However, improved tools and equipment serve same purpose for both genders (Fig. 11).

Fig. 11: Drudgery reducing implement for farm women

- Reduce drudgery
- Increase inputs utilization efficiency
- Ensure timeliness in field operations and reduce turnaround time for next crop
- Increase productivity of man- machine system
- Conserve energy
- Improve quality of work and also quality of produce
- Enhance the quality of life of agricultural workers

Drudgery of farm women in various field operations could be reduced by providing improved farm tools and equipment. The improved tools and equipment are primarily developed keeping men workers in consideration while farm women in the country are also involved in most of the operations. Hence, already developed equipment may not be suitable to farm women as such because ergonomical characteristics are different than men workers. The result is that women workers have to carry out the operation with their hands, and there is a lot of drudgery involved in it in addition to occupational health problems. The posture adopted during the operation are also not proper and lead to occupational health problems, if not given due attention. This may also result unemployment of women workers, more over it also appears that improved technology is being kept away from them. The importance of developing farming technologies relevant to farm women has only recently been recognized as an extensive participation of farm women in the field of agriculture, food security, horticulture, processing, nutrition, sericulture, fisheries, and other allied sectors has been gradually realized in coming years. The suitability of equipment to farm women can be judged in better way using ergonomical studies as ergonomics cover all aspects that deal with anthropometry, assessment of workload, working environment and safety features/mechanism to optimise human-machine environment system. This helps in increasing their working efficiency with reduced drudgery by fitting to the capabilities and limit of human operators/ workers.

Ergonomical characteristics of women farm workers

Ergonomics is the scientific study of the relationship between a person & his/ her working environment, which includes working environment, ambient conditions, tools & materials, methods of work & organization. The performance of a tool/ equipment not only depends on the constructional features but also on the workers operating it. Important ergonomical data suitable for design of equipment and work methods are:

a) **Anthropometric data:** It includes data on various body dimensions of workers. Seventy nine body dimensions useful for farm equipment design have been identified and data collected for 4500 women workers (Sunderum, 2013). The mean height and weight of Indian female agricultural workers are 151.5 cm and 46.3 kg as against 163.3 cm and 54.7 kg for male workers. The equipment needs to be designed keeping in view the limiting dimensions of women workers in consideration. It will help to make the equipment women friendly and safe for operation.

b) **Muscular strength data:** It is generally considered that a woman has about 2/3 strength as that of a man.

c) **Maximum aerobic capacity:** The maximum aerobic capacity, also called as maximum oxygen consumption rate sets the limit for maximum physical work capacity of a person. For women, this value is generally 75% of that of men. As per the data available for Indian workers, this value for women workers is about 1.5 litres/min.

d) **Physiological cost of operation:** Physiological cost of any operation is expressed in terms of heart rate and oxygen consumption rate. For an 8 hour work period for women workers a work load requiring oxygen at a rate of 0.6 l/min is considered as the maximum limit for acceptable work load. The heart rate for such a work load will be about 110 to 120 beats/min.

e) **Posture:** A good working posture requires minimum static muscular effort. If a work can be done in a standing posture instead of bending or squatting posture, it should be preferred for long duration jobs. Also for long duration work, a sitting posture may be better than the standing posture.

f) **Load carrying capacity:** Load to be carried by a woman worker should not exceed 15.0 kg (about 40% of body weight). The mode of load carrying should be such that the static loading of hands and arms is avoided. For hilly terrains, the limit will be lower depending on slope and the terrain. Women have different ergonomical characteristics and therefore, due attention needs to be given to their capabilities and limitations while designing various equipment (Chandra and Gite, 2012)

Extension services

It is important to ensure greater access for farm women to various inputs (including farm tools and equipment) needed by them to carry out their work more efficiently and with minimal drudgery. Farm women have little access to non-formal education and trainings. It is a fact that the agricultural extension services are mainly composed of male agents and as such they tend to channelize knowledge and training on improved technology to male farmers/workers only. There is also lack of infrastructural facilities for women in relation to technical training, accommodation and transport provisions. It is imperative that Central/ State government departments, R&D institutions and NGOs promote improved technology to enhance labour productivity of women workers and to reduce their drudgery. They should also recruit the women extension staff in their extension wing for effective transfer of women specific technologies to farm women. The CIAE Bhopal is providing necessary training on various tools and equipment to women facilitators from different states. These facilitators can act as resource persons for their own state to propagate the technologies.

Nutritional, health and livelihood security

Several horticultural crops, especially tuber crops are used as staple food in the world. Fruits and vegetables are also rich source of vitamins, minerals, proteins and carbohydrates, etc., which are essential in human nutrition as protective foods, and have importance for nutritional security of the people. Fruits and vegetables provide substantial amount of nutrients important for human health they are particularly the important source of micronutrients, pro vitamin–A vitamin-B6, vitamin C and vitamin E as well as folic acid, iron and magnesium. With the development of new improved varieties of horticultural crops. The demand for seed and genuine planting material has increased manifold across the country. This offers unique scope for

Fig. 12: Exposure visit of women group to vegetable field in Kalahandi

development of high- tech nursery which further generates employment opportunities for rural women.

Recycling of farm renewable resources by women

Women entrepreneurs can play powerful role in confidence building and creating awareness in other women to promote self-reliance. With majority of woman in rural India, it is important to develop entrepreneurship skills in rural environment. Entrepreneurship development among rural women helps to enhance their personal capabilities and increase decision making status in the family and society as a whole. They are engaged in starting individual or collective income generation programme with the help of self-help group. This will not only generate income for them but also improve the decision-making capabilities that led to overall empowerment.

Harnessing the nutrient energy of within farm renewable resources include crop residues, farm wastes. They are valuable sources of plant nutrients and humus. In tropical and subtropical soils found in India, there is a general deficiency of organic carbon and plant nutrients due to rapid loss of these components by biodegradation. To make up for these losses, extensive utilization of organic residues in agriculture is essential. In addition they also protect the soil from erosion. The manurial value and quality of these wastes could be improved by composting and enriching these organic sources along with inexpensive materials such as rock phosphate. In India, there is great potential for utilization of crop residues / straw of some of the major cereals and pulses. Approximate availability of straw is to the tune of 141.2 m t, which contributes about 0.7, 0.84 and 2.1 m t of N, P_2O_5 and K_2O respectively, after deducting 50% quantity utilized as animal feed. (Sharma, 2009). Farm manures should be returned to the fields. As a rule, organic matter fit for soil application should not be burned. In India, the estimated production of dung and urine from bovine population works out to 1002 and 658 m t respectively. They contribute about 5.71 m t of N, P, & K. Cow dung is an important input for biogas plants having dual advantage of providing both fuel (gas) and fertilizer (slurry) (Sharma, 2009). Agro-forestry systems can lead to more nearer to the nutrient cycling than agriculture.

Almost all the women in villages prefer to sell cow dung cakes, but it is high time to train them make available alternate sources of firewood so as to prevent use of cow dung directly for burning. Subabul tree can be used as an alternate in place of cow dung cakes as fire wood in household usage. Being a fast growing tree, Subabul, should be advocated for plantation on their North-West sides of field boundaries to avoid shading on their farm. FYM also contains plant material (often straw), which is used as bedding for animals absorbing the feces and urine. Agricultural manure in liquid form, known as slurry, is produced by more intensive livestock rearing systems where concrete is used, instead of straw bedding. Manure from different animals has different qualities and requires different application rates. For example horses, cattle, pigs, sheep, chickens, turkeys, rabbits, all have different properties. For instance, sheep manure is high in nitrogen and potash, while pig manure is relatively low in both. Horses mainly eat grass and a few weeds so horse manure can contain grass and weed seeds, as horses do not digest seeds the way that cattle do. Chicken litter, coming from a bird, is very concentrated in nitrogen and protein and is prized for both properties. Processing of farm yard manure is also not up to mark by the rural people. Women being more receptive than men should also be trained on scientific method of preparing farm yard manure. This will help in propagating the right usage of FYM.

Similarly women may be trained for production, utilization and marketing of vermi wash (a liquid foliar spray), Panchagavya, bio pesticides and herbal pest repellents using locally available plant materials/ from farm wastes. Production and distribution of efficient biological inputs such as biofertilizer with indigenous inoculums are produced with a network of trained small and marginal farmers for improved farm productivity and income generation. Introduction and testing of post harvest technologies for value addition of organic products with aim to enhance livelihood options for poor rural tribal women through introduction of post harvest energy saving technologies in processing of local fruit is an important approach. Alternative vocation for income generation, small-scale (household level) scientific rearing of small animals such as goat, poultry and pig, sustainable utilization of natural resources and value addition, mechanized processing of bamboo furniture and product applications, diversified cropping systems, agro-technology for improving the land use, fisheries, cultivation of horticultural produce by using organic/bio-fertilizers, value addition in banana fibers, etc. are very lucrative enterprises. The women of Nagla Banjara (Rajasthan) now weave the bamboo baskets and other woven bamboo structures and earn a small amount of Rs.1000/-per woman. Spinning and weaving of cotton, silk and woolen textiles, Kauna Grass mat manufacturing, bee keeping etc. also account for income generation to the women in India.

Future strategy for technology development and promotion

In the changing scenario, the participation of women workforce in agriculture is going to increase to 50% by 2020 i.e. out of the total estimated agricultural workforce of 240 million, about 120 million will be the women workers. This is expected to happen mainly because male workers will either get involved in other non-farm activities or migrate to towns and cities for other jobs. To meet this situation, it is necessary to take the following steps.

- Design the tools/equipment keeping in view the anthropometric data of women workers.
- Organize demonstrations and trainings to rural women on various modern tools/equipment in proper and safe operation. This will help in reducing their drudgery and increasing productivity.
- Encourage manufacturers/entrepreneurs to fabricate improved tools and equipment
- Assist farm women, after being duly trained to get loans from banks/ others
- Building up of linkages with central/ state depts., NGOs, banks, and other stakeholders to promote these improved tools and equipment.

References

Bhatt, B.P., 2013 Gender perspective an agricultural strategy for gender mainstreaming. Lectured delivered during training on gender perspective on IFS. ICAR research complex, Patna, Bihar pp.-1

Chandra, P. and Gite, L.P. 2012. Technologies for Women in Agriculture-Experience and Achievements of CIAE. Proceedings: First Global Conference on Women in Agriculture.

Das D.N. 2002. Fish farming in rice environment of North-Eastern India. Aquaculture Asia: T (11):43.

Dehadrai P.V.1992. Opportunities for women in Rice-Fish culture, in C.C.R.De Aa cruz, BA light foot et.al) eds. Rice Fish Research and Development on Asia 24 Manila ICLARM .

FAO, 2012 The state of world fishery and Aquaculture 2012. Part-II – selected issues in fisheries and aquaculture.

Jahara Y.1998. Women in small scale fisheries in Malasia in proceedings of the symposium on women in Asia fisheries, cheang-mai, Thailand.

Kumar.R. and FulZale R.M.1998 – Factors affecting the time utilization of farm women in household, agriculture and dairying activities. Dairying, Foods, Home Science 17(1) : 55-59.

Kusakabe K 2003. Women's involvement in small scale aquaculture in North-East Thailand, Development in practice, 13(4:333-345)

Meena M S, Singh K M, Bhatt B P and Ujjwal Kumar. 2013. Model Training Course on Gender Perspective in Integrated Farming System 17-24 January 2013 at ICAR Research Complex for Eastern Region, Patna, Bihar.

Nandeesha, M. C., 1994. Aquaculture in Cambodia. Info fish International 2:42–48.

Nataraju, M.S. and Lovely, R.S., 1993, Extent of participation of rural women in crop and animal production activities. An Analysis. *Indian J. Adult Edu.*, 54 (3): 52-57.

Nanda, S.S. Behera, M. Sahoo, S.C. and Jena, D. 2009. Gender issues and prospectives in Agriculture. Directorate of Extension Education OUAT, Bhubaneswar.

Sadika Sharmin, M. Serajul Islam* and Md. Kamrul Hasan.2012. Socioeconomic Analysis of Alternative Farming Systems in Improving Livelihood Security of Small Farmers in Selected Areas of Bangladesh. The Agriculturists 10 (1): 51-63.

Sanjeev Kumar, 2013. Lecture Delivered during model training course on gender perspective in integrated farming system, 17-24 January 2013 at ICAR Research Complex for Eastern Region, Patna, Bihar.

Sharma, A.K. 2009. Bio – fertilizers for sustainable Agriculture. Published by Agro bios (India), Agro House, Jodhpur. Pp 407.

Shivani, 2013. Role of women in integrated farming System, Some experiences from eastern region, Lecture delivered during model training course on gender prospective in IFS, 17-24. January, 2013. ICAR Research Complex for Eastern Region, Patna, Pp. 56.61.

Singh, S.K. 2013. Lecture Delivered during model training course on gender perspective in integrated farming system, 17-24 January 2013 at ICAR Research Complex for Eastern Region, Patna, Bihar.

Singh, S.S.2013 Women empowerment through crop based farming system. Lectured delivered during training on gender perspective on IFS. ICAR research complex, Patna, Bihar Pp -25.

Sunderum, P.K . 2013. Women friendly agricultural engineering technology for reduced dregeny. Lactone delivered during model training course on gender prospective in IFS. 17-24 January 2013 at ICAR Research Complex for eastern region. Patna. Pp. 158-172.

Integrated Farming System Practices: Challenges & Opportunities, pp 297-314

New India Publishing Agency, New Delhi, India

12

Family Farming and Gender Participation in Integrated Farming Systems

D. Jena

Introduction

Agriculture has been the joint venture of both men and women since the time immemorial. Women play a significant role in agriculture and are responsible for half of the world's food production in most of the developing countries. According to International Food Policy Research Institute (IFPRI), women play a central role as producer of food, managers of natural resources, income earners and care takers of household food and nutrition security. Food and Agriculture Organization (FAO) studies also confirm that women are the mainstay of small scale agriculture, the farm labour force and day to day family subsistence. About 70 per cent of agricultural workers, 80 per cent of food producers and 10 per cent of those who process basic foodstuffs are women and thus making up more than two-third of the workforce in agriculture production. They currently account for 60-70 % of all food production in all the developing countries. Other than agriculture and allied sectors like dairy, poultry, fishery, goatery and sericulture, women are also skilled in handicrafts, wooden works and various other types of small enterprises which boost the rural economy. They are also involved in small scale enterprises like rope making, broom making,

---

College of Home Science, Orissa University of Agriculture & Technology, Bhubaneswar-751003, Odisha, India bamboo works. All women in rural areas, irrespective to their age, size of their family, size of land holding, caste and community, perform major agricultural tasks. The rural women form the most important productive work force in the economy of majority of the developing nations like India. Agriculture sector employs 4/5th of all economically active women in the country and 48 per cent of India's self employed farmers are women (Kumar, 2013). In present scenario, nearly 84 per cent of economically active women in India are engaged in agriculture and allied activities. In a study undertaken by DRWA (Directorate of Research on Women in Agriculture) in low land and rainfed rice in Odisha revealed that under low lying situation, women spent nearly 20 days in uprooting and transplanting, 10 days in intercultural operations and 20 days in harvesting and transportation. In up land rainfed rice women were involved for 40 days mainly in gap filling, weeding, harvesting threshing etc. (Arya, 2008). The rural women of low-income category and lower caste perform more work in farm related field activities while the women of higher income groups perform more work in post harvest processing like storing of food grains, parboiling, winnowing, preparing puffed rice etc. In tribal areas, women participate in all agricultural operations except ploughing. They are seen to do more hard work in the fields than men. Depending upon the socio-cultural variations among different tribal communities, Non Timber Forest Products (NTFPs) are collected by tribal women. They collect food, fodder, fuel, medicinal plants, building materials and other household materials. Every day during off seasons, they go to forest for collecting NTFPs and sell it in the local hats (market) for day to day expenditure of their family. Sal and kendu leaves are also collected by them for their livelihood support. Ladies are the decision makers in farming sector as well as on other domestic affairs because of their greater contribution of household income. Marketing of their produce and also other materials collected from forest by them are exclusively done by the women group. Tribes mostly depend on forest and forest products during the dry period of the year. Now-a-days non-availability of NTFPs has shifted women from self employed to wage employment.

Family farming

The United Nations declared 2014 as the "International Year of Family Farming" (IYFF) to recognize the importance of family farming in reducing poverty and improving global food security. According to the UN, it aims to promote new development policies at national and regional levels that will help small holders and family farmers to eradicate hunger through small scale sustainable agricultural production. The IYFF has been linked in many countries to the launching of the zero hunger challenge. Since overcoming hunger involves concurrent attention to calorie deprivation, protein hunger and hidden hunger caused by the deficiency of micronutrients in the diet, IYFF offers an opportunity for achieving a shift from food security to nutrition security. Unlike corporate farming which involves monoculture, family farming tends to be based on crop, livestock, fish, agro-forestry, and mixed farming systems. Therefore, it can otherwise be called "Integrated Farming System in every Household". That system will provide nutrition security to each and every household in a sustainable manner.

Further, by 2050, global population will be about nine billion. The IYFF can create awareness among all stakeholders [family farmers, Government, civil society, researchers and financial institutions and private sectors] about the potentials of Family Farming and challenges they face in this regard and take opportunity to strengthen institutional infrastructure to develop sustainable agriculture based on family farms. During the IYFF media have a responsibility to highlight specific farm production models that our research institutes and State Agricultural Universities have evolved for successful implementation by family farmers to minimize incidence of hunger, poverty and environmental degradation.

Food and nutrition security

Food and nutrition security to every individual through sustainable livelihood support is most essential in rural sectors. Food security means access by all members at all times to enough food for an active and healthy life. Nutrition security can be defined as physical and economical access to foods which provide all nutrients as per requirement of the body according to age, sex, physical activity and physiological status of every individual to maintain disease free healthy life. Nutrition security is the essential component of food security.

The term "food security" was first used in the international development literature of the 1960's and 1970's and referred to the ability of a country or region to assure adequate food supply for its current and projected population. The focus of international and national efforts was to grow more food and reduce population growth rates to sustainable levels. Food security was measured by food grain production to ward off famine, improving availability and access to food at affordable cost, to meet the energy requirements and prevent chronic under nutrition among the ever growing population. The World Health Organization (WHO) defines Food Security as having three facets: (I) Food availability (II) Food access and (III) Food use. Food availability is having available sufficient quantities of food on a consistent basis. Food access is having sufficient resources, both economic and physical to obtain appropriate foods for a nutritious diet. Food use is the appropriate use based on knowledge of basic nutrition and care as well as adequate water and sanitation. The FAO adds the fourth facet: the stability of the first three dimensions of food security over time. Although national food security is important as providing a foundation for country as well as state level, food security for each and every household is to provide adequate food to every member of the family in adequate quality, quantity, safety and cultural acceptability.

The population now faces the dual burden of mal nutrition with persistent inadequate dietary intake and under nutrition on one side and low physical activity above requirements and over nutrition on the other side. Over 40 per cent of preschool children in India are underweight. India is rated very poor in terms of food security, Available evidence suggests that in countries with high stunting rates, under five underweight rates are directly linked to household food insecurity and are not good indicators of nutrition security. Under nutrition among women increases reproductive and maternal health risks and lowers productivity. This situation contributes to women's diminished ability to gain access to other assets later in life and undermines attempts to eliminate gender inequalities. Sudden erupt in prices of important food items in domestic and global market and its consequences for poor people has once again brought to the front the fragility of food situation and vulnerability of poor including women and children. Therefore improving the performance of agricultural sector through gender focused agricultural programmes worldwide is the need of the hour. Therefore, sustainable agriculture growth is important to check the hunger and poverty in the vulnerable population of the country as for 1 per cent growth in agriculture sector there would be a 2-3 per cent reduction in poverty.

The National Food Security Bill 2013 aims to provide 5kg. of food grains / head per month at subsidized prices from State Governments under the Targeted Public Distribution System (TPDS).Indian Council of Medical Research (ICMR) recommends that an adult requires 14 kg of food grains per month and children 7kg. While providing food grains will help increase access to calories, it will not ensure to get all important nutrients. Pulses are to meet protein needs in vegetarian diets. Vegetables intake which is essential to provide the needed micronutrients are low dietary intake is the major factor responsible for under nutrition. Therefore family farming is the most important enterprise for getting nutrition security from different components of farming in rural areas.

Concept of gender

Gender means the socially constructed differences in roles and responsibilities assigned to women and men in a given culture or location and the societal structures that support them. Every society has different 'scripts' for male and female members to follow. Thus members learn to act out their feminine or masculine role, much in the same way as every society has its own language. The term gender was first used by Ann Oakley and others in 1970s as analytical tools to understand the characteristics of men and women which are socially determined in contrasts to biological differences. Gender is also defined (World Bank) as economic, social, political and cultural attributes and opportunities associated with men and women (Srinath.K, 2008).But **Sex** means the biological difference between male and female, which are universal, obvious and generally permanent. In a family farming the role and responsibility are more important for its sustainability.

Gender role

1. The role refers to the activities performed by men and women in different situations, times and within the different cultures, classes, castes, ethnic groups etc. The roles of men and women are shaped by various forces such as social, cultural, economic, environmental, religious and political. The gender roles may change depending on the socio-cultural dynamics of the society. Men and women have different roles, different needs and constraints. Gender role differs from biological role of men and women.

The differentiation in task and responsibilities are as follows

Dual role of Male	Triple role of Female
Productive roles : Refers to activities carried out by men in order to produce goods and services to meet the subsistence needs of the family, i.e. in agriculture, productive activities include planting, irrigation, weeding, harvesting, selling etc.	**Productive roles** : Refers to activities carried out by women in order to produce goods and services to meet the subsistence needs of the family, i.e. in agriculture, productive activities include transplanting, weeding, harvesting, parboiling, storing etc.
-	**Reproductive role**: Reproduction of human resources (the child bearing and rearing responsibilities maintenance and reproduction of labour force)
Community management/ community politics (producing community goods, collective consumption of water health care, education and well beings)	**Community management** / community politics (producing and utilizing community goods, collective consumption of water health care, education in a common plat form)

Employment is an important condition for food security, economic security and sustainable livelihoods among the families. According to 2011 census, the population of Odisha was reported 4.20 cores i.e. about 3.47 per cent of the population of the country. Out of total number of workers in Odisha, 67.9 per cent was male and 32.1 per cent was female workers. The main male workers constituted 61.0 % where as female main workers were 17.87 % of the total workers. Male Cultivators were reported 23.4 %where as male marginal workers were reported 39.0 %. Female agricultural workers were reported 70.70 %.

Female agricultural labourers were 76.2 % whereas male labourers were reported 38.4 %. It has been observed that percentage of female participation as agricultural labourers and marginal workers were more as compared to male in Odisha according to 2011 census.(Fig.1)

The women population constitute nearly half of the human resource. The literacy rate of women (64.36%) in Odisha is also at par with the literacy rate (65.46 %) of India Table 1. The literacy gap in gender has been observed 18.04 in Odisha and 16.6 in India. According to census, 2011 the highest employment of women has been observed in agricultural sectors as agricultural labourers of 76.20 % followed by 70.70 % as agricultural workers, 54.52 % as marginal workers, 17.87% as main workers and 16.5%, 17.7 % and 11.2% women employed in organized, public and private sectors respectively. To understand the totality of women's working life, it is necessary to understand the nature of their tasks, the time involved, their role and the benefit they gain.

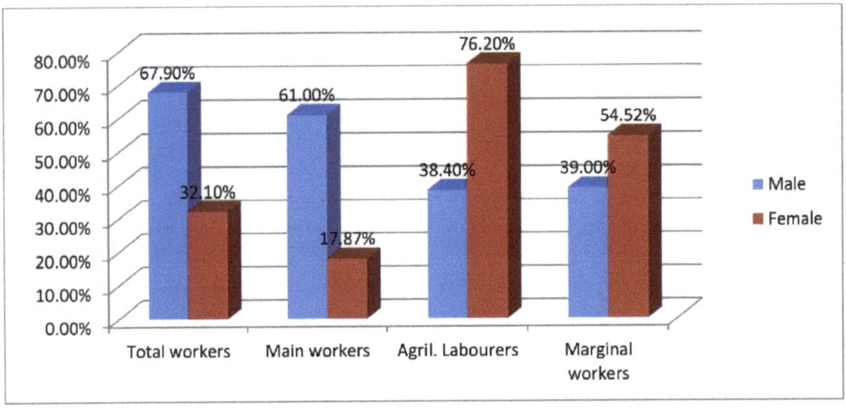

Fig. 1: Gender participation in Odisha (2011 census)
*: Odisha Economic Survey, 2013-14

Table 1: Status of Women in Odisha

Status of women	Odisha	India
Females per '000, male	978	940
Females per '000 male (SC)	987	945
Females per '000, male (ST)	1029	990
Women literacy rate (%)	64.36	65.46
Literacy gap in gender(M/F)	18.04	16.6

Source: Census 2011

Gender role in family farming

Fig. 2: Women transporting rice bundles from field

According to the Food and Agricultural Organization (FAO), family farms are engaged in all family-based agricultural and allied activities that contribute to production in vital sectors, viz. agriculture, forestry, fisheries, animal husbandry and aquaculture. More specifically, the FAO Director-General Graziano da Silva, describes *"a family farm is managed and operated by a family and predominantly reliant on family labor, including that of both women and men"* (Fig. 2). The practice of inheritance of land/ farm to the next generation is an incentive for the children to manage the farm efficiently and improve its productivity from generation to generation by investing in it and adopting scientific techniques. According to Agricultural Census 2010-11 out of 138 million farm holdings in the country, 117 million are small and marginal holdings. Small and marginal landholdings together shot up from 62% of total in 1960-61 to 85% in 2010-11 and own nearly 44% of the cultivated area. Marginal farmers share 67% in total number of holdings and 22% of cultivated area and farming for them becomes a significant source of livelihood. Small and marginal farms need higher costs for adopting modern technology and accessing production inputs and extension, credit and marketing services. Declining size of holdings without any alternate source of income has resulted in fall of farm income, thus causing discourage of future generations. A large number of smallholders have to move to non-farm activities to expand their incomes.

In India, the average holding size is estimated to be 0.32 and 0.24 ha in 2030 and 2050, respectively. (ICAR Vision Document - 2050). At present, 63 percent holdings are below 1 ha accounting for 19 percent of the operated area while over 86 percent of holdings are less than 2 ha account for nearly 40 percent of the area (APCAS, 2010).

The small land holders are better contributors to the total production (78%) but weak in terms of generating adequate income and sustaining their own livelihood. Small holdings (below 0.8 ha) does not generate enough income to keep a farm family out of poverty despite high productivity (Chand et al., 2011). Continuous decline in average size of land holding has implications for financial institutions since financial investment to create assets is often not viable on marginal farms (4). In Odisha, the average land holding has declined from 1. 30 ha in 1995-96 to 1.15 ha in 2005-06. About 84 % of farmers in the state are small and marginal with average holding size of < 0.8 ha. From the question-answer session of Odisha Lagislative Assembly on 29.11.14, the total number of farmers in the state has been recorded increased to 46,67, 466 in 2010-11 over 40,67, 135 in 2001-02. In the state , the total marginal workers are 33,68,296, small farmers are 9, 18,647, medium farmers are 63, 688, large farmers are 55, 74 and agriculture labourers are 24, 20.540 in numbers. Since women constitute half of the adult population their involvement in farming sectors are very much significant. Without substantial participation by them agriculture would not go ahead. Therefore gender analysis is more essential for its sustainability as well as future strategic measures.

i) Gender role in crop production

Among all cereal crops rice is the predominant crop in Odisha. Maize, ragi, bajra and other minor millets are the food grains grown in irrigated and rain fed areas. It is observed that the farm women are involved in different farming operations like transplanting, (Fig. 3) weeding, cutting and bundling of crop, parboiling and storing in rice based cropping system while men perform maximum percentage of work in land preparation, sowing, fertilizer and pesticide application, cutting, bundling, marketing. In intercultural operations the lower caste women are engaged in field operations with their male counterpart as agricultural labourers. In case of higher caste, large and small farmers, the women are involved mainly in post harvest processing of cereal crops like parboiling, drying, storing and keeping seed materials properly. In India women carry out as much as 80 % of the work in rice production (Pandey. H, 2001). in pulse and oil seed crops, women are mainly involved in harvesting and post harvesting activities in Odisha.

Fig. 3: Transplanting of paddy by women

ii) Horticultural crops

Women play an active role in various activities in horticultural crops and their participation has increased substantially. In vegetable cultivation, the women in Odisha are involved in stubble collection, nursery raising, care and maintenance of crops, harvesting, storage of seeds properly, post harvest handling and value addition and marketing of vegetables (Fig 4). Harvesting of vegetables like beans, pea, chili & okra, tomato, brinjal etc. are mainly done by women. After harvesting of produce cleaning and grading of the produce are mainly done by women. The post harvest management, processing, storage and utilization of vegetables are generally the domain of women in family farming. They are actively involved in winter and summer vegetable cultivation like tomato, cauliflower, cabbage, chili, brinjal and green leafy vegetables mainly for the purpose of income generation. The operations performed by women in fruit orchards are digging, filling of pits, application of FYM, watering, watch and ward, cleaning of orchard, picking, assembling and transporting etc. Farmwomen were found to participate in all the activities of coconut products. Sixty per cent of women were involved in broom making and 45 % in oil extraction from coconut. (Das, 1990). Value addition of fruits and vegetables are exclusively done by women with their traditional knowledge. Developing kitchen garden and flower cultivation traditionally in backyard is the female domain in most of the household irrespective to caste and economic status.

Fig. 4: Women in western Odisha selling vegetables

iii) Livestock production

In animal husbandry, women are considered as the owner of livestock. They play multiple roles ranging from care of animals, grazing, fodder collection, cleaning of animal shed, feeding animals, milking and value addition of milk by preparing curd, cheese, ghee etc. to get more profit from milk processing (Fig. 5). In all category of farming, women perform the tasks of collecting cow dung and making cow dung cake as fuel by mixing dung with twigs and crop residues in rural areas. Dung composting and carrying to the fields is undertaken by both men and women. In most of the cases women are the custodians of sheep, goats and poultry in the household and often children also actively take part in their management (Fig. 6). Men are involved in decision making regarding

Fig. 5: Grazing of goats by women

Fig. 6: Feeding of cattle by women

sales and purchase of animals, purchase of feed & medicines, taking animal to the L.I centre etc which are mainly done by men. Livestock rearing is an assured source of income to meet the emergency needs of a household and to ensure nutrition security in rural areas. In tribal pockets poultry keeping in every household is a tradition which contributes nutritional security and livelihood support to the tribal farm women.

iv) Fish production

In Odisha, nature and extent of women's participation in fishery varies across the state. In general, women belong to the fishermen category are mostly involved in catching, selling of fish, drying/ curing & marketing of dried fish. It has been observed that, the Coastal Gram Panchayats (GPs) of Rajnagar and Mahakalapada block of Kendrapara district, most of the women are involved in catching of fish, crabs and lobsters from the creeks and other natural bodies (Fig. 7). In coastal blocks of Ganjam district women are involved in fish catching and processing particularly making dry fish and marketing done by women (Fig. 8). They work as wage labourers in large scale fish processing industries. In other districts, men cast nets while women and children catch fish with hands in case of lower caste marginal and landless households. However, marine fish capture is a men's domain. Women's participation under Self Help Group (SHG) in all most all districts in small-scale fisheries is increasing day by day.

Fig. 7: Women catching fish from river in Puri dist.

Fig. 8: Fish catching by family members from creek of Kendrapara district

v) Forestry

Forest has been the store house of food materials like fruits, nuts, honey, leafy vegetables, mushrooms, flowers, roots and tubers etc. of tribal households. All these items are mainly collected by tribal women and also they play a dominant role in forestry. In developing world a significant portion of forest industry work force is made up of women (FAO, 1998). Besides these they collect non Timber

Forest Products (NTFPs) like medicinal plants, fuel wood, wild grasses for broom making, house construction materials from forest areas even if from distant places resulting high drudgery and low work efficiency of women.

Keeping women participation in all sectors in view it can be proposed to reorient them more technologically competent for family farming/ integrated farming so that there should not be gender bias and discrimination among them.

Technologies suitable for family farming

In order to minimize the risk of the farmers from climatic variations, integrated farming or family farming system approach will be encouraged in the State. A proper combination of different farm production systems namely, agriculture, horticulture, livestock, poultry, agro-forestry, sericulture and pisciculture will be promoted. Due to low agricultural productivity, the small and marginal farmers as well as about 15 to 18% landless families living in the rural areas are unable to generate remunerative employment and about 40% families are forced to live in poverty. With lack of food and income security, poor families are compelled to migrate to cities in distress, keeping their agricultural lands fallow, may become a major national challenge. Family farming offers the potential scope to solve the technology development problems. Research organizations in many countries are shifting towards farming system approach with emphasis on participatory on-farm research. Orientation to integrated farming with suitable and judicious utilization of resources by both men and women is more essential for nutrition safety.

Agriculture from ancient times has been the family based rural enterprise. The rural household gets maximum employment opportunities and their bread and butter from it. Therefore, family farming is an age old practice in a new concept with modern technologies to revive the integration of crop with animal component and allied off-farm activities like mushroom cultivation, vermi composting, fodder cultivation, azolla cultivation and value addition of agricultural produce etc for income generation and alternate livelihood option for them. The entire farming system revolves around better utilization of time, money, resources and family labour. The farm family gets scope for gainful employment round the year, thereby ensuring good income and higher standard of living.

Several integrated farming systems with different combinations have been proved to be the most profitable venture in agriculture. Crop components like cereals, pulses, oilseeds, fruits and vegetables , animal components like dairy, poultry, duckery, Goatery, fishery and off farm activities like mushroom cultivation, vermi composting, value addition of horticultural products, handicrafts from bamboo, sabai grass, golden grass, applique work, jute, coconut coir etc. can be suitable combinations for integrated farming system. These systems of enterprises provide stable income, household food security and employment opportunities both for male and female. There is no scope for horizontal expansion in production due to shrinkage of cultivable land verses ever increasing population. This can be possible through adoption of Family farming / integrated farming system based on the resources and labour force available within the family.

For landless farm family, the scope of integrated farming is considerably less extensive. The homestead area can be utilized judiciously for this farming. In backyard, nutritional gardening, vermi compost and mushroom can be grown. Cattle, goat/ sheep and poultry can be reared with minimum space. The system will provide meat, milk, eggs, fruits, vegetables and mushroom for consumption as well as income generation. Since all the above enterprises are mostly done by women, food and nutrition insecurity would not be a constraint for landless households. Ducks can be raised in ponds, and pond dykes can be used for vegetables, vermi composting and poultry raising. The following integrated farming systems are the key examples of successful gender sensitized enterprises. Attempts have been made for implementing income generating enterprises in addition to their conventional practices by the farmers and farm women in Puri district.

1. Pond based IFS: Pond based farming system is a common practice of coastal district like Puri, having different Land holding size and category of farmers. A marginal WSHG having holding size of 0.97 ha could manage to get net return of Rs 89,235.00 involving 173 mea days, where 106 man days were contributed finale workers
    - In case of small farmer having 1.45 ha Land holding, the contribution of male worker was more than 50% (223 male man days) of the female workers (107 man days).
    - In case of medium group of farmer the male mandays was more than 48% (524 mandays) than female (265 mandays) with land holding size g 2.45. ha.
2. Dairy based IFS: Interestingly, it was observed to that 50% man days was contributed from women workers (101 man days) as compared to male (199 man days) by small farmers.
3. Vegetables based IFS: In vegetable based farming system covering baby corn, cowpea, tomato, okra and chili, the contribution of babycorn to the net income was higher (Rs167,900.) in the Land holding of 2.00 ha and employment generation between male and female was more or less same.

Table 2: Gender analysis and income opportunities of Integrated Farming System / Family Farming conducted in different locations in Puri district

Si.No.	Type of Integrated farming system	Area/category (ha)	Components taken	Gross Return (Rs)	Net return (Rs.)	B:C ratio	Employment generation (man days/Year)		
							M	F	T
1	Pond based IFS	0.97(Marginal*)	Vegetables-fish-duckery-vermi compost-azolla	1,64,700	89,235	2.2	67	106	173
4	Dairy based IFS	1.46(small)	Paddy-ground nut- green gram- betel vine-fodder-azolla- dairy-goatrey-vermi compost	3, 72, 280	2,35,370	2.7	99	101	300
5	Vegetable Crop based	2.0 (Small)	Chili- baby corn-cowpea-tomato-okra	2,62,900	1,67,900	2.7	295	310	605
2	Pond based	1.45 (Small)	Paddy-banana-coconut-fish-poultry-dairy-duckery	2, 50,700	1,40,200	2.7	223	107	330
3	Pond based	2.54 (Medium)	Paddy-banana-vegetables-betel vine-fish-dairy-goatrey -vermi compost-mushroom-honey bee	7,47, 030	4,65, 210	3.3	524	265	789
6	Crop based	4.0(Medium)	Rice-banana-coconut-vege tables-fish-duckery-dairy-honey bee	5,84,180	3,40,480	2.4	320	147	467

*Women SHG at Khirikhia village taken area, 0.972 ha. on leased basis

4. Crop based (IFS: In crop F.S. system with land holding of 4.0 ha, the employment of male workers were more than double (320 man days) as compared to the female (147 mandays), Nanda et al 2011.

   - Integrated Farming Systems are proved to be a profitable family farming with combination of crop-fish-vegetables- dairy- fruits orchard-vermi compost etc. involving members of the family. It gives sustainable livelihood support to the farm family even under adverse climatic situations. A marginal and small farm family will get employment throughout the year. Agro by-products can be utilized judiciously in the farming system which fetches more profit than conventional farming. Every inch of land is utilized systematically in the system to get maximum profit and ensure household food and nutritional security. More emphasis is still required to generate model suited to various farm size holdings in different agro climatic conditions involving family labour.

The above analysis revealed that in all types of IFS, both men and women have got equal share in shaping their household economy and food security by introducing women friendly enterprises like mushroom cultivation, duckery, vermi compost, azolla and poultry keeping. Mushroom cultivation at household level was exclusively done by women where as in commercial cultivation from bed preparation to marketing done by men. In dairy based farming system, activities such as caring, feeding, grazing and cleaning of shed were done by women. In poultry keeping feeding, cleaning of shed and watering were done by women where as vaccination marketing are the job of men. Similarly in vermi composting women had major role in putting half decomposed cow dung and watering and care of the vermi unit while application and marketing were mostly done by men.

Fig. 9(a) Visit of Dean, Extn. Edn at IFS site of Khirikhia Village, Puri

Fig. 9(b) Pond based IFS developed by Women SHG

Fig. 9(c) Feeding to ducks by Women SHG

Fig. 9(d) Vermicompost by Women SHG

Fig. 9(e) Azolla feeding to ducks in IFS by SHG

Fig. 9(f) Azolla cultivation in IFS by SHG

Fig. 9: Pond based IFS of Maa Basuli women SHG at Khirikhia village, Puri district.

Suitable combinations of enterprises for family farming in gender prospective

Gender prospective and location specific suitable combinations of enterprises are recommended for different land holdings and farming situations for sustainable livelihood of rural households. Women friendly enterprises are :

- Post harvest processing of cereals and pulses
- Vegetable and fruit cultivation in nutritional garden, flower cultivation and value addition of these products
- Value addition of nuts and oilseeds
- Mushroom cultivation and its value addition
- Handicrafts from bamboo, sabai grass, coir, golden grass etc.
- Dairy with value addition of milk and milk products
- Fodder and azolla cultivation
- Poultry keeping, goatery among WSHGs
- Fish cultivation and value addition of fish
- Value addition of Non Timber Forest Products (NTFPs)

There is a vast scope to improve the household profitability by judiciously utilizing family labour using innovative practices and ensuring multiple uses of various household resources. Small and medium size water bodies can be brought under multi-component production systems like fish-duck and horticultural crops in the pond dyke will ultimately lead to ensuring nutrition security, enhancing income and sustainable livelihood of small farm holdings. Women can only be empowered if they have enough opportunities in these systems. This could be possible through location specific trainings and critical need based support to them. Developing women-centric farming system models will be a real challenge as men are migrating to rural non-farm sectors. A proper combination of different farm production systems namely, agriculture, horticulture, livestock, poultry, agroforestry, sericulture, pisciculture and value addition of produce will be promoted. As stated above, these systems of farming utilized wastes from different components, livestock, poultry and aquaculture byproducts in the production. For example, ducks were raised in ponds, and pond dykes were used for horticultural crops like banana and vegetables, vermi compost, azolla for duck feeding and animal keeping. The system provides meat, milk, eggs, fruits, vegetables, grains and fish for nutrition security and income.

Future strategies for efficient involvement in integrated farming system

- Technologies are to be developed for farm women in the areas of maximum participation including harvesting and post harvest processing, livestock production, processing of milk, mushroom cultivation, vermi compost, value addition of different farm produce. Community food processing units should be established in each village for facilitating village level processing and value addition by farm women.
- Low cost and women friendly drudgery reducing implements with ergonomically suitable should be designed for the agricultural operations. They should be trained to use farm machineries at their farm level.
- Women from marginal and small holding should secure on time quality seeds, other inputs and technical knowledge of using them in right time.

Majority youths in rural areas are losing their interest in agriculture and are looking for jobs or even betel shop in urban areas. The young generation should involve in agriculture and allied sectors actively so that they can carry out family farming from generation to generation. Therefore, Policy and Programmes need to be designed targeting **rural youths s**pecifically to make them fully aware of their role in family farming.

- Involvement of farm women in different entrepreneurial activities would be beneficial to the Women Self Help Groups (WSHGs) in their capacity building for income generation.
- Successful interventions of production oriented farm technologies for farm women through Krishi Vigyan Kendras would be helpful to enhance agricultural productivity and to ensure food and nutrition security for the growing population.
- Providing family farm inputs with adequate financial and scientific support to ensure food security should be the bottom line of all food and agriculture policies of developing countries.
- Needs of rural women should be studied regularly and effective measures should be taken to solve their strategic needs.

Conclusion

Women represent half of the human resources of our country. The socio–economic empowerment of women is essential for sustainable development of the nation. Thus, the active involvement will help the farm women to become equal partners in the process of rural development. Both men and women have equal share in shaping their household economy and food security from existing farming systems. Sustainability of family farming depends upon the adoption of Integrated Farming System which should be gender equality and gender sensitivity. The existing cropping systems are unsustainable due to decreased cultivable land, degradation of natural resources, yield stagnation, increase in cost of cultivation, environmental pollution and erratic rain fall due to climate change. Today our farming community faces a great challenge of food security such that it is required to make more involvement in farming. While integration of components like poultry, duck and sheep/goat/pig with fish has been proved to be economically viable, several other components like dairy and horticultural crops can also be effectively integrated .Further, two or more compatible components can also be integrated with fish for making the system economically sustainable. Farming system holds a great promise and potential for enhancing production, betterment of rural economy, employment generation and also improving the socio-economic status of weaker rural community. Recycling of household resources and modification of existing farming need immediate attention for food and nutrition security as well as livelihood support to the family.

References

APCAS. 2010. Asia and Pacific Commission on Agricultural Statistics, 23rd Session (APCAS 10-28), 26-30 April 2010, Siem Reap, Cambodia.

Chand, R., Prasanna, P. A. L. and A. Singh. 2011. Farm size and productivity: Understanding the strengths of smallholders and improving their livelihoods, Economic and Political Weekly, 46 (26 & 27).

Das, P.K. 1990. Rural women: their role and status in coir and cashew processing sectors of Kerala. NIRD. Workshop on Women in agriculture background papers, Oct. 29th -1st Nov.

FAO. 1998. Women Feed the World, Rome.pp.2-19

Kumar, P. 2013.Empowering women for food security, *Kurukshetra, A Jr. of Rural Development*. 61(10):28-31

Nanda, S.S. and Jena, D 2011. Integrated farming system: A profitable livelyhood support to farmers in Puri district. Technical bulletin 1. KVK, Puri

Odisha Economic Survey.(2013-14), Government of Odisha,

Pandey. H. 2001.Understanding farm Women, NRCWA, pp: 19-27

Srinath.K, 2009. Priorities for gender research in agriculture. Report of winter school on Participatory research for Mainstreaming Gender Concerns in Agriculture, NRCWA: 1-8

M.P.S.Arya, 2008. Production and Appropriate Technologies for farm women. Report of Winter School on Participatory research for Mainstreaming Gender Concerns in Agriculture: 26-34

http://www.mydigitalfc.com/op-ed/strengthening-family-farming-india-469

http://indiamicrofinance.com/2014-year-of-family-farming.html

http://www.fao.org/fsnforum/forum/discussions/family-farming

Integrated Farming System Practices: Challenges & Opportunities, pp 315-342
© 2016, Editor, Prof. Sankarsana Nanda,
New India Publishing Agency, New Delhi, India

13

Agribusiness through Mushroom Production in Integrated Farming Systems

K.B. Mohapatra

Mushrooms have been part of fungal diversity for around 300 million years. A mushroom is a macro fungus with a distinctive fruiting body, which can either be epigeous (above ground) or hypogeous (underground) and large enough to be seen by naked eye and to be picked up by hands. They lack the green matter (chlorophyll) present in plants and grow on dead and decaying organic materials, as phytoparasite or in mycorrhizal association with plants. From these decaying substrate, they absorb their nutrition with the help of fine thread like structures (mycelium), which penetrate into the substratum and are generally not visible on the surface. After the mycelium has grown profusely and absorbed sufficient food materials, it forms the reproductive structure which generally comes out of the substrate and forms fruiting body, commonly known as mushroom. Mushrooms can be divided into four categories (i) Those that are fleshy and edible fall into the edible mushroom category i.e., *Agaricus bisporus*; (ii) Mushroom that are considered to have medicinal applications are referred to as medicinal mushrooms i.e., *Ganoderma lucidum*; (iii) Those that are proved to be or poisonous or suspected of being poisonous are named poisonous

Centre of Tropical Mushroom Research and Training, Department of Plant Pathology, College of Agriculture, Orissa University of Agriculture & Technology, Bhubaneswar-751003, Odisha, India mushrooms i.e., *Amanita phalloides*; (iv) those in a miscellaneous category, which include a large number of mushrooms whose properties remains less defined. The toxic/poisonous species commonly referred to as 'toadstools' are objects of fear and distrust and have led to certain amount of inhibition and taboo towards mushroom consumption.

Use of mushrooms as food by man has been practiced since time immemorial and probably predates any historical records. The early civilizations of Greeks, Egyptians, Romans, Chinese and Mexican appreciated mushroom as delicacy, knew something about their therapeutic value, often used them in religious ceremonies. Greeks regarded them as strength food for warriors while Romans considered them as 'Food of Gods' and Chinese regarded mushrooms as 'Elixir of life'. Much referred 'Somrus' in ancient Indian literature is actually a decoction of mushroom. Archeological and other accounts reveal that ancient man used mushroom in religious ceremonies. The Aztecs of South America referred mushroom as 'teononacte' (God's flesh) and worshipped a group of mushrooms as being divine. The high esteem in which mushrooms were held in different civilizations is also indicated by the restriction of the use of many species to the rulers.

Currently 14000 mushroom species are known to exist. Of these about 50 % or 7000 species are considered to posses varying degrees of edibility and almost 3000 species from 31 genera are regarded as prime edible mushrooms. To date only 200 of them are experimentally grown, 100 of them economically cultivated, approximately 60 commercially cultivated and about 10 have reached to industrial scale production in many countries. Furthermore, about 2000 are medicinal mushrooms with varieties of health attributes. The number of poisonous mushroom is usually reported to be relatively small (approximately 1 %), but there is an estimate that about 10 % may have poisonous attributes while 30 species are considered to be lethal.

Importance

Agriculture continues to be the main strength of Indian economy. With the variety of agricultural crops grown today the country has achieved food security by producing about 260 million tonnes of food grains. However, the struggle to achieve nutritional security is still on. In future, the ever increasing population, depleting agricultural land, climate changes, water shortage and need for quality food products at competitive rates are going to be the vital issues. Hence, it is imperative to diversify our agricultural activities in areas like horticulture to meet these challenges and to provide food and nutritional security to our people. Mushrooms are one such component that not only use vertical space but also help in addressing the issues of quality food, health and environmental sustainability. There is need to promote both mushroom production as well as consumption for meeting the changing needs of food items. Fortunately, mushroom trade has gained importance in the recent years possibly for the global shift towards vegetarian food and recognition of mushroom as a functional food. Mushroom cultivation offers scope to recycle agro-wastes as carbon pool into good quality protein, much of which is otherwise wasted in the field. This unique horticultural venture has tremendous scope to meet the food shortage without undue pressure on land.

Integrated farming system and mushroom cultivation

Integrated farming is a farming system where high quality food, feed, fibre and renewable energy are produced by using resources such as soil, water, air and nature as well as regulating factors to farm sustainably and with as little polluting inputs as possible. Being bound to sustainable development, the underlying three dimensions; 'economic development', 'social development' and 'environmental protection' are thoroughly considered in the practical implementation of integrated farming. However, the need for profitability is a decisive pre-requisite. To be sustainable, the system must be profitable, as profits generate the possibility to support all activities outlined in the integrated farming network. Mushrooms are one such component that not only impart diversification but also help in addressing the problems of quality food, health and environment related issues. One of the major areas that contribute towards goal of conservation of natural resources as well as increased productivity is recycling of agrowastes. Such kind of activity can enhance income and impart higher level of sustainability. Unlike most of the development efforts, in mushroom cultivation, the emphasis has been on economic development with improvement in quality of life of the rural poor through an integrated effort. Labour intensive nature of cultivation, significant value addition, dependence on locally available, abundant and cheap raw material, low initial investment, simple technology and limited space requirement make it a worthwhile proposition. Mushroom, being an indoor crop, happens to be women friendly in nature. This brings about removal of poverty and social development in harmony and therefore, fits well in any farming system. Integrating mushroom cultivation in existing farming systems will supplement the income of rural poor, provide gainful employment and will lead to inclusive growth.

Mushroom: a functional food

It is well recognized that mushroom contains all essential components of a balance food (Table 1 and 2). In fact, being rich in highly digestible lysine rich protein, vitamins and minerals, mushroom lacks fats and are low in carbohydrate (low calorie food). They are rich in folic acid, phosphorus, potassium, calcium, copper, iron and vitamin-B complex. In place of starch, mushroom contains sorbitol and linolenic acid. They are excellent sources of thiamine, riboflavin, niacin, pantothenic acid, biotin, folic acid and vitamin B-12 (Table 3). Further, mushrooms are also reported to have high medicinal attributes. A large number of mushrooms contain biological active polysaccharide protein complex having anti-tumour, immuno modulating and antioxidant properties. *Ganoderma lucidum* showed hypo-glycemic, hypo-lipidemic, endurance enhancing, immune-modulatory, anti-hypertensive, antiplatelet aggregation, hepato-protective, anti-HIV and anti-atherosclerotic effects. Many medicinal mushrooms also showed hematological, antiviral, anti-tumour, antioxidant, cardiovascular and renal effects. The interesting fact is that the K: Na ratio is very high and it is suitable for people who suffer from hypertension. Besides its nutritional and medicinal values, mushrooms are relished as a source of food due to the pleasant aroma, taste and its fleshy nature.

Table 1: Proximate composition (per unit fresh weight) of the cultivated mushrooms (Rai and Sohi, 1988).

Mushroom/ vegetable	Moisture	Protein	Fat	Carbohydrate	Fibre	Ash	Calorie
Button mushroom (*Agaricus bisporus*)	90.1	2.9	0.3	5.0	0.9	0.8	36
Oyster mushroom (*Pleurotus sajor-caju*)	90.2	2.5	0.2	5.2	1.3	0.6	35
Paddy straw mushroom (*Volvariella volvacea*)	90.1	2.1	1.0	4.7	1.1	1.0	36
Cabbage	91.9	1.8	0.1	4.6	1.0	0.6	27
Cauliflower	90.8	2.6	0.4	4.0	1.2	1.0	30
Potato	74.7	1.6	0.1	22.6	0.4	0.6	97

Table 2: Protein content in food stuffs (Kaul, 1983)

Food stuff	Protein in edible portion (per unit fresh weight)
Button mushroom (*Agaricus bisporus*)	3.7
Paddy straw mushroom (*Volvariella volvacea*)	3.9
Peas	2.6
Cabbage	1.5
Cauliflower	2.7
Potato	2.1
Apple	0.3
Banana	1.1
Chicken	20.01
Fish	10.0
Egg	12.8
Milk (cow)	3.2

Table 3: Essential amino acids (per cent crude protein) in edible mushrooms (Bano and Rajarathnam, 1982; Li and Chang, 1982)

Amino acid	Button mushroom (*Agaricus bisporus*)	Oyster mushroom (*Pleurotus sajor-caju*)	Paddy straw mushroom (*Volvariella volvacea*)
Leucine	7.5	7.0	4.5
Isoleucine	4.5	4.4	3.4
Valine	2.5	5.3	5.4
Tryptophan	2.0	1.2	1.5
Lysine	9.1	5.7	7.1
Threonine	5.5	5.0	3.5
Phenyl alanine	4.2	5.0	2.6
Methionine	0.9	1.8	1.1
Histidine	2.7	2.2	3.8

Background information

It is estimated that around 200 billion tonnes of organic matter are generated annually through the process of photosynthesis (Zhang, 2008). A chunk of this organic matter and many agro-industrial wastes furnish large volumes of solid wastes, residues and byproducts which pose serious environmental pollution (Lal, 2005). In fact, the physico-chemical properties of such lingo-cellulosic materials offer great scope for their enormous biotechnological importance. Global efforts are being made involving innovative technologies and disposal methods for the utilization of these lingo-cellulosic materials into more profitable materials by solid state fermentation (Chang, 2006). Among various methods, mushroom cultivation is the most economical and relatively short biological process for the biotransformation of such materials into protein rich food (Chiu and Moore, 2001). The production of tropical mushrooms like oyster mushroom (*Pleurotus* spp.), paddy straw mushroom (*Volvariella volvacea*) and milky mushroom (*Calocybe indica*) utilizing locally available substrates such as paddy straw, wheat, soybean, cotton wastes, coffee waste, water hyacinth, saw dust, sugar cane bagasse, wild grasses and various categories of refuse and ligno-cellulosic wastes have great potential to exploit and transform it into a highly nutritious food within short period of time (Singh et al., 2011). Cotton wastes and sunflower stalks are examples of agro-industrial wastes which are of limited/ no economic use. In recent years, cotton wastes have become popular as substrates for straw mushroom (Chang, 1982) and oyster mushroom (Hamlyn, 1989) cultivation. Oyster mushroom (*Pleurotus ostreatus*) is the third mushroom of the world produced by China (Li, 2012). Its production in India is also becoming popular in almost all the states having temperate, sub-tropical and tropical climate. Paddy straw mushroom (*Volvariella volvacea*) is another mushroom dominating whole of Odisha and available in every nook and corners of the state. It is the sixth mushroom of the world in production (Ahlawat and Tewari, 2007). Paddy straw mushroom is a fast growing fungus with a crop duration of 14-15 days. Outdoor cultivation of paddy straw mushroom under trees like coconut, casurina, cashewnut etc. is very much successful during summer (Thakur and Mohapatra, 2013). Odisha is the only state where paddy straw mushroom is grown commercially for eight months a year involving common people. The paddy straw mushroom is successfully cultivated in the plains of Kerala throughout the year where the temperature ranges between 28-32°C (Prakasam, 2012). In India, four mushroom varieties such as *Agaricus bisporus*, *Pleurotus* spp., *Volvariella* spp. and *Calocybe indica* have been recommended for the year round cultivation. The Indian subcontinent is known worldwide for its varied agro-climatic zones with a variety of habitats that favour rich mushroom biodiversity (Verma, 2013). Among tropical mushrooms, oyster, paddy straw and milky mushrooms are very much in cultivation in India. Besides, cold loving white button mushroom is also in cultivation in suitable agro-climatic situations of the country.

Global scenario of mushroom production

China is the largest producer, consumer and exporter of mushrooms in the world followed by USA and Netherlands (Table 4). China grows over 60 mushroom species in a small to commercial scale. Ten most widely cultivated mushrooms in China include *Pleurotus ostreatus, Lentinus edodes, Auricularia auricula, Agaricus bisporus, Flammulina velutipes, A. polytricha, P. cornucopiae, Coprinus comatus, Agrocybe chaxinggu* and *Volvariella volvacea* (Li, 2012). The world production of mushroom in China in 2010 was 21,524,473 tonnes, of which the major share goes to oyster mushroom. Oyster mushroom (*Pleurotus ostreatus*) is the third largest mushroom of the world produced by China with a production of 4,929,000 tonnes (Chinese Edible Fungi Association, 2011).

Table 4: World production of mushrooms (metric tonnes)

Country	1997	2007
China	562194	1568523
USA	366810	359630
Netherlands	240000	240000
Poland	100000	160000
Spain	81304	140000
France	173000	125000
Italy	57646	85900
Ireland	57800	75000

Contd.

Country	1997	2007
Canada	68020	73257
UK	107359	72000
Japan	74782	67000
Germany	60000	55000
Indonesia	19000	48247
India	9000	48000

Source: World mushroom and truffles : Production, 1961-2007; United Nations, FAO, FAO Stat (8/28/2009)

The three major mushroom producing countries as per FAO data namely China, USA and Netherland account for more than 60 % of the world production; however, share of China itself is 46 % which is about half of the world mushroom production. According to current Indian estimates, mushroom production of India is about 1,20,000 metric tonnes, which is 3 % of the world mushroom production.

National scenario

Mushroom production in the country started in the 70s but growth rate, both in terms of productivity as well as production has been phenomenal. In seventies and eighties button mushroom was grown as a seasonal crop in hills, but with the development of technologies for environmental controls and increased understanding of the cropping systems, mushroom production shot up from mere 5000 tonnes in 1990 to over 1,00,000 tonnes in 2010. Today, commercially grown species are button and oyster mushrooms, followed by other tropical mushrooms like paddy straw mushroom, milky mushroom etc. Two to three crops of button mushroom are grown seasonally in temperate regions with minor adjustments of temperature in the growing rooms; while one crop of button mushroom is raised in North Western plains of India seasonally. Oyster, paddy straw and milky mushrooms are grown seasonally in the tropical/ subtropical areas from March to October. The areas where these mushrooms are popularly grown are Odisha, Maharashtra, Tamil Nadu, Kerala, Andhra Pradesh, Karnataka and North Eastern region of India. Some commercial units are already in operation located in different regions of our country and producing the quality mushrooms for export. The present production of white button mushroom is about 80 % of the total production of mushrooms in the country. The state wise production of mushroom in India is presented in Table 5.

Table 5: State wise mushroom production in India (tonnes)

Sl.No.	State	Button	Oyster	Milky	Others	Total production
1.	Punjab	58000	2000	0	0	60000
2.	Uttarakhand	8000	0	0	0	8000
3.	Haryana	7175	0	3	0	7178
4.	Uttar Pradesh	7000	0	0	0	7000
5.	Tamil Nadu	4000	2000	500	0	6500
6.	Himachal Pradesh	5864	110	17	2	5993
7.	Odisha	36	810	0	5000	5846
8.	Delhi	3000	50	20	0	3070
9.	Andhra Pradesh	2992	15	15	0	3022
10.	Maharashtra	2725	200	50	0	2975
11.	Other states	1891	1214	315	311	3731
	Total	100683	6399	920	5313	113315

Source: RMCU, DMR, Solan, 2010.

Status of mushroom production in Odisha

The research on edible mushroom in Odisha made its humble beginning in the Department of Plant Pathology, College of Agriculture, Orissa University of Agriculture and Technology, Bhubaneswar in 1972 with a view to generate profitable and sustainable production technology. Having achieved success in developing mushroom cultivation and spawn production technology, research efforts were further strengthened and transfer of technology was initiated with the establishment of Centre of Tropical Mushroom Research and Training in the University with the financial support of Government of Odisha in the year, 1991-92. This research organization paved the way for initiation of commercial mushroom cultivation in the state within two years of its establishment.

In depth study of production of spawn and mushroom cultivation particularly paddy straw mushroom, oyster mushroom, milky mushroom and button mushroom were undertaken. Farmer training programmes and demonstrations on spawn production and mushroom cultivation were then extended to all over the state. Technical assistance was provided for development of individual/ group/ private sectors in establishing spawn production units and mushroom cultivation centres. The overall activities of the centre further gained momentum with the establishment of All India Coordinated Research Project on Mushroom in the University during 2009-10 with the financial assistance from Indian Council of Agricultural Research, New Delhi. At present the total mushroom production of the state has reached an all time high of 12,334 tonnes/annum contributing to over 10 per cent of the country's production (Table 6).

Table 6: Mushroom production in Odisha (tonnes)

Sl.No.	Mushroom	2009	2014	Annual increase (%)
1.	Paddy straw	5000	8129	12.51
2.	Oyster	810	4095	81.11
3.	Milky	0	0	0
4.	Button	36	110	41.11
	Total	5,846	12,334	22.19

Fig. 1: Paddy straw mushroom.

Paddy straw mushroom has been a popular vegetarian diet of the people of Odisha since long. People in rural areas are in the habit of collecting this mushroom grown naturally during the rainy season on straw piles. However, it could not find a regular place in the diet due to its non-availability in other seasons. Now the scenario has changed altogether and straw mushroom is being cultivated 8-10 months a year (Fig. 1). Besides, methods for off-season cultivation of paddy straw mushroom in the coastal ecosystem of the state are in the offing. The conditions in most of the agroclimatic situations of Odisha are favourable for taking up cultivation of a number of edible mushrooms. Diverse agro-wastes suitable for mushroom production are also available in abundance. Paddy straw, the main input for mushroom cultivation is plentily available at cheaper rates. Besides, cultivation is comparatively easier, as it involves minimum investment, labour and space. Straw mushroom is cultivated largely as an intercrop in the coconut plantations in the coastal agro-ecological situation. However, in the inland districts it is cultivated under thatched roof. Large number of farmers rely on the cultivation of *V. volvacea* as a secondary source of income, making use of their waste paddy straw. The enterprise has assumed the proportion of a cottage industry among the rice farmers in the hot, steamy climate in the East and South-Eastern coastal plain zone of Odisha (Mohapatra *et al.*, 2013).

Odisha produces about 3.05 million tonnes of paddy straw per annum and a major part of it is left out to decompose naturally or burnt *in situ*. Therefore, farmers utilize paddy straw for straw mushroom cultivation owing to its availability at cheaper rates throughout the state. In most cases, cultivation is done in non-pasteurized substrate with incorporation of pulse powder or wheat bran. The acidic medium in hot and humid climate is more prone to competitor moulds and diseases. Therefore, the biological efficiency hovers at 10 per cent which is low. However, there is scope for improvement in biological efficiency by using more productive strains of the species, producing good quality spawn and effective substrate management. Further, higher and more stable yields (30-45 %) could be obtained through adoption of indoor method of farming (Ahlawat and Tewari, 2007).

Fig. 2: Straw mushroom in coconut plantation

Besides, pasteurized paddy straw substrate supplemented with cotton seed hulls could give better productivity (Fig. 3).

Outdoor cultivation of paddy straw mushroom

The outdoor cultivation method was introduced in the middle of the 1980's. This method is ideal in rural areas where paddy straw is abundant after each paddy harvest. Low cost house for growing paddy straw mushroom has been developed. Besides, outdoor cultivation under trees like coconut, cashewnut, casurina etc. is very much successful during summer (Fig. 3). However, outdoor cultivation leads to uncertainty in production with irregular and low yields due to difficulties in controlling temperature, relative humidity and insect pest problems. Further, yield deteriorates owing to site contamination out of moulds and mushroom diseases. The traditional outdoor method uses the bed type approach and utilizes a number of agricultural wastes like dried paddy straw, rice stubbles, water hyacinth, banana leaves and stalks (Reyes and Abella, 1997).

Fig. 3: Straw mushroom in bamboo plantation

Good quality paddy straw bundles are collected, made to 45 cm length by trimming both the ends and soaked in 2 % solution of calcium carbonate for six hours. Bundles are taken out, excess water drained off to 65 % substrate moisture and beds are raised having dimension of 1.5^1 x 1.5^1 x 1.5^1 (length x breadth x height). Requirement of spawn and organic additive has been standardized at 3 % each of the dry weight of the substrate. Three layered raised beds are prepared for spawning and organic supplementation at 1:1:2 proportion. Beds are topped with a thin layer of straw followed by covering with polythene sheet that maintains appropriate temperature for mycelial ramification (30-35°C) and fruit body formation (28-30°C). Polythene sheet is removed after the emergence of pin heads, watering done as and when required and the fruits of the first flush are harvested at 14-15 days followed by providing the polythene cover once again. The second harvest from the second flush is harvested at 21-22 days. One bed thus prepared, may yield approximately one kilogram of fruits with a biological efficiency of 15 % with appropriate aftercare.

Milky mushroom

About two decades ago, milky mushroom (*Calocybe indica*) was identified as a wild edible mushroom in India. However, complete commercial production techniques were developed for the first time in Tamil Nadu (Krishnamoorthy, 1995). Milky mushroom is a potentially new species to the world mushroom portfolio. It is a robust, fleshy, milky white, umbrella like mushroom which resembles button mushroom (Pani, 2012). The species is suitable for hot humid climate and can be cultivated indoor in high temperature and high humidity areas. It grows well in the temperature range of 25-35°C and relative humidity at more than 80 %. The cultivation technology is very simple, involves less cost and no special compost is needed for the cultivation (Fig. 4).

Fig. 4: Milk mushroom

The cultivation process resembles that of oyster mushroom with the additional process of casing. The mushroom can be harvested at 40 days after spawning with the total crop cycle of 60 days. The milk mushroom has an extended shelf life of 3-5 days compared to other cultivated species, making it more amenable to handling, transportation and storage. Its robust size, sustainable yield, attractive colour, delicacy, long shelf life and lucrative market value attracted the attention of both mushroom consumers and prospective growers. It is a low fat mushroom with good amount of essential amino acids (Ruhul *et al.,* 2010).

Oyster mushroom

Fig. 5: Oyster mushroom

Oyster mushroom (*Pleurotus* spp.) has species suitable for both temperate and sub-tropical regions. It is the third largest cultivated mushroom of the world. In Odisha, cultivation is limited to winter months (November-February) and the production stands at 4095 tonnes/annum contributing to 33 per cent of total mushroom production of the state. The mushroom can be cultivated in a wide range of temperature (22-28°C) employing most cellulosic farm wastes. Oyster mushroom can be grown using a variety of containers such as polythene bags, nylon nets, baskets, shelves, trays etc. The spawn run is rapid and the first crop of mushroom can be harvested in about three weeks time. Gray oyster mushroom (*Pleurotus sajor-caju*), white oyster mushroom (*P. florida*), pink oyster mushroom (*P. eous*) and blue oyster mushroom (*Hypsizygous ulmarius*) are the ruling species of the state. Pink oyster is gaining popularity owing to its attractive colour along with good taste and flavour for small scale semi-urban and urban units. The biological efficiency of oyster mushroom is very high (100 per cent) and the shelf life is better (24 hours) than straw mushroom. Production cost is low with little longer cropping cycle (45 days). Further, it is suitable for post-harvest processing. However, the consumer demand is limited in the state (Fig. 5).

Button mushroom

Fig. 6: Button mushroom

The button mushroom (*Agaricus bisporus*) is most popular variety of mushroom in the country. At global level, it ranks first in terms of production. Punjab is the leading state contributing to 60 per cent of the total production of the country. Being a temperate mushroom, production can be taken up year round in controlled environment or seasonally during winter months. Odisha has just entered into commercial production with 110 tonnes/annum and it is likely to grow further in near future (Fig. 6).

The commercial cultivation of button mushroom is done on wheat or rice straw which may be supplemented with wheat or rice bran, calcium carbonate (chalk powder), calcium sulphate (gypsum) and chemical fertilizers. All these things are mixed thoroughly and then aerobically fermented. This process is called as composting.

Compost is suitable for the growth of mushroom mycelium and restrict the growth of other microorganisms. After composting, nutrients of the raw material used for making compost becomes readily available for the growth and development of the mushroom mycelium. On the other hand, it does not favour the growth of other microorganisms. Essential requirements for the cultivation of button mushroom are mushroom house, composting yard, ingredients and raw material for compost, mushroom spawn, casing, equipment, accessories and pesticides. Compost can be prepared employing long method or short method. Fully prepared compost is dark brown in colour, has no trace of ammonia, no unpleasant odour but smells like fresh hay. The desired quantity of spawn (1-1.5 % of the dry straw) is mixed thoroughly with the required quantity of compost which is then filled in racks, trays or bags. The trays or bags are stacked in the growing room at 22-26°C temperature and 80-85 % RH for mycelial growth. Under favourable environmental conditions spawn run is completed in 15-20 days. Spawn run compost is covered with 4-5 cm thick layer of casing soil for induction of fruit bodies. Within 4-5 days mycelium spreads in the casing soil. Thereafter, temperature of the mushroom house is lowered down to 18°C and maintained between 14-18°C during rest of the fruiting period. In ideal conditions, fruits are observed after 15-20 days of casing. Harvesting is done when the cap size is 3-4.5 cm in diameter. The yield in bag cultivation method is estimated to be 15-20 kg/100 kg compost in about 8 weeks time.

Economics of mushroom cultivation

Mushroom cultivation is a profitable enterprise. The cost involved in raising one bed of straw mushroom of 1.5' x 1.5' x 1.5' size comes to Rs.50 /- with a production of one kilogram mushroom within a crop cycle of 21 days. The net return is Rs.50/- per bed assuming the market rate at Rs.100/- per kilogram. Likewise, the cost for raising one bag of oyster mushroom is Rs.30/- per bag with a production of 1.5 kilogram mushroom within a crop cycle of 45 days. The net return is Rs.30/- per bag assuming the market rate at Rs.40/- per kilogram. A model small mushroom production unit (300 sq.ft.) with the investment of Rs.25,000/- accommodating 120 beds of paddy straw mushroom per month during summer and rainy season and 225 bags of oyster mushroom per two months during winter season gives an estimated net income of Rs.6,000/- per month.

Protected (indoor) cultivation of mushroom

It is imperative to say that mushrooms can not be grown year after year with full commercial access, unless proper growing conditions are provided and adequate facilities are available for the control of competitor moulds, diseases and insect pests. Possibly, such conditions can be fulfilled in shelf growing, by the construction of properly insulated and ventilated mushroom houses accommodating store room, spawn running room, cropping room as well as packing and preservation room.

In Odisha, raising of simplified and low cost thatched mushroom houses are being encouraged for seasonal cultivation of mushrooms round the year with greater precision. The houses are appropriately designed to maintain required temperature and humidity inside, besides having access to ventilation. House of variable sizes may be raised as per the requirement of a grower. However, bigger houses are often difficult to maintain in terms of mushroom crop requirement and therefore, are not desirable. The vertical space in the mushroom house can be utilized effectively by raising three-tiered structures (shelves), mandatory for indoor cultivation. The thatched roof can be supported by a wooden or bamboo frame with insect proof net protecting the sides. During the hot summer months, humidity of mushroom house can be raised by spreading sand on the floor up to a thickness of 4-6 inches, wetting it sufficiently. Experiments have shown that these low cost houses perform better than outdoor cultivation in terms of productivity. Further, infestation of beds out of competitor moulds and incidence of diseases and insect pests are found to be low in low cost houses (Fig. 7a). Various modifications of the thatched houses are being designed now-a-days in order to make it more permanent and mushroom friendly. Shade net houses and houses having asbestos roof are therefore viable alternatives to the thatched sheds. In order to maintain appropriate temperature and humidity, foggers or micro-sprinklers can be fitted inside so as to make the summer cultivation of paddy straw mushroom more easier and simpler. The houses can be disinfected at regular intervals by applying germicidal sprays for raising healthy crops. By and large, higher productivity is obtained from these growing rooms in comparison to the outdoor units (Fig. 7b).

In view of the higher sale price of the produce, off season cultivation or winter cultivation of paddy straw mushroom is gaining popularity now-a-days in the state. Hence, poly house cultivation of paddy straw mushroom is being popularized during winter season wherever growers are interested. It is possible to maintain required temperature (34-35°C) and humidity (80 %) in the polyhouse through provision of foggers or microsprinklers and therefore, satisfactory yields can be obtained from the tropical mushrooms even during winter months. Low cost polyhouses with 200 micron ultraviolet proof polythene sheets erected even on bamboo frame can be raised by the growers for the purpose (Fig. 7c). Length wise, on the rear wall, a vent with an exhaust fan has to be provided for evacuation of hot air. Poly mushroom growing houses can be of variable dimensions. However, the appropriate dimension for maintaining suitable microclimate and obtaining good yields could be 20[1] x 10[1] x 9[1] (length x breadth x height).

Fig. 7a: Straw mushroom in thached house

Fig. 7b: Straw mushroom in as bestos house

Fig. 7c: Straw mushroom in poly house

Indoor cultivation using partially composted substrate

Indoor cultivation of paddy straw mushroom with partially composted substrate can be a boon for mushroom growers in view of its higher productivity (Quimio, 1993). This gives 30-45 % yield in comparison to 10 % obtained from outdoor farming. Semi-industrialization of paddy straw mushroom cultivation can be made possible utilizing this technology. The substrate with paddy straw and cotton waste in equal proportions each partially composted for a period of six days with addition of calcium carbonate and rice bran in appropriate amounts. The compost is spread on shelves in the growing room and steam pasteurized at 60°C for six hours. After natural cooling, spawning is done and left covered with polythene sheet. After the spawn run is over, light and intermittent fresh air are introduced into the growing room for induction of fruit bodies. Suitable temperature and humidity are maintained inside the growing room. A quantity of 30-45 kg of fresh mushrooms can be harvested in two flushes within the crop cycle of 21 days per each 100 kg of compost. Hence, this system is far more productive than the outdoor/indoor cultivation. This system has successfully been initiated in Odisha and efforts are on to popularize it among the farmers.

Fig. 8: Straw mushroom in composted substrate

Mushroom spawn production

In nature, mushroom spores help in survival of a species from one generation to the next. These spores are extremely small and therefore, difficult to handle as seed. Further, spores need time and specific conditions to germinate and during this time the competitor fungi grow faster to exhaust the available substrate. Therefore, a pure culture of the desired mushroom mycelium is first raised on a convenient artificial culture medium and then added to a substrate to give it an advantage. Spawn is a vegetative mycelium from a selected mushroom grown on a convenient medium. It comprises mycelium of the mushroom and a supporting medium, which provides nutrition to the fungus during its growth. Spawn is used as seed in mushroom cultivation. Appropriate kind and quality of spawn is very important for mushroom cultivation (Fig 9). The isolation, purification and maintenance of mushroom cultures requires technical expertise and aseptic laboratory facilities. It is difficult to maintain pure cultures and/or spawn for small mushroom growers. They have to rely entirely on commercial spawn producers, reliable government and non-government organizations that play a vital role in supplying quality spawn of a desired mushroom strain or variety. At present Odisha is ahead of other states in establishment of commercial spawn production centres. It has been estimated that a quantity of over 2318 tonnes of spawn is produced per annum from 210 spawn production units present

Fig. 9: Mushroom Spawn

across the state in both private and public sectors. However, mushroom productivity often declines owing to use of poor quality spawn supplied by some spawn producers. In this context, it is imperative to produce quality spawn in order to exploit the yield potential of different mushroom species.

Maintenance of spawn quality

A major source of contamination for the growing mushroom mycelium is the grain seed or the substrate used to prepare the spawn. Modern equipment and facilities for sterilization and the maintenance of sterile conditions, however, are capable of reducing fungal and bacterial contamination to 0.1 % or less of the number of spawn units. Quality control in spawn making consists of constant inspection to eliminate those units visibly contaminated or those units exhibiting unacceptable differences in appearance and growth.

Mushroom spawn, whether prepared as a family home project or on an industrial scale using modern equipment, should be in excellent condition when delivered to the growers.

Spawn of most mushrooms can be refrigerated, but it should be warmed to normal room temperature before it is used as inoculum or as planting spawn. *Volvariella* spawn, in particular, should not be kept either at refrigerated or room temperatures for more than four weeks after mycelial growth is complete. The presence of pink chlamydospores in *Volvariella* spawn indicates a highly fertile and active growth, although their time of appearance and density depends upon the strain and substrate used.

Vigorous growth of the planting spawn is a pre-requisite to good growth and yield. In outdoor cultivation of *Volvariella,* substrates used for production are not pre-sterilized. The spawn will therefore, compete with organisms naturally present in the substrates used for production. If the spawn is not vigorous, the mushroom mycelium will be overgrown by competitor organisms. If it is vigorous it will overcome many of the competitive organisms and produce more mushrooms. The same is true for other substrates which are not pre-sterilized.

Contaminated, old or no-growth spawn should never be sold to growers. Spawn makers should maintain a testing facility where they can test each batch of spawn for production characteristics. Sales personnel should visit growers using their spawn so that they can observe problems at first hand. Problems related to spawn production must be corrected quickly.

Mushroom value addition

Mushrooms are living entities and are governed by a number of factors that lead to post-harvest spoilage and losses. These include infection with microorganisms, spoilage by enzymes and unfavourable conditions of humidity, temperature, atmosphere during picking/harvesting, transportation, packaging, storage and marketing. Mushrooms contain about 90 % moisture, are highly perishable and cannot be stored for more than 24 hours at the ambient temperature. Once the fruiting body matures and harvested, degradation process starts and it becomes unconsumable after sometime, if not properly handled. Development of brown colour is the first sign of deterioration and is a major factor contributing to quality losses. It is suggested that spoilage might be caused by the action of bacteria on the mushroom tissue and browning of mushroom is due to a combination of auto-enzymatic and microbial action on the tissue. Sound post-harvest practices have since been developed to extend the shelf life of fresh mushrooms.

Harvesting of mushrooms at optimum stage of maturity is of great significance. Harvesting of under matured or over matured fruit bodies results in poor texture, flavor and immediate degradation. White button mushroom are harvested when the cap size reaches 30-45 mm. In general these should be harvested at a stage when the cap diameter is twice the length of stem. Pre-harvest spray of 2 % ascorbic acid improves the colour by inhibiting the polyphenol oxidase activity. Oyster mushrooms are harvested when the fruit body has curled under edges because fully matured fruit body is fragile and difficult to handle. As the maturity advances, the fruit body looses water and thus shrinkage occurs. The loss in turgidity is followed by darkening of colour and liquescent margins. Paddy straw mushrooms should be harvested at button or egg stage. The button stage occurs when the mushroom has begun to differentiate into the cap and stem inside the universal veil which continues until the differentiation is complete and the veil is broken. Once the stem is visible, the elongation starts which is considered undesirable. The deterioration is less in the mushrooms harvested at egg stage because the polyphenol oxidase activity is slow. Due to the fast post-harvest maturity, many changes occur as the caps open and it gets the appearance of flattened umbrella. This is accompanied by loss in weight due to loss of moisture and release of spores. These spores cause pink colouration of fruit bodies and thus causing off-flavour and rendering the produce unmarketable.

Freshly harvested mushrooms are highly perishable and require immediate processing for stable preservation. Pickling and sun-drying are economically viable methods for preserving mushrooms during the glut period to meet the domestic requirement in India. However, freezing and freeze-drying gives a product of excellent quality and can be explored commercially for export market (Fig. 10a&b).

Fig. 10a: Mushroom canning

Fig. 10b: Pickles from straw mushroom

Recycling of spent mushroom substrate for beneficial purposes

Natural plant wastes are the basic substrates for production of all species of edible mushrooms, capable of utilizing lignin, cellulose and hemicellulose and producing fruit bodies with excellent nutritional and medicinal attributes (Rajarathnam *et al.,* 1997). Mushroom growing is an eco-friendly activity as it utilizes the wastes from agriculture, animal husbandry, poultry, brewery etc. But dumping of 'spent mushroom substrate' after crop harvest may cause serious environmental problems including ground water contamination (Beyer, 1996).

Compost is considered as 'Spent' when one full crop of mushroom has been taken and further extension of cropping become unremunerative. Mushroom industry needs to dispose off more than 5,00,000 tonnes of used mushroom composts each year called 'spent mushroom substrate' (SMS) (Fox and Chorover, 1999). The large dumped piles of spent mushroom substrate become anaerobic and give off offensive odours. The run off from such piles may contaminate nearby water sources and pollute them. Hence, research activities are being carried out on aspects like traits of SMS, effect on surrounding environment and its diversified uses.

Traits of spent mushroom substrate

Spent mushroom substrate (SMS) normally contains 1.9:0.4:2.4 % (N-P-K) before weathering and 1.9:0.6:1.0 (N-P-K) after weathering for 8-16 months. Nitrogen and phosphorus do not leach out during weathering but potassium being more leachable is lost in significant amount during weathering. SMS contains much less heavy metals than sewage sludge and therefore, non-hazardous. Weathering causes a slow decrease in the organic matter contents (volatile solids) and led to different characteristics of weathered SMS because of ongoing microbial activity in the field. The samples obtained from 6,12, 24, 36 and 48 months old decomposed SMS do not show significant variations in pH, bulk density, particle density and water holding capacity. However, other parameters like total dissolved solids, conductivity, N, P, K, calcium chloride and nitrate content decrease with increasing age of the SMS decomposed by natural weathering process (Ahlawat *et al*, 2004) (Table 7).

Table 7: Analysis of spent mushroom substrate (SMS) and comparative figures for horse and chicken manure

Contents	Unit	Average fresh SMS	Average weathered (8-16 months SMS)	Horse manure	Chicken manure
Sodium, Na	% dry weight	0.72	0.22	0.3	0.13
Potassium, K	% dry weight	2.35	1.03	1.2	2.25
Magnesium, Mg	% dry weight	0.71	0.91	0.25	1.2
Calcium, Ca	% dry weight	4.93	6.16	0.85	0.8
Alluminium, Al	% dry weight	0.40	0.80	0.06	0.2
Iron, Fe	% dry weight	0.44	0.92	0.06	0.25
Phosphorus, P	% dry weight	0.36	0.55	0.9	2.5
Ammonia N, NH_4	% dry weight	0.11	0.03	NT	NT
Organic nitrogen	% dry weight	1.83	1.89	3.5	6.11
Total nitrogen	% dry weight	1.93	1.92	3.5	6.11
Solids	% dry weight	43.39	49.43	24	20
Volatile solids	% dry weight	62.78	44.29	NT	NT
pH	Standard units	7.28	8.05	7.2	6.5
Manganese, Mn	ppm dry weight	332.92	438.62	115	300
Copper, Cu	ppm dry weight	46.26	61.68	25	20
Zinc, Zn	ppm dry weight	103.88	136.41	130	175
Lead, Pb	ppm dry weight	14.89	18.17	NT	NT
Chromium, Cr	ppm dry weight	8.53	11.31	NT	NT
Mercury, Hg	ppm dry weight	0.07	0.19	NT	NT
Nickel, Ni	ppm dry weight	11.93	15.74	24	20
Cadmium, Cd	ppm dry weight	0.43	0.32	0.1	0.4
NPK ratio	ppm dry weight	1.9-0.9-2.4	1.9-0.6-10	1.2-0.3-0.4	3.0-1.3-1.2

% x 10,000 = ppm, NT = Not tested

(*Source:* Wuest and Fahy, 1991)

Recomposting of SMS

Spent mushroom substrate being rich in organic matter adds nutrients to the soils, helps in neutralizing acidic soils, facilitates plant growth in barren areas and also it improves water quality along with bio-remediation of contaminated soils. In order to improve the physico-chemical characteristics of SMS for its use as manure, recomposting methods are employed.

Recomposting methods

Natural

In natural weathering process, the Kachha pit of 5^1 x 5^1 x 5^1 is filled up to the top with SMS and left for natural recomposting process for next two years.

Aerobic

A perforated base is first prepared in the bottom of a Kachha pit of 5^1 x 5^1 x 5^1 with the help of waste wooden material. The perforated bottom of the pit is connected perpendicularly with 2^{11} diameter hollow plastic pipes placed at a distance of one square feet. Holes of 0.5^{11} diameter are provided in the plastic pipes at a distance of 15 cm each in a circular fashion. The pit is filled with SMS and left for recomposting for a period of two years.

Anaerobic

In anaerobic process, the Kachha pit of 5^1 x 5^1 x 5^1 is filled with SMS and covered with one foot thick layer of normal soil to create anaerobic conditions and left as such for two years.

Physico-chemical properties of recomposted SMS

The paddy straw mushroom spent substrate has pH in the range of 8.82 to 9.16; while oyster mushroom spent substrate has pH between 6.51 to 7.69. The oyster mushroom SMS contains higher nitrogen (1.82 %) as compared to paddy straw mushroom (1.06 to 1.46 %). The other nutrients like organic carbon and phosphate are higher in paddy straw mushroom spent substrate than in oyster mushroom spent substrate. However, contents of potassium, magnesium and calcium are at par in two types of SMS. The ingredients used and method of substrate preparation employed always have bearing on the properties of two types of SMS. With the increasing recomposting period, pH decreases in SMS of all mushrooms, while conductivity increases in oyster and paddy straw mushroom spent substrates. Nitrogen content, porosity, water holding capacity, total dissolved oxygen and microbial count increase during recomposting by all the methods in all the mushrooms. The contents of phosphorus, potassium, sodium, calcium, carbon, nitrate, total dissolved solids, particle density and bulk density decrease with time in SMS of all mushrooms. Recomposting of SMS plays a vital role in reducing the residual level of various fungicides, insecticides and heavy metals.

Recycling of SMS

The addition of SMS in nutrient poor soil improves its health by improving the texture, water holding capacity and nutrient status. Incorporation of SMS leads to an increase in both pH and the organic carbon content. Experiments have shown that the dry matter content of plants increases with the incorporation of increasing amount of weathered or unweathered SMS in soil (Chong et al., 1987). The mixing of spent mushroom substrate in soil has shown plant growth promoting activity in several plant species (Table 8).

Use of spent mushroom compost in horticulture

Spent mushroom substrate makes the substrate suitable for raising vegetables. The spent mushroom substrate being rich in N, P and K acts as a good growing medium for vegetables like cucumber, tomato, broccoli, cauliflower, peppers, spinach etc., but the response of the plants varies at different levels of SMS incorporation.

Weathering in open for two to three years make SMS more suitable for either ornamental shrubs or other horticultural plants of economic importance (Beyer, 1996).

Cereal crops

Studies have been conducted on several crops out of which maize has shown promising results. SMS incorporation at 100, 200 and 400 t ac^{-1} in nutrient poor soil shows positive effect on the silage and grain yield of corn crops (Wuest and Fahy, 1981).

Mushrooms

Use of SMS in mushroom cultivation is as casing material for button mushroom. The use of aerobically fermented spent mushroom substrate as casing material gives mushroom yield at par with peat based casing material with an additional advantage of less bacterial blotch (*Pseudomonas tolaasii*) infection on fruit bodies (Szmidt, 1994).

Bio-remediation of contaminated soil

The release of huge industrial wastes and poor availability of pre-treatment facilities contribute towards the increased level of contaminants in the soil. In addition to being a rich nutrient source for various crops, spent mushroom substrate possesses unique physico-chemical and biological properties, which make it an ideal bio-remediative agent for various environmental protection activities. SMS adsorbs the organic and inorganic pollutants and harbours diverse

Table 8: Response of different plant species towards different doses and age of SMS

Crop	Age of SMS	Quantity (q ha^{-1})	Impact			Source of information
			Yield	Quality	Diseases/ insect pests	
Tomato	Button mushroom SMS	NR	NR	Improves firmness and ascorbic acid	Restricted root knot infestation	Dundar et al., 1995, Verma, 1986
Capsi-cum	12 months aerobically recomposted	250	Enhanced plant growth and fruit yield	Superior quality	2-20 % reduction in insect pests and diseases	NRCM Solan
Cauli-flower	12 months anaerobically recomposted SMS + chemical fertilizers	250	Enhanced growth and curd yield	Superior quality	40-60 % reduction in insect pests and diseases	NRCM, Solan
Pea	12 months old anaerobically recomposted SMS	200	Enhanced growth and pod yield	Superior quality	Reduction in Fusarium wilt and powdery mildew	NRCM, Solan
Brinjal	12-24 months anaerobically recomposted substrate	250	Enhanced growth and bulb yield	Superior quality	NR	NRCM, Solan
Maize	Fresh	1000-4000	Enhanced silage and grain yield	Sueprior protein content	NR	Wuest and Fahy, 1991.

category of microbes, which have the capability of biological breakdown of the organic xenobiotic compounds.

Bio-remediation of pesticides

The mixing of spent mushroom substrate at 10, 20 and 30 % w/w in soil results in faster degradation of decis, malathion, bavistin and mancozeb in comparison to soil without any amendment of SMS. Mixing of soil at 30 % gives best results and the effect is at par with 20 % SMS amendment. The concentration of insecticides reduces to half after one month of SMS treatment and to negligible level after six months of SMS treatment. In case of fungicides also maximum degradation occurs at 30 %.

Bio-remediation of heavy metals

The mixing of spent mushroom substrate also results in faster degradation of lead and cadmium in soil in comparison to soil without any amendment of SMS.

Vermi composting

Decomposition of SMS takes a minimum of six months and quality manure from SMS can be obtained from long term decomposition. The duration of SMS decomposition can be shortened to few weeks by vermin composting of SMS. Fresh as well as 15-20 days old rotten spent mushroom substrate is an acceptable material for the worms to multiply and to convert it into manure for field crops.

Spent mushroom substrate as organic fertilizer

Spent mushroom substrate is nutrient rich. However, N release from the SMS is very slow. Therefore, plant nutrition may be insufficient. On the other hand, phosphorus, potassium, micronutrient and the growth substances in SMS are present in sufficient quantity. Availability of nutrients from SMS can be enhanced by preparing an organic mineral fertilizer from the spent mushroom substrate.

Spent mushroom substrate for disease management

The actinomycetes, bacteria and fungi inhabiting the compost, not only play role in its further decomposition but also exert antagonism to the normal pathogens surviving and multiplying in the soil ecosystem. *Pseudomonas* and *Bacillus* present in the SMS exert antagonism to a number of pathogens affecting fruits and vegetables.

SMS has the potential of solving several agriculture related problems. However, it has to be suitably processed before using it as a feeding material for vermicomposting, plant disease management, preparation of organic-mineral fertilizer and bio-remediation of contaminated soils besides the purpose of raising horticultural and several other crops.

Harnessing science for mushroom improvement

i) Conservation and characterization of Biodiversity

The mushroom biodiversity is fast depleting due to deforestation, urbanization, climate change and unsystematic exploitation through collection of wild mushrooms and hence, there is an urgent need to collect and conserve this biodiversity. Some species particularly the mycorrhyzal mushrooms are facing extinction. Periodic collection of indigenous mushroom diversity from the various agroclimatic zones of the country and their genetic characterization for wider gene pool and novelty will be necessary for strategic and anticipatory research in genetic improvement of cultivated mushrooms.

ii) Commercializing technologies for diversified farming

Majority of the farmers are cultivating mushrooms only during particular season. It is also paradoxical to note that India is largely a tropical country and we mainly cultivate temperate mushrooms. The tropical and sub-tropical mushrooms like oyster (*Pleurotus* spp.) , paddy straw (*Volvariella* spp.), milky (*Calocybe indica*), reishi (*Ganoderma lucidum*), wood ear (*Auricularia* spp.) etc. are not cultivated on a larger scale. Hence, the continuous cultivation of different mushrooms depending on the season is certain to increase the economic returns of the mushroom growers. Round the year cultivation assumes much significance especially for rural livelihood security.

iii) Propagating genetic resources

Spawn (mushroom seed) is the most crucial input for successful cultivation of mushrooms. Therefore, its genetic and physical purity and quality need to be maintained at the highest level. To maintain these very conditions, spawn can only be produced in highly sophisticated laboratories. However, in India, the spawn industry is an un-organized venture and needs research support in the years to come so that it may attain quality standards and competitiveness comparable to multi-national companies.

iv) Utilizing spent mushroom substrate

Mushroom growing is an ecofriendly activity and it utilizes the byproducts from agriculture, poultry, brewery etc. and in turn produces a quality food with excellent and unique nutritional as well as medicinal attributes. The spent mushroom substrate (SMS) left after final crop harvest is a matter of concern as it creates various environmental problems including ground water contamination. As mushroom production is increasing, so is the SMS generation, which calls for alternative management of this waste. Fortunately, SMS has many positive attributes still left for its potential uses. Studies on natural resource management in terms of complete recycling of agro-waste can be initiated so as to contain pollution problems.

v) Plant health and bio-risk management

Diseases caused by biotic and abiotic agents are quite common in seasonal farms especially those using unpasteurized substrate. Changing climatic patterns further aggravate these problems. However, the incidence of diseases and competitor moulds is relatively less in environmentally controlled units. Epidemiological studies on new competitors and parasitic moulds, bacterial and viral diseases need to be undertaken and integrated disease management packages for major mushroom diseases and competitor moulds be developed. Parallely, basic information on identification, biology, behavior and status of common insect pests of mushrooms can be recorded for planning their effective control strategies.

vi) Post-harvest technology

Increased productivity demands proper post-harvest infrastructure to enhance shelf life and marketability. Mushrooms are highly perishable in nature and therefore, spoilage starts after 24 hours of harvest. Short shelf life poses unique problems in packaging, marketing and preservation of mushrooms. In India, the retail packaging for fresh marketing is highly crude and primitive and is done in hand sealed polypropylene bags. Improved technology as is adopted in developed countries has to be commercialized for value addition in the harvested produce.

vii) Technology transfer approaches

The apex body, Directorate of Mushroom Research at Solan takes the responsibility to impart training to the trainers from various state and central organizations. However, the Krishi Vigyan Kendras, Agricultural Universities and State Development Departments can directly train the growers. Emphasis from TOT programmes should be given on publication and distribution of literature, adoption of model village, close association with self-help groups and NGOs, holding of mushroom melas, creation of awareness through mass media, preparation of documentary films and video clips and dissemination of information through internet, mobile and other ICT tools.

Conclusion

The Centre of Tropical Mushroom Research and Training along with All India Coordinated Research Project on Mushroom are making concerted efforts in pushing Odisha ahead of other states in mushroom production. This would probably be the appropriate way to search for alternative nutritional sources for our huge population and help achieve non-green revolution.

Mushrooms are truly health foods and promising neutraceuticals. Odisha has tremendous potential for mushroom production owing to the availability of agricultural wastes in abundance, manpower and suitable climate. Further, there is increasing demand for quality products both in domestic and export markets. Mushroom being a women friendly crop, could be facilitated well with a strong Mission Sakti existing in the state. To be successful in both domestic and export market, it is essential to produce quality fresh mushrooms and processed products devoid of pesticide residues at competitive rates. It is also important to commercially utilize the spent mushroom substrate left after cultivation for making manure or vermin-compost for additional income and total recycling of agro-wastes. It is worthwhile to mention here that few of our entrepreneurs have got recognition at the national and international levels owing to their excellence in mushroom production and allied activities. With the untiring efforts of all concerned, possibly Odisha mushroom industry will see a new dawn in the near future.

References

Ahlawat, O.P. and Tewari, R.P. 2007. Cultivation technology of paddy straw mushroom (*Volvariella volvacea*). Technical Bulletin, National Research Centre for Mushroom, Chambaghat, Solan (H.P.). p.36.

Ahlawat, O.P. and Vijay, B. 2004. Effect of casing material fermented with thermophillic fungi on yield of *Agaricus bisporus*. *Indian J. Microbiol.* 44(1): 31-35.

Bano, Z. and Rajarathnam, S. 1982. *Pleurotus* mushrooms as a nutritious food. In: S.T. Chang and T.H. Quimio (eds.) Tropical Mushrooms: Their Biological Nature and Cultivation Methods. The Chinese University Press, Hong Kong, 493p.

Beyer, M. 1996. The impact of the mushroom industry on the environment. *Mushroom News* 44(11): 6-13.

Chang, S.T. 1982. Cultivation of *Volvariella* mushrooms in South East Asia. In: Tropical Mushrooms: Their Biological Nature and Cultivation Methods (Chang and Quimio eds.). The Chinese University Press, Hong Kong, 493p.

Chang, S.T. 2006. The world mushroom industry: Trends and technological development. *Int. J. Med. Mush.* 8(4): 297-314.

Chiu, S.W. and Moore, D. 2001. Threats to biodiversity caused by traditional cultivation in China. In: Fungal conservation (Moore, Nauta and Rotheroe eds.), The 21st Century Issue, Cambridge.

Chong, C., Cline, R.A. and Rinker, D.L. 1987. Spent mushroom compost and paper mill sludge as soil amendments for containerized nursery crops. Combined proceedings. *International Plant Propagators Society* 37:347-353.

Fox, R. and Chorover, J. 1999. Seeking a cash crop in spent substrate. http:/aginfo, psu. edu./pea/ sgg/mushrooms, html.

Hamlyn, P.F. 1989. *The Mycologist* 3 (4): 171-173.

Kaul, T.N. 1983. Cultivated Edible Mushrooms. Regional Research Laboratory, Council of Scientific and Industrial Research, Srinagar.

Krishnamoorthy, A.S. 1995. Studies on the cultivation of Milky mushroom, *Calocybe indica* P. and C., Ph.D. thesis, Tamil Nadu Agricultural University, Coimbatore, India, p.124.

Lal, R. 2005. World crop residue production and implications of its use as a biofuel. *Environment Int.* 31(4): 575-584.

Li, G.S.F.and Chang, S.T. 1982. Nutritive value of *Volvariella volvacea*. In: S.T. Chang and T.H. Quimio (eds.) Tropical Mushrooms: Their Biological Nature and Cultivation Methods. The Chinese University Press, Hongkong. 493p.

Li, Yu, 2012. Present development situation and tendency of edible mushrooms industry in China.

Mohapatra,K.B., Chinara, N. and Behera, B. 2013. Low cost cultivation techniques for *Volvariella volvacea* grown in Odisha. Proceedings of National Symposium on 'Mushrooms for Medicinal Value and Nutritional Security under Changing Climatic Conditions', 27-28 December, 2013, UHF, Nuani, Solan (HP), pp.10.

Pani, B.K. 2012. Sporophore production of milky mushroom (*Calocybe indica*) as influenced by depth and time of casing. *Int. J. Advanced Bot. Res. (IJABR)* 2(1): 168-170.

Prakasam, V. 2012. Mundkur Memorial Lecture: Current Scenario of Mushroom Research in India. *Indian Phytopath*. 65(1): 1-11.

Quimio, T.H. 1993. Indoor cultivation of straw mushroom *Volvariella volvacea. Mushroom Research* 2(2): 87-90.

Rai, R.D. and Sohi, H.S. 1988. How protein rich are mushrooms. *Indian Horticulture*, 33:2-3.

Rajarathnam, S., Shashirekha, M.N., Zakia, Bano and Ghosh, P.K. 1997. Renewable lingo-cellulosic wastes, the growth substrates for mushroom production: National strategies. In: Advances in Mushroom Biology and Production (R.D. Rai, B.L. Dhar and R. N. Verma., eds.), pp.291-304. Mushroom Society of India, NRCM, Solan, India.

Reyes, R.G. and Abella, E.A. 1997. Mycelial and basidiocarp performance of *Pleurotus sajor-caju* on the mushroom spent of *Volvariella volvacea*. Proceedings of International seminar on the "Development of Agri-business and its impact on Agricultural Production in South East Asia, Tokyo NODAI Press, pp.491-497.

Ruhul, A., Abul, K. Nuhu, A. and Tae, S.L. 2010. Effect of different substrates and casing materials on the growth and yield of *Calcite indicia. Mycobiol.* 38(2): 97-101.

Singh, M. 2011. Mushroom production: An agribusiness Activity. In: Mushroom Cultivation, Marketing and Consumption (eds. Singh, M., Vijay, B., Kamal, S. and Wakchaure, G.C.), Directorate of Mushroom Research, ICAR, Solan (HP), pp.1-10.

Szmidt, R.A.K. 1994. Recycling of spent mushroom substrates by aerobic composting to produce novel horticultural substrates. *Compost Science and Utilization* 2(3): 63-72.

Thakur, M.P. and Mohapatra, K.B. 2013. Tropical Mushrooms: Present Status, Constraints and Success Story. Proceedings of Indian Mushroom Conference, 16-17 April, 2013. PAU, Ludhiana (Punjab), pp.42-43.

Verma, R.N. 2013. Indian Mushroom Industry-Past and Present. Bulletin 8 of World Society for Mushroom Biology and Mushroom Products. P.16.

Wuest, P.J. and Fahy, H.K. 1991. Spent mushroom compost traits and uses. *Mushroom News* 39(12):9-15.

Zhang, Y.H.P. 2008. Reviving the carbohydrate economy via multi-product lignocelluloses bioefficiencies. *J. Ind. Microbial. Biotechnol.* 367-375.

Integrated Farming System Practices: Challenges & Opportunities, pp 343-381

New India Publishing Agency, New Delhi, India

14

Apiculture Practice in Integrated Farming Systems

C.R.Satapathy* and Madhusmita Panda**

Introduction

Honey bees are the most fascinating creature found in the nature and are very closely associated with our cultural heritage. They have been evolved with structural perfection and behavioral instinct to render their service as the best pollinator and enable to realize the potential yields of several cross pollinated crops under various agro-ecological situations. They are indispensable for human survival. The words of the great scientist Albert Einstein quoted below is very pertinent in this context *"If the bee disappeared off the surface of the globe, then man would only have four years of life left. No more pollination, no more plants, no more animals, no more man"*.(Thakur, 2011). Thus, starting from honey hunting since time immemorial to traditional bee keeping and its present transformation to scientific bee keeping has been a continuous endeavor of human for harnessing both direct and indirect benefits from bee keeping activities. Bee keeping has been a very popular agro based enterprise which does not compete with any resource of farming system rather acts as an integral component of Integrated farming system. Now-a-days it has not only attracted the attention of a diverse group of people, administrators and institutions for various purposes but also it has been the point of attraction of the rural poor and youth for their livelihood improvement.

*Department of Agronomy

**Department of Entomolgy, College of Agriculture, Orissa University of Agriculture & Technology, Bhubaneswar-751003, Odisha, India

Presently the practice of bee keeping has switched over in its objective from honey mode to pollination mode. In this context production of large number of honey bee colonies for managed bee pollination along with conservation and augmentation of other pollinators has been the need of the hour in one end, and in the other end nearly one-third of the country's population lives below poverty line, who can take it as an avocation for their livelihood support. India possesses substantial biodiversity and regarded as a country of mega bio-diversity. Nearly 8 % of the world's documented animal and plant species are found in our country with more than 750 species of bee friendly plants (Lakra, 2006). Since some time, agriculture sector, as a whole, has been confronted with numerous challenges linked to food and energy crisis coupled with climate change and degradation of natural resources threatening the food security and livelihood security of poor and vulnerable population. Food and Agriculture organization (FAO) in this context defines food security as "A situation that exists when all people, at all time, have physical, social and economic access to sufficient, safe and nutritious food that meets their dietary needs and food preferences for an active and a healthy life." Therefore, conservation of natural resources, maintenance of biological wealth and acceleration of agricultural growth are considered of paramount importance in the present context as well as of the future. In Brundtland commission report (1987) "OUR COMMON FUTURE", the sustainable development is defined as "Development that meets the needs of the present without compromising the ability of future generation to meet their own needs" (WCED, 1987).

Agriculture was, has been and will always remain as the most important sector of human intervention at both national and global level. This is obviously because; food is the prime requirement for any living organism next to oxygen and water. Sustainability is a pre-requisite to keep the intervention ongoing acceptably. Changing scenario in the field of agriculture, particularly the ever expected Sustainable Agriculture has brought the eco-friendly integrated farming system approach to forefront over monoculture in spite of several advantages of the latter.

Thus, the Integrating farming in present context relies upon raising various crop and other allied enterprises in an unit area in a coordinated framework in as a compactable manner as possible with optimum utilization of all available inputs, reducing the use of off-farm inputs and enhancing the use of on-farm inputs, as well as effective waste recycling with an objective for obtaining enhanced crop production and productivity keeping in view the socio economic acceptability, environmental safety and sustainability of the farming system. Integrated farming system are often less risky. If managed efficiently, they benefit from the synergism among enterprises. The components of Integrated farming system has broadened, integrated and revolutionized the production system over the conventional sole farming like that of crops, livestock, aquaculture, horticulture, agro-industry or any other allied activities. It could be combinations of crop with one or more of other allied components. The most common examples of Integrated Farming System models are crop-livestock, agri- horticultural systems, paddy-fish culture, agri-silvicultural systems, crop production integrated with piggery, goatery, duckery etc. But importance of apiculture as a component of Integrated Farming System (IFS) has hardly been realized. Apiculture is undoubtedly one of the ideal components in IFS but lack of understanding the bee language, the science and arts of bee keeping and skill involvement in handling of bee colonies are the bottlenecks in its inclusion in IFS by the farmers. Apiculture otherwise colloquially called beekeeping is defined as an *art and science of collecting/ procuring colonies of desired honeybee species, hiving them in the standard and specified bee boxes, installing them in appropriate sites, managing optimum number of colonies scientifically round the year and harnessing both direct and indirect benefits of the activities*. (Satapathy and Jena 2009).It can well fit in the system both as an allied enterprise as well as an agro inputs.

Bee keeping: It's attributes for sustaining livelihood & ecosystem

- It provides food and cash income without ownership of land .It does not require very specific and fertile cultivable land. Hives can be placed anywhere convenient but should be suitable for bees too. Land where bee boxes are installed normally remains unused and so beekeeping does not use up valuable land. Wild, cultivated as well as wasteland areas all have value for beekeeping because bees collect nectar and pollen from any floral species. Beekeeping fits in very well to small-scale farming systems.

- Bees do not compete for resources needed by livestocks and crops, as they forage on unharnessed resources viz. nectar and pollen. Bees essentially require nectar, pollen and fresh water for their growth and development other than fresh air.

- Bee keeping can be adopted as a spare time, part time, and full time occupation by any person irrespective of age, sex, caste, creed, and profession, educational & economic status and can be adopted as an enterprise. Rural youths, resource poor farmers as well as other rural and non-rural people can adopt this as a profession. This is because bees do not need daily care and beekeeping can be done when other work allows. Thus, anybody willing can be a sole beekeeper or have beekeeping as a subsidiary avocation.

- It serves as a source of regular income for farm families mostly in the honey flow season without much recurring investment. The bee keepers those who have expanded the enterprise for production of bee boxes and other bee equipments can generate more family income.
- Bee keeping can be undertaken for complementary services, such as crop pollination and enhances productivity levels of various agricultural, horticultural and fodder crops. Providing colony on rent to orchardists can render dual benefits i.e. benefit of income obtained through rent and higher honey production. Further, the amount that would have been spent towards watch and ward by the bee keeper is also saved.
- Hive product are of low volume but high value with good shelf life. Honey bees provide a plethora of products including honey, wax, pollen, royal jelly, propolis, venom, etc. that a small-scale farmer can sell from a single farm enterprise. However, for products like pollen, royal jelly, propolis, venom, etc. little more expertise and added facilities are required. The hive product propolis can be obtained, if bee keeping is practiced with *Apis mellifera* which is a good propoliser but not from the Indian hive bee *Apis cerana indica*.
- Honey bee products have good market value and earn foreign exchange.
- The hive products can also be transformed with minimal processing into value added products, for example wax can be processed into candles. The comb foundation sheets widely used in beekeeping with *Apis mellifera* are prepared from bee wax only in a comb foundation mill. This again generates employment and added income.
- Tools and equipments required for bee keeping as well as the bee boxes, frames etc can be made locally which is another source of employment for local craftsman other than the bee keepers to earn extra wages.
- Beekeeping is benign and environmentally friendly. Beekeeping generates income without destroying habitat. Beekeeping encourages ecological awareness due to the fact that beekeepers for their own interest need to conserve the environment.
- Bee keeping assists in overcoming the problem of malnutrition and human health disorders. Honey is not only considered and regarded as *amrut* but also widely used either as medicine or carrier of the ayurvedic medicine. Scientifically it is an energy rich high calorific product and can be treated as a complete food.

- Bee pollination is vital for life on the earth in terms of agricultural produce obtained through enhanced crop production and role played for bio-diversity conservation.
- Bee keeping provide effective linkages to other farming system with positive ecological consequences.
- The technology is simple and low cost but essentially needs interest, patience, acquiring preliminary knowledge relating to bee behavior for beginning the enterprise and utilization of gained knowledge for achieving further success.

Bee keeping an ideal component of IFS

Honey bees by default, very much associated with different crop ecosystem including agro ecosystem for their own survival and in return contribute immensely for safe and secured environment. But human intervention following modernizations without realizing the importance of pollinators has endangered the pollinators *vis-α-vis* our food security. Scientific bee keeping, a non land based income and employment generating activity has become very popular now a days and has a great role to envisage in this context. Besides, the direct benefits of beekeeping, it is more alluring to raise bees for the purpose of harnessing indirect benefits like biological security through crop pollination contributing additional yield of crops valued at 15-20 times more than the value of hive products. Unlike other components of the production system viz seeds, fertilizers, pesticides, water etc each contributing to yield, honeybees though not applied but occurring naturally in the crop ecosystem contribute the major share to the yield. In absence of honey bees the yield of bee friendly plants would not have been realized how important the input may be. A farmer may make full exploitation of all inputs he can like seeds, fertilizers and pesticides etc., but in absence of a successful pollinator like honey bee the sustainability of the production system can only remain as a dream. Thus, honey bees are considered as the most important agri inputs and bee keeping renders a real value to the integrated farming system. In IFS, bee keeping plays the role more as input in contrast to as an enterprise; though the latter can't be underestimated from the beekeepers point of view. The honey production will continue to remain as the prime objective of beekeeping for the bee keeper in addition to the other direct benefits derived out of the enterprise. It is matter of great irony that we can survive any length of time without using any of the bee based hive products obtained directly from the bees but in contrast it is really not possible at all to live long if bees don't render their pollination service as agriinput either in nature or under the conditions of managed bee pollination. Besides producing an elite class of super foods and around-the-clock pollination efforts that touch nearly every corner of the planet, the influence of honeybees is directly linked to mankind's ability to live in abundance and, perhaps, even survive. This is because about 1/3rd of the total food crops of the world require to be pollinated by the insects. When we talk of pollinators we cannot skip the title of "Best Pollinator" that has been given to the honey bees because of the facts that the body parts of the honey bees are especially modified to pick up pollen grains, e.g., pollen baskets (Corbicula) in the hind legs, presence of body hairs, exhibiting flower constancy and fidelity, potentiality for long working hours, special communication system because of their social behavior, ability to maintain high populations when and where necessary and adaptability to different climates and niches (Satapathy et. al 2010). Contrast to the naturally occurring bees, the apiary bees viz, *Apis cerana indica* and *Apis mellifera* are still advantageous as they are quite amenable for shifting near to crop ecosystem and directing them to desired crops for harnessing pollination services. Therefore, including apiculture or bee keeping as a component of Integrated Farming System is definitely wise enough that would enable us to protect the ecosystem by putting the bees at work. Compromise for land, inputs, time will not be necessary by including bee keeping in IFS, rather it will support and add to make the system more viable and profitable.

Asian bees and bee keeping has also been linked since ages with our socio-economic, natural, cultural and religious heritage although, so far no statistical data is available about the direct and indirect benefits of honeybees in terms of value of their hive products and cross pollination services (Verma and Satapathy, 2012). It is estimated that about 15-20 thousand species of bees are found in the world. These include honey bees, bumble bees, stingless bees and solitary bees. Insects are predominant among the biotic pollinating agents and honey bees are the most efficient and best pollinators among all the insects.

The importance of bee keeping in this context can be summed up by the statement of father of Green Revolution in India, Dr. M.S. Swaminathan who in his forwarding note in the FAO service bulletin on tropical and sub-tropical apiculture quoted that "*inspite of all global resolution on food security, several 100 million children , women and men are going to bed hungry every day, particularly in the countries of the 'South'. Since prospects for global food security system appears to be small at the present moment, it will be for the developing countries characterized by poverty and under / nutrition, to build their own national food security system. In this task, apiculture can play a useful role. At present with little expenditure, honeybees will not only provide food and income, but will also enhance the productivity of horticultural and other field crops by their pollination activities.*"

Fig.1: Body structure attributing to efficient pollination by bees

Need of bee keeping for crop pollination

The crops particularly the angiosperms commonly grown or found in nature are broadly either self fertilized with their own pollen or self infertile and need to be cross pollinated with pollen from other plants of the same species. It is estimated that, in nature 5 per cent of flowers are self pollinated and 95 per cent are cross pollinated. Three important types of pollination generally occurs from physical and genetically viewed point (Verma and Satapathy, 2012). They are;-

1. **Autogamy**: It is the process of transfer of Pollen grains from anther to the stigma of the same flower. It is also called as self fertilization or self pollination.
2. **Xenogamy**: It is the process of transfer of pollen grains from the anther of a flower to the stigma of a flower borne on another plant. It is also called as cross pollination.
3. **Geitonogamy**: In this process of pollination, pollen grains from anther of a flower are transferred to the stigma of another flower borne on same plant. Genetically it is self pollination but physically it is cross pollination.

Xenogamic and geitonogamic plants depend essentially upon pollinating agents to set seeds or fruits. Self fertile plants when cross pollinated in presence of pollinating agents produce quantitatively more and qualitatively better seeds and fruits/capsules.

Ten per cent of cross-pollinated crops depend upon wind and water and 85 per cent upon insects. In the course of evolution of angiosperms, enotomophily more specifically pollination by the bees has played a vital role. Both angiosperms and bees have been co evolved since 60-100 million years ago for mutual benefit. Species of bees in world though estimated up to 15-20 thousand but in the Indian subcontinent there is relatively very less number of species of bees visiting the flowers. Among these species bees of seven families viz., Apidae, Halictidae, Colletidae, Megachillidae, Andrenidae, Stenotritidae and Mellitidae belong to super family Apoidea of order Hymenoptera pollinate the crops widely. Members of apidae are the most reliable agent for pollination and about 80 per cent insect pollination is done by the bees alone. Other insects normally switch over from one type of flower to other types during successive visit, so fail to pollinate the flowers as efficiently as honey bees. Bees are reported to pollinate over 70 per cent of the world's cultivated crops. Among bees following five species (Table -1) of honey bees are most important.

Table 1: Major species of honeybees

Sl.No	Common Name	Scientific name	Sub family	Family
1.	Rock bee -	*Apis dorsata F.* -	Apinae	Apidae
2.	Little bee -	*Apis florae F.* -	Apinae	Apidae
3.	Indian hive bee -	*Apis cerana indica F.* -	Apinae	Apidae
4.	Italian honey bee -	*Apis mellifera L.* -	Apinae	Apidae
5.	Sting less bee -	*Tetragonulla iridipennis Smith* -	Melliponinae	Apidae

Insect pollinated flowers develop certain adaptations to attract insects. They are large sized flowers, bright colored petals, scent and odors, special nectar secreting glands, edible nectar, edible pollen, special mechanisms of advantage to plant itself. Indian hive bee and Italian bees are hived in wooden boxes and extensively used for managed pollination of field and orchard crops. It has also been reported that about 15 per cent of the 100 principal crops are pollinated by the domestic bees. The association of crops and bees can be realized from the fact that *A. cerana indica* and *A. mellifera* collect nectar load of 15-30 and 30-40mg respectively in each trip. A bee brings about 10-20 full loads of nectar/day and for each load it visits 50-500 flowers. Similarly the pollen load in *A. cerana indica* and *A. mellifera* varies from 7-14 mg and 12-29 mg respectively which they collect from 50-200 flowers. Each pollen load is collected in 10-20 min. time and as many as 47 loads are collected / day. This long period and intimate association of bees for nectar and pollen gathering facilitate high rate of pollination. The sting less bees is very small and known to pollinate over six dozen of crops in Odisha itself. They produce very little honey but the quality of honey is much superior to others. These bees are also hived in small wooden box of 25 x 15 x13 cm size (Fig-2) for pollination of crops in back yard.

'A' 'B'

Fig. 2: Stingless bees in wooden hive (A- Hive, B- Bee colony)

Many of our important crops (Table-2) need bee pollination (Satapathy 2009), Pollination service rendered by bee improve our national income through enhanced yield of agricultural and horticultural crops. The yield increase due to bee pollination in different crops experimented is presented below in Table 3 & 4 (Saraswata 2011).

Table 2: Principal crops in different States needing pollination service by insects

State	Principal Crops
Andhra Pradesh	Sesames, Sunflower, Safflower, Red gram, Horse gram, Cotton, Chili, Coriander, Cumin, Coco palm, Coffee, Cucumber, Ivy gourd, Melons, Cashew, Citrus fruits.
Assam	Rapeseed and Mustards, Areca palm, citrus spp., Loquat, Litchi.
Bihar	Rapeseed and Mustards, Linseed, Niger, Redgram, Guava, Litchi, Cucurbits.
Gujarat	Sesame, Cotton, Lucerne, Vegetable crops, Red gram.
Haryana	Rapeseed and Mustards, Cotton, Lucerne.
Himachal Pradesh	Apple, Cherry, Cole crops (seed crops), Buckwheat, Rapeseed and Mustards, Plums, Persimmon.
Jammu & Kashmir	Rapeseed and Mustards, Apple, Cherry, Cole crops (seed crops), Apricots, Buckwheat.
Karnataka	Sesame sunflower, Niger, Red gram, Cotton, Horse gram, Coco palm, Areca palm, Citrus spp., Avocado, Cardamom, Coffee.
Kerala	Areca palm, Coco palm, cardamom, Coffee
Madhya Pradesh	Sesame, Rapeseed and Mustards, Linseed, Niger, Red gram, Cotton, Coriander, Vegetables, Citrus spp.
Maharashtra	Sesame, Sunflower, Safflower, Niger, Linseed, Red gram, Cotton, Lucerne, Chilies, Coco palm, Cashew, Mandarin, Orange, Grape fruit, Guava, Sweet Orange, Lime, Melons, Cucumber, Onion, Strawberry.
Odisha	Sesame, Niger, Rapeseed and Mustards, Sunflower, Mango, Coconut, Cucurbits,
Punjab	Rapeseed and Mustards, Red gram, Cotton
Tamil Nadu	Sesame, Sunflower, Niger Red gram, Cotton, Chilies, Coriander, Coco palm, Areca Palm, Coffee, Cardamom, Mandarin Orange, Lime Lemon, Cashew, Cucurbits, Pear, Plum, Peach.
Uttar Pradesh	Rapeseed and Mustards, sesame, linseed, Almond, Apple, Apricot, Cherry, Litchi, Peach, Plum, Citrus sp., Loquat, Berseem, Lucerne, Red gram, Chilies, Buck wheat, cucurbits and Cole crops.
West Bengal	Rapeseed and Mustards, Litchi

Table 3: Yield increase in different crops due to bee pollination

Sl.No.	Oil seed Crop	Yield increase (%)	Sl. No.	Pulse Crop	Yield increase (%)	Sl. No.	Vegetable Crop	Yield increase (%)
1	Mustard	128.1- 159.8	10	Alfalfa	23.4-19733	19	Radish	22-100
2	Rai	18.4	11	Berseem & Clovers	23.4-33150	20	Cabbage	100-300
3	Rape seed	12.8-139.3	12	Vetches	39-20000	21	Turnip	100-125
4	Toria	66 -220	13	Broad bean	6.8-90.1	22	Carrot	9.1-135.4
5	Sarson	222	14	Dwarf bean	2.8-20.7	23	Onion	353.5-9878
6	Safflower	4.2-114.3	15	Kidney bean	500-600	24	Brinjal	35-67
7	Linseed	1.7-40	16	Runner bean	20.6-1100	25	Cucumber	21.1-411
8	Niger	260.7	17	Arhar	21-30	26	Radish	22-100
9	Sunflower	20-3400	18	Other pulses	28.7-73.8			

Table 4: Yield increase in different fruit crops due to bee pollination

Sl.No.	Oilseed Crop	Yield increase (%)
1.	Apple varieties	180-6950
2.	Pear	240-6014
3.	Plum	6.7-2739
4.	Cherry	56.1-1000
5.	Strawberry	17.4-91.9
6.	Raspberry	291.3 – 462.5
7.	Persimmon	20.8
8.	Litchi	4538-10246
9.	Citrus	7-233.3
10.	Grapes	756.4-6700
11.	Squash	771.4-800
12.	Guava	70-140
13.	Papaya	22.4-88.9
14.	Mosambi	36-750
15.	Orange	471-900

Other benefits

- Bee pollination not only increases the yield but also improves the quality of seed and fruits. Increases number and sizes of seeds and yield of crops, more nutritive and aromatic fruits are produced.
- Stimulate germination of pollen on stigma.
- Adequate pollination ensures early seed set, uniform maturity and early harvest.
- Increases fruit set and reduces fruit drop.
- Increases viability of seeds, embryos and plants.
- Oil content in oil seed crop increases due to bee pollination.
- Enhances resistance to diseases and other advance environmental conditions.
- Increases nectar production in the nectarines

Managed bee pollination in IFS

India has about 160 million hectare of cropped area of which more than 55 million hectare is under bee dependent crops other than forest, pastures, waste lands and non cultivated lands. It has been discussed earlier that, honey bees are the best pollinator and can be made available in considerable number whenever and wherever needed. IFS is a well designed and planned crop ecosystem which can be made more productive through managed bee pollination Planned or managed bee pollination is of paramount importance to improve production of agricultural, horticultural and other useful crops.

Important consideration for managed bee pollination

Productions of honey and pollination service to crop are two important gifts honey bees have offered to mankind. Although honey production remains as the prime interest of the bee keeper but pollination service is more important. Therefore, managed pollination is practiced. It should be based on the following considerations.

- The most important consideration to the bee keeper is the ultimate benefit he is getting by extending pollination service to the farmers.
- Migration of colonies is essential for effective managed bee pollination. The type of crop needing pollination service, area under the crop, number of colonies required for crop area are the points to be considered for managed bee pollination. Normally 3-5 bee colonies of good strength are required per hectare. More number of colonies are required if crop is not attractive, however overcrowding of colony should be avoided in the interest of the bee keeper. Optimum number of colonies required per hectare in different crops (Verma and Satpathy, 2012) is furnished in Table - 5
- Prosperous or strong colonies i.e. colony with young laying queen ,healthy brood in all stages new brood comb, comb with adequate honey and pollen store and ideal bee strength covering at least six brood comb should be utilized for the purpose.
- Appropriate management practices should be adopted to keep the colonies in peak foraging activity and with predominant pollen gatherers.
- Visit to the crop by forager can be enhanced by providing stimulant food during night hours comprising of sugar solution in which few flowers and pollen of the target crop are soaked for some hour.
- Farmers requiring pollination service should have planned cropping system to avoid competition between crops for bee pollination and safety measures against the honeybees for better pollination.
- Colonies should be safely migrated during night hours when bees are calm and quiet and should be placed near crops at 10-20 per cent flowering stage. Hives should be placed in the vicinity of the crop field.

Table 5: Optimum number of bee colonies required per hectare for adequate pollination.

Sl.No.	Crop	Colonies/ha(No.)	Sl.No.	Crop	Colonies/ha(No.)
1.	Almond	2	11.	Mango	8-15
2.	Apple	2	12.	Onion	12-36
3.	Apricot	2-5	13.	Orange	4
4.	Brassica	2-5	14.	Peach	1-25
5.	Cauliflower	2-3	15.	Pear	1-5
6.	Carrot	2-4	16.	Radish	5
7.	Citrus	3-4	17.	Sesame	5
8.	Cotton	1-12	18.	Sunflower	1-4
9.	Litchi	5	19.	Safflower	2
10.	Lucerne	4-8	20.	Watermelon	1-30

Income generation through pollination service

Scientific bee keeping provides multifacet income to the bee keeper. Apart from income generation through sale of hive products, bee colony, beekeeping equipment, sharing intellectual skill of bee keeping, beekeeper can earn by renting his boxes for pollination service. Besides rent amount, the farmer also harvests more honey from the hive that adds to his income. Himachal Pradesh is the leading state in India and South East-Asia in renting honeybee colonies in a scientific way for pollination of apple orchards .The State department of horticulture in the state is providing bee colonies to the orchardists at a very nominal rate of rent for pollination. Due to increased awareness among the orchardists about the importance of bee pollination in enhancing apple production, there is an increase in demand and the orchardists are unable to get required number of bee colonies for their orchard .This practice has been initiated in Bihar to harness the honey from litchi .In many parts of the world beekeepers are charging rent per colony for pollination. Increased awareness among farmers and beekeeper for mutual benefit will increase the income of both deriving pollination service of the hived honey bees.

Ample of information is available worldwide on effect of honey bee pollination on yield of different crops. Similar studies conducted in Odisha revealed yield increase in different oilseed crops from 10-79 per cent (Table -6) (Satapathy and Padhi, 2012) besides increase in oil content in mustard and sesamum.

Table 6: Effect of bee pollination on yield of oilseed and litchi crops in Odisha.

Sl.No.	Crop	Yield increase (%)
1.	Mustard	10.6
2.	sesame	25.0
3.	Niger	33.0
4.	Safflower	64.0
5.	Sunflower	79.0
6.	Litchi	5.3 times

Fig. 3: Pollination studies in Odisha

Bee keeping

In India, the knowledge about the honey bees and use of the naturally occurring sweet i.e. honey dates back to ancient times .The then people used to hunt/ rob honey from feral colonies of native bee species viz., *Apis dorsata* Fab., *Apis florae* Fab. and *Apis cerana indica* Fab. Attracting the Indian hive bee, *Apis cerana indica* to keep them in various indigenous feral hives such as log or wall hives and gathering honey from these colonies was an obvious practice till 19th Century. Discovery of bee space, construction of wooden hives with movable hive frames in the western countries gave impetus to the scientific beekeeping. Adopting the concept of bee space and movable frame wooden hive, Father Newton in 1910 designed a hive for Indian hive bees which was later named after him as Newton hive in which Indian hive bee could be successfully kept. Thus, it is imperative to mention that the scientific bee keeping in India is only about a century old practice. During course of time practice of beekeeping has undergone a lot of transformation. The higher honey yield and other desirable attributes of exotic honey bee, *Apis mellifera* Linn. Lured / gratified the scientists to introduce this species into India. Several attempts were made from 1920s to 1950s to introduce and establish *Apis mellifera* into the various parts of the country but all these attempts went in vain. The first successful introduction and establishment of *Apis mellifera* in the erstwhile Punjab was made possible by Dr A. S. Atwal, a renowned bee scientist of Punjab Agricultural University, Ludhiana, through "Interspecific Queen Introduction Technique" and later on through the import of its disease free nuclei. The stocks of A. *mellifera presently* existing in our country has descended from the queens and disease free nuclei of exotic honey bee imported from 1962-1966. Introduction and establishment of this *Apis mellifera* brought about the amber revolution in the country(Gatoria et. al.,2007). After successful introduction, A. *mellifera* and A. *cerana* are coexisting since last 30-35 years in India. Among these two species *Apis mellifera* is undoubtedly much superior to *Apis cerana* in many respect and very amenable for commercial bee keeping but has got its own limitation with respect to its adoptability in new location and necessity of migration to fulfill the foraging satiation. Similarly, other than more swarming and absconding tendency, *A c. indica* has a limitation that it does not thrive well in Indo gangetic plains where summers are too hot. In spite of many good attributes of the exotic species when viewed from its adaptability to a particular agroclimatic location the native bee *Apis cerana indica* is no way less. Temperature has a lot of bearing on various biological activities (Table 7) and success of the honey bee species. Therefore beekeeping in any locality can be initiated with the species looking into its suitability as bees never relay or depend upon the beekeeper for food, colony activities or caring for their brood etc. The responsibility of the bee keeper is to provide the conditions and amenities required by the bees by understanding their language and social behavior for enabling them to work normally. Therefore, the scientific beekeeping can be undertaken for pleasure or as an agro enterprise or component of IFS or as a sole avenue for food and livelihood security. A thorough understanding of beekeeping which is defined as an art and science of collecting/ procuring colonies of desired honeybee species, hiving them in the standard and specified bee boxes, installing them in appropriate sites, managing optimum number of colonies scientifically round the year and harnessing both direct and indirect benefits of the activities is essential.

Table 7: Effect of temperature on bee activity

Temperature	Activity
33-36°C	Most favorable for colony build up, comb construction and brood rearing
33-34°C	Favourable for queen to lay eggs
> 34°C	Try to control hive temperature through fanning, reduce inner temperature of hive
> 33°C	Some bee seal the entrance being congregated to reduce hive temperature
< 27 °C	Cover the entire brood frame to keep the brood frame warm
20°C	New queen stop nuptial flight for mating
16°C	Drone fails to move around the box
14°C	Workers remain in group for sustain low temperature
10°C	Workers cannot fly
< 8°C	Lethal to the honey bees

Bee keeping as a practice in IFS

There is vast potential for beekeeping in the country and ample scope for including it as a component in IFS. However, due to lack of knowledge, scientific beekeeping is not being practiced by the beekeepers. It is necessary for beekeepers to participate in the trainings / other capacity building programmes on the subject to gain scientific knowledge on the subject. Success or failure of bee keeping depends upon the conditions where in this activity is practiced and the knowledge and skill of the bee keeper. Under natural and more bee friendly conditions, the chance of success is easier and greater because of the availability of bee pasturages in plenty and for a longer period. Duration of major honey flow, minor honey flow, dearth and miner dearth periods exhibit direct bearing on growth and development of the honey bees and are dictated by regional conditions. Based on the availability of natural bee flora and the cropping situations honey flow period differ greatly in different parts of the country. Deforestation for modernization, repeated natural calamities in coastal India devastating huge foraging plants, changed environmental happenings and many other adverse conditions in one end and the intensive agricultural activities by man in the other end have changed the honey flow conditions. Further, honey bees perform their entire duty viz., foraging, brood rearing, nest cleaning, protection against natural enemies etc themselves and hardly depend on bee keepers for any of its essential colony activities other than installing the colony in suitable place and providing conditions amenable for their growth and development. Keeping all these in mind planning should be made to take up the activity to harness the direct and indirect benefit from this component under IFS. However, irrespective of the purpose the following guidelines are to be adhere to initiate bee keeping which can be suitably modified based on the purpose or requirements.

Guidelines for starting apiary or bee keeping

1. Training

- Successful beekeeping essentially requires strong interest, patience, preliminary knowledge about bee keeping and utilization of gained knowledge subsequently for which training is a must for a new person.
- The person interested for starting commercial bee keeping must undergo both theory and practical training on handling of bee colonies from any recognized institution(s) since it involves lot of special skills and clear cut understanding of bee behaviour.
- Central Bee Research and Training Institute (CBRTI), Pune regularly offers training to the interested persons. Besides, training is imparted by KVIB or KVIC or AICRP centers on Honey bees and Pollinators (ICAR) in SAU's.

2. Suitable site

- The selection of ideal site is essential for successful bee keeping .The area should be neat and clean, well ventilated, and well lighted.
- Bees are known to fly as far as 12 km (8 miles), but usually foraging is limited to food sources within 3 km. approximately. About 75 percent of the bees from colony forage within one kilometre radius while young field bees only fly within the first few hundred meters.
- Foraging requires energy and the honeybee's evaluation as to where, what and how long to forage is all related to the economics of energy consumption and the net gain of food to the colony. For example, foraging bees may not access a high quality food source because its distance requires energy expenditure exceeding the energy value of the food source. Generally bees fly only as far as necessary to secure an acceptable food source from which there is a net-gain. Factors that influence foraging behaviour and determine profitability includes weather e.g. wind, temperature, and sunlight ; distance of the food source from the hive (including differences in elevation); food quality (concentration of sugar, protein content of the pollen), and quantity of nectar or pollen.
- Presence of bee flora within a radius of one kilometer capable of providing sufficient pollen and nectar to bees at least for 6-8 months is ideal for bee keeping. For this preparation of floral calendar that depicts the source and duration of nectar and pollen availability over calendar months is a prerequisite.

- Cultivation of bee friendly seasonal crops like mustard, sunflower, sesamum, niger etc under unprotected or protected in a planned way meets the requirement partially and make the site more amenable for bee keeping.
- There should be no serious environmental problems nearby i.e. the crops in the locality should either be free from insecticidal application or protection schedule should be compatibly adjusted to beekeeping. Besides, a commercial bee keeper should develop a floral calendar of his locality.
- There must be a source of fresh and clean water nearby or alternative arrangements may be made for the purpose. In absence of clean water they tend to collect water from any source which is likely contaminating the hive products.
- Hives must be secured from wind and shaded from strong sunlight.
- There should be no water dripping from overhead branches or shade over the bee hives.
- Site should be well communicated but away from main road, railway tracks, power station, brick kilns etc. Swampy, smoky, dusty and polluted area must be avoided. There should not be any source of stagnant / dirty water, chemical industry/ sugar mill, etc. nearby the apiary
- Bee boxes should not be installed at a height higher than the normal foraging height.

3. Bee species

- Beekeeping can be done by using two species of honey bees viz. *Apis cerana* and *Apis mellifera* depending upon floral conditions and capability of investments. However, success in both the cases depends on quality of bees, management practices and the support of natural condition, particularly availability of the bee flora.
- Indigenous species, *Apis cerana indica* is amenable for beekeeping in many parts of the country and suitable for stationary bee keeping.
- The Italian honey bees, *Apis mellifera* is a better honey gatherer and docile species with many other desirable traits suitable for commercial bee keeping and performs well if migratory bee keeping is practiced. It gives more return when practiced under condition of migratory bee keeping with diversification.
- Besides these two, many other apis bees, colletid, andrenid, mellitid, halictid and megachillid bees visit our crops under IFS as potential pollinators for pollination which the farmer should know and take steps for their augmentation and conservation.

4. Bee colony

- The selection and procurement of good bee colony is very vital for initial establishment of a colony in new area. In case of *A. mellifera* normally divided colony is purchased or procured from the bee keepers while bee keeping with *A.c.indica* may be started by collecting natural colonies or swarm colonies or divided colonies. When source and quality of the mother colony is not known, the quality of swarm colony also remains doubtful and close vigilance is required to record performance of such colony.
- The person having fair knowledge about quality of bee colony should purchase/procure the bee colony either from any recognized organization or from bee keepers.
- The ideal bee colony should be of at least of four frames strength with young queen (< one year) for higher reproductive efficiency, sufficient broods, adequate pollen and nectar store and free from diseases and pests.
- Bees of the colony should be with fair adoption to floral resources, easily managed and with no or little swarming or absconding instinct.
- Selection and multiplication of honey bee colonies should be done from disease resistant, high honey yielding colony. The queen should be new one with high fecundity.
- Bees of the colony should be good honey and pollen gathers.

5. Starting time

- Starting time for establishment of new apiary depends upon the nature. Colony availability is a major constraint in this context. Normally it should be initiated coinciding with breeding season of bees i.e., preferably in early spring or in case of *A.c.indica* when swarms take off from the hives commences.
- During this period bee keeping may be initiated either by capturing swarms or natural colony or dividing any populous colony, the later being more preferred.
- If apiary establishment is done early in the season, colony development accomplished in a couple of months and such colonies are capable of withstanding the dearth period. In case it is delayed, the colony development declines coinciding with dearth period.
- Rainy season is the dearth period for bees and very unsuitable period to start establishing the apiary. During this period colonies are also not available.

6. Colony density /unit area

- Colony density /Unit area for the purpose of effective pollination under managed pollination has been standardized (Table 5) which varies depending upon the crop concerned and the purpose. Large number of colonies (36/ha) are required for effective seed production in onion.
- However, for establishment of an apiary or bee nursery, the number of colony /unit area depends upon the availability of bee foraging plants in the vicinity. In an apiary it is very important to keep optimum number of colonies (50-100) for better establishment and management However, over- stocking or overcrowding of colonies in the apiary must be avoided.
- During installation of the Indian hive bee, *Apis cerana indica* colonies should be placed well scattered with a minimum distance of 8 to 10 feet apart between two hives where as in the exotic species colonies can be kept relatively at a closer proximity.

7. Bee keeping tools

- Prior to start an apiary, the essential bee keeping tools should be procured well in ahead which includes bee hive , hive stand, ant well, nucleus box, capturing hive, smoker, bee veil, queen gate, queen excluder sheets, hive tool and the honey extractor etc. as per the desired species of bee to be used for beekeeping. The native *A. cerana* is kept in BIS or ISI. Standard "A" & "B" type boxes while Longstroth hive is required for exotic species, *A.mellifera*.
- Besides, other small tools may be collected as and when required depending upon the management operation.
- The bee keeping tools in our country are available for purchase from KVIB or progressive beekeepers who have taken another step ahead to derive income from sale of bee keeping equipments.

8. Consumable inputs

- Sugar (to meet the stimulating or supplementary or dearth feeding), formic acids and other medicines for bee health protection should be kept ready and timely utilized as per prescription.
- In case of *A. mellifera* comb foundation sheets should be kept ready before onset of monsoon.
- Use of tetracycline, however should be avoided strictly particularly when honey is to be exported.

9. Colony inspection

- Colony inspection should be done at regular interval of 7-10 days depending upon the season in order to know the colony performance and decide the management required
- Colony inspection should be done on clear sunny days preferably at temperatures between 20^0 and 30^0 C. It should never be done in cold, windy and cloudy days. Use smoker whenever needed to subdue the bees.
- Colony inspection should be done by standing on any side and when maximum bees are outside on foraging trip.
- Prior to colony inspection one should go with all preparedness with protective clothing, bee veil, smoker and ancillary items.
- Normally, in rainy season hive should be inspected at weekly interval and bottom board should be made clean by hard brush to damage wax moth egg and larvae. Inspection is to be done in accordance to the safety procedure of hive handling with least disturbance to the bees.

10. Seasonal and specific management:

A. Seasonal management

Good beekeeping logically seeks to take advantage of the bee flora of the area with the least number of well-managed strong colonies than large number of weak colonies. Because stronger colonies produce more surplus honey and such a strategy minimizes the cost of equipment and reduces labor while increasing honey yield. Management strategies in beekeeping need to enhance or improve space for brood and honey with queen's ability to lay more eggs to raise worker bee population in a colony. It should be kept in mind that the temptation to remove all of the honey at the end of the honey flow period becomes too expensive later on. Removing and selling honey often gives immediate gain but leaving it on the colony for the bees' for use during the dearth periods is an investment for the future, or a deferred gain. Insufficient honey stores may lead to death of bees of starvation or the colony may become so weak that it succumbs to predators. A sufficient amount of honey left on a colony for the dearth period assures that it will survive the period and be in good shape at the beginning of the next build-up period. Thus, good beekeeping practice requires good understanding of bee life cycle, good timing of bee management operation and maintenance of healthy and strong colonies before nectar and honey flow. The essential management practices to be taken up during different calendar months are furnished below in brief. However, the practices may vary slightly depending upon the climatic situation in different agro-ecological conditions and the species of bee used for bee keeping.

Jan-March (pre summer season)

This is the period when population of bee increases inside bee boxes and honey flow commences coinciding with blooming of field, horticultural crops and forest plants. Therefore the following operations are to be carried out.

- Place along with superframes super fitted with old side combs removing them from side frames of brood chamber gradually over succession of time.
- Place empty frames between two strong brood frame for comb construction in *Apis cerana indica* and frames fitted with comb foundation sheet in *A. mellifera.*
- Extract honey scientifically using honey extractor when 70-80 per cent frames are sealed.
- It is ideal time for colony division or colony multiplication. Initiate colonies production for sale.
- Provide conditions to improve weak colonies to become stronger by providing new comb in brood chamber or brood s from stronger colony.
- Favourable time for mite infestation and TSB disease. The disease prevention may be achieved by maintaining populous healthy colony.
- Bottom board needs to be kept clean periodically
- Provide supplementary feeding to the colony if necessary. Adequate precaution must be taken while providing the feeding to the colony.

April-May (hot summer)

- When temperature in the apiary increases beyond 37°C, water is used by bees to evaporate and cool the colony. Thus, water sources nearby bee hives should be ensured and space between bottom board and brood chamber need to be created for increasing ventilation.
- By this time the new queen after successful mating starts egg laying and there remain no need of the drone. So remove the drones if present by using drone trap.
- Remove combs uncovered by bees if any.
- Time for wasp attack, preventive measures must be ensured particularly in hilly areas.

June-September (rainy season)

Rainy season is the worst and dearth period for bees. Bees can't forage normally and adequate foraging plants are not available limiting the availability of nectar and pollen. So queen reduces egg laying, bee population in the colony declines and colony become weak. This is the period for more wax moth infestation. So the following management practices are to be adopted.

- Bottom board cleaning at weekly interval regularly must be practiced to keep the colony free from infestation by wax moth and other natural enemies.
- Area around bee hive must be kept clean and dry to avoid growth of grasses and bushes covering hive entrance.
- Proper cross ventilation should be ensured in apiary site
- Excessive or continuous rains affect bee foraging and lead to food scarcity in hives. Dearth feeding is a must to manage colony during food shortage. During feeding care should be taken to avoid bee fighting.
- It is advisable to keep fewer stronger colonies rather than more number of weaker colonies. So weak colonies may be united to form stronger colonies is recommended to surpass the dearth period.

October-December (winter)

- Practice of uniting the weak colonies to form stronger colonies need to be continued.
- Measures must be initiated to make the colony stronger
- The bee frames are to be kept closer to each other in compact manner. Using dummy board or gunny bag covers warmth within the hive is created.
- It is the time to remove old combs and provide new frames for comb building.
- Development of drone cells may be curbed to avoid earlier swarming.
- Provide stimulant feeding i e., (sugar: water:: 2:1) to encourage population built up prior to honey flow season to harness more yield.

B. Specific management

Specific seasonal management practices for both the species i.e. *A. mellifera* and *A.c.indica* have already been developed and plenty of information is available for successful bee keeping which must be adhered to by the bee keeper soon after installation of the colony. The information and skill can be acquired during the course of training. For specific problem, specialists in the field of beekeeping should be consulted. Some specific managements viz. colony division, colony union, swarm control, curbing drone population, management of absconding, drifting, robbing, laying worker management etc are to be undertaken in appropriate time.

a) Swarming

- Swarming is the natural process of multiplication of honey bees.
- In this process the old queen bee accompanies the swarming bees leaving behind a cross section of old population having drones and a few queen cell from which new queen emerges.
- If swarming colony is not managed timely and properly it may issue secondary and tertiary swarms and causes great loss to the bee keeper.
- Sometimes when swarm is issued in the end of the breeding season queen of the mother colony either remains unmated and bee strength declines or if mates then it fails to raise the population to the desired strength for exploitation of ensuing honey flow season.

Indication

- Over-crowded condition in the brood nest followed by the appearance of huge number of drones and few queen cells are indication of a colony ready for issuing swarm.

Swarm control

Preventive measures

- Removal of congestion
- Reversal of brood and super chamber (applicable in A. *mellifea*)

Remedial measures

- Clipping of wings of queen bees.
- Caging queen by putting queen gate.
- Placing queen excluder sheet between bottom board and brood chamber.
- Destruction or queen cells.
- Division of colony is the best solution.

b) Absconding

- It is a condition in which all the bees of a colony leave the hive due to unfavourable internal or external conditions.
- *Apis cerana indica* also has got more tendency to abscond as compared to with *A. mellifera* among apiary bees while *Apis dorsata* and *Apis florea* are nomadic and abscond quite frequently in search of suitable site and bee flora.

Causes

- Extremely unfavourable weather conditions
- Scarcity of food
- Excessive smoking
- Severe attack of bee enemies.
- Human interventions

Consequences

- Absconding results in deserted combs
- Multiplication of wax moths on deserted combs

Preventive measure

- Provide feeding sugar syrup to bees during dearth period.
- Check attack of wax moth and other bee enemies.
- Avoid pesticides hazards.
- Avoid excessive smoking of honey bee colonies.
- Avoid excessive swampy conditions or thick vegetation around honey bee colonies
- Keep colonies in shade during severe summer season.

c) Robbing

- Robbing in the apiary may be intra-specific, during which workers of stronger colonies start robbing the food from the weaker colonies of the same species.
- It may be inter-specific during which worker bees of one species rob colonies of other species.

- During rainy season, this menace becomes more serious because of scanty availability of nectar in the field.
- During robbing fight takes place between the robbers and the guard bees at the entrance of the colony being robbed.
- The intruders try to kill the queen bee so that the remaining bees get demoralized and surrender themselves completely to the robbers.

Preventive measure

- The hives should be made bee proof, except the main entrance, using mud.
- The main entrance of the colony should also be narrowed down to single bee space so that only one bee enters or leaves the hive at a time.
- Feeding should be given only late in the evening in some suitable feeders inside the hive. Care should be taken that during this process the sugar syrup does not get spilt over the hive and the ground.

d) Queenless colony

- Queen is the most important member of the honey bee colony.
- It is the only perfect female which is capable of laying both fertilized and unfertilized eggs and is thus very important factor in augmenting colony population
- It not only maintains cohesion among the colony members but also helps in maintaining normal activities in the colony.
- The pheromones emitted by this single organism in the colony are responsible for maintaining normal conditions and activities in the colony as a whole.
- Its prolonged absence creates unnatural conditions in the colony.
- Queen less colonies, if not properly and timely managed, develop into laying worker colonies particularly during non-breeding seasons and finally get perish.
- Problem of queenlessness may also occur even during normal breeding seasons.

Reasons of queenlessness

- Accidental Death of the Queen bee due to
  - i) Improper handling
  - ii) Death during transportation
  - iii) Bee Enemies
  - iv) Weather conditions
  - v) Use of pesticides
- Robbing
- Swarming
- Theft of the Queen bee

Symptoms of queenlessness

- On becoming queen less, the colony tends to build queen cells provided the eggs or young larvae are present in the colony.
- The presence of queen cells anywhere in the middle of the combs rather than on the margins might be one sign of missing queen bee.
- The queenless colony is easily differentiated from a queen-right colony as the former becomes more aggressive than the latter
- Absence of eggs in the comb cells may also be one of the signs of queenlessness.
- The number of guard bees at the hive entrance of such colonies also increases.
- Since brood rearing in queen less colonies get depleted and may finally cease, the pollen incoming in the colony accordingly decreases and finally stops.
- Honey collection activity also dwindles.

Managing the queenless colonies

- Provision of eggs, young larvae or gyne cells can be undertaken only when there is breeding season and there is good drone population for successful mating of newly emerged queen bee.
- Provision of mated queen or uniting of colonies could be the only options during off season i.e. when drones are not available

e) Laying worker colonies

- The laying workers are imperfect females which develop from the normal eggs of the colony in absence of the queen for a long period.
- Their ovaries get a little developed in the absence of queen substance and these bees start laying infertile eggs which give rise to drones of poor quality.
- The eggs by laying worker bees are laid in worker brood cells and are placed in a haphazard manner.
- In one cell, usually more than one egg is seen because many workers lay eggs in the same cell irrespective of the fact whether egg is already present or not.
- The eggs are also mostly seen attached to the cell wall rather than at the base as is in the normal queen right colony.
- The pattern of brood also looks uneven, irregular and scattered and the honey and pollen is stored below the brood in such a colony

f) Drone management

Drones or male bees in a colony appear coinciding with the initiation of honey flow and when population in a colony increases to get ready for swarming. Excepting mating they don't perform any other work of the colony and in other hand consume a lot of honey store. So drone management is required for higher honey production

- Drones are tolerated by workers till queen is mated.
- Once queen is mated they are driven out after tearing their wings.
- Drone trap can be used to collect and remove the drones from the colony.

11. Protection from pests sand diseases:

- Adequate care should be taken while maintaining bee colonies. The colony may be made populous to remain free from pests and diseases and can counter the attack of natural enemies.
- Technologies like equalizing the colony, uniting the colonies and related operations are to be exploited to keep the colony stronger. The specialist in the field of beekeeping should be consulted to mitigate any type of specific problem.

- In spite of all possible care, bees sometimes succumb to the infestation by over dozen of natural enemies

a) Wax moth management:

- The basic rule of a wax moth control is to maintain strong queen right colonies.
- Close all cracks and crevices of the hive and reduce entrance size.
- Keep the bottom board clean. Collect and burn the debris periodically
- Avoid pesticides poisoning which otherwise weaken the colonies
- Remove extra combs or comb remaining uncovered by bees from the hive, especially during dearth period.
- Destroy severely attacked combs and boil them indirectly in water and strain to harvest bees wax.
- Fumigation with sulphur @ 250-300 g/m^3 space or Aluminium phosphide @ 0.75 g/ m^3 space is very much effective in killing the larvae in stored combs in air tight rooms/ containers/ chambers.

b) Ant management

- Keep the apiary clean by removing the dead logs, rotten woods, stones and cut the grass regularly.
- Do not keep the honey bee boxes without stands. It has been observed in many apiaries that bee keepers place honey bee boxes directly on the earth surface.
- Use ant well in *Apis cerana* or similar alternatives for *Apis mellifera* to prevent climbing of ant to bee hives.
- Searching ant nests in the vicinity of the apiaries and driving away the ants by using repellents such as ethanol, sodium fluoride, sulphur, Borax, kerosene oil etc. is effective in reducing their attack.
- Hive stands wrapped with cloth dipped in burnt mobil oil is also very effective in reducing ant attack.

c) Wasp management

- Ensure food availability in the colonies, if necessary artificial feeding may be given.
- Locate nests of wasps and destroy them

- Eliminate alighting board and reduce hive entrance
- Flapping the visitor wasps at the hive entrance or foraging in and around an apiary at early stag e of predation is more effective in disrupting the hunting phase and preventing the predation process from reaching the more destructive phase.
- In the month of July-August, wasp traps using honey, sugar or rotten fruit or meat mixed with small quantity of any contact insecticide can be used as bait for effectively trapping the hunting wasps and reducing the predation of honey bees.
- Lastly, if above measures fail, migrate the honey bee colonies to safer places or search out the

d) Mites management

i) *Acarapis woodi the e*ndoparasitic mite causes Acarine disease in honey bees. It infests 3-4 days old workers. This disease spreads from one colony to another colony by drones or by robbing. This mite blocks the tracheal passage resulting in death due to respiratory failure.

- Wings become 'K' shaped
- The adult bees cannot fly and drop down on top cover or by the side of hive. Sometime they congregate near the hive entrance. The abdomen of bees swollen and appear shiny.
- Using Folbex (Fluvalinate) strip inside hive for 30 min under air tight condition reduces the mite attack. This is used at an interval of 7 days for 6-7 times. Nuvan can be used very carefully. Spray solution of Nuvan is prepared by using 0.5 ml in 10 liters of water. 10 ml of this solution is sprayed on news paper or blotting paper of size 20x6 cm. This paper is inserted in perforated polythene & the polythene is hanged in the hive. Care should be taken that the bees should not come in contact with the insecticide.

ii) *Varroa jacobsoni* the ectoparasitic mite usually enters the brood and infest the 4-5 days old brood. Drone broods are more vulnerable. It can cause death of infested brood.

- Emerged adult workers are weakened by this mite that remains clinging the bees at inter segmental membrane and are reddish brown in colour with pin head size.
- Dead larval and pupal broods are seen near hive entrance and underdeveloped worker and drones too congregate at hive entrance.

- Broods occur in irregular pattern.
- Cotton or cloth soaked with 85 % Formic acid @ 5 ml kept inside a perforated match box or inside a glass vial which is placed inside the hive. This is applied continuously for 21 days.
- Destruction of sealed drone brood.
- Application of finely sieved wheat flour on the bees during evening hours @ 10-15 g at an interval of 4 days for 4 times reduces *Varroa* attack.
- Thymol + Formic acid at 1:1 and garlic paste at 80 g/dose found effective against *Varroa* infestation in *A. mellifera.*

iii) *Tropilaelaps clareae* - Ectoparasitic mite

- This mite clings to the bees and appears like a white spot. Usually it spreads though *Apis dorsata.* The sealed broods are cut open due to their infestation resulting in reduction of population.
- Irregular brood pattern is seen.
- Infested bees have disjointed wings
- Fumigation with Formic acid as in case of *Varroa* mite

e) Viral disease

- Thai sac brood is a viral disease infecting *Apis cerana* colonies and it created havoc in India during 1991. The virus is transmitted through mites.
- The brood get rotten and become sac like.
- Infection is more during spring season.
- Infected combs have punctured cappings with holes.
- Infected larvae die prior to pupation.
- Colour of dead brood become straw coloured, then brown and finally greyish black or black.
- The head end of larva becomes black.
- Destruction of infected combs by burning reduces the infection
- Keeping the colony strong and checking on robbing and absconding.
- Avoid capturing and hiving swarms of unknown source.
- Avoid exchange of hive parts for other management
- Disinfect beekeeping equipments by soaking in a solution of soap-formalin

for 24 hours or by immersing 3-4 times in $KMnO_4$ solution (@50 g in 1 liter water) and then 20-30 min in boiled water.

- Colonies showing tolerance or some resistance should be multiplied and screened in successive generations

f) Protozoan disease

- Nosema is caused by a spore forming protozoan, *Nosema apis*.
- The infected bees become dysenteric with distended and swollen abdomen.
- Affected bees have disjointed wings and are found crawling in front of the hive.
- Large number of spores can be observed in the midgut contents of infected bees.
- The disease is particularly severe during winter and spring.
- Sterilization of infected combs and equipments by fumigation with 80% acetic acid for one week.
- Fumagillin @ 10 mg per liter of sugar syrup is best known chemical for the control of this disease.
- Feeding Metron syrup (Metranidazole) with 50% honey solution (1 : 1) reduces the disease.

12. Apiary size

- Ordinarily, a beginner can start a small scale apiary (Apiary with five colonies) or a medium scale apiary with 10 (Ten) colonies and subsequently go on adding colonies based on the bee flora available and the specific requirement.
- In order to earn an amount of Rs.1.0 Lakh /Annum, the bee keeper should maintain minimum 70 - 80 colonies of the *A.c.indica.*

13. Man day required

- Unlike other enterprises, bee keeping does not require continuous labour and in the other hand certain operations are there which can't be left unattended due to unavailability of labourer. Normally, a man day /week/ ten hives is enough to take up the routine operations easily.

14. Production of honey and bee colony

- The honey from the hive is the main attraction of most of the new bee keepers. The surplus honey stored in the hive by the honey bees during the honey flow season is normally extracted by the bee keepers using the honey extractor.

- Honey should be extracted when 70-80% combs in the super chamber is filled with honey and are capped.
- The extraction should be done preferably in the afternoon hour and in a protected area following scientific instructions. Honey should not be kept exposed to air for a longer period.
- The honey extracted should be processed and stored in glass bottles.
- Apart from honey, production of colonies for sale is another important attraction of the experienced bee keeper. Production of bee colonies is equally or even more remunerative than the price obtained from honey production.

15. Value of direct benefit

- Net income from 100 bee colonies is minimum Rs. 1.70 lakhs /annum which can be more if proper scientific practices are adopted.
- It helps in rural development and promotes small scale industry in village.
- Unemployed youth can start this business with minimal funds. This enterprise provides good employment opportunity to produce bee colonies, hives and to manufacture appliances / equipments.

16. Value of indirect benefit

The report from some western countries indicate that, the annual income through augmentation of crop yield due to bee pollination is 15-20 times more than that of the values of honey and beeswax. Such case cannot be different for India. Thus the value accrued owing to enhanced crop yield due to cross pollination is quite high. Besides the socio economic conditions of the farmers can be increased by renting the honey extractor, renting the bee colony for crop pollination, sharing charged intellectual skill for promotion of bee keeping.

IFS, Honey bees and insecticides

Crop protection system is an inevitable component of crop production system. Realization of maximum return from a production system is made possible through compatible blending of the protection system keeping in view the associated environment and socioeconomic situations.IFS being a multi component production system must depend upon the protection system wherein the chance of compatibility of the later is a challenge. Conventional insecticides such as chlorinated hydrocarbons, organophosphates, carbamates and pyrethroids were successful in controlling insect pests during the last few decades. But the rapidly developing consequences were alarming. So, People have continually sought new solutions for controlling insect pests and the search for new, safe and selective insect control agents is a continuing effort made by the manufacturer and the researchers of different research institutions throughout the world particularly for replacement of available insecticides and develop insecticides with new target site of action which are more ecologically acceptable as a part of integrated pest management (IPM) programs. The neonicotinoid is one such group of insecticides (viz. Imidacloprid, Acetamiprid, Thiomethoxam, Thiachloprid, Chlothianidin and Dinotefuran) that are highly active for foliar application, soil and seed treatment. Neonicotinoid insecticides are successfully applied to control sucking pests in a variety of agricultural crops including pests on oil seed rape and other crops. They have combining systemic properties with relatively low application rates and act agonistically at the postsynaptic nicotinic acetylcholine receptors in the central nervous system of insects and also have particularly strong activity against sucking insects such as aphids, jassids, whiteflies, and thrips. They show good compatibility with all relevant crops with no phytotoxicity. However, they may not only affect pest insects but also non-target organisms such as pollinators. Hazardous effect of neonicotinoids to bees including honey bees (Table -8 and 9), bumble bees and solitary bees have been reported from various corner of the globe. In India neonicotinoid have been widely used and efficacy has been reported against the sucking insect pests while for their uses has been banned in France and Germany. Many pesticides that have been classified as dangerous to the bees are registered under the insecticide Act, 1968, but continue to be used in India. FAO has provided a list of 77 toxic pesticides that are highly dangerous to bees, of which 31 are sold in India. Some other pesticides need to be applied either late in the night or when the bees are not in flight.

Table 8: Toxicity of neonicotinoid to Honey bee, *Apis mellifera*

Sl.No	Insecticides	LD_{50}
1.	Imidacloprid	18 ng
2.	Clothianidin	22 ng
3.	Thiomethoxam	30 ng
4.	Dinotefuron	75 ng
5.	Nytenpyran	130 ng
Cyanosubstituted Neonicotinoid		
6.	Acetamiprid	7.1 µg/bee
7.	Thiacloprid	14.6 µg/bee

The crops under IFS need protection from pest and pollination service of honey bees. Under such circumstances it is essential on part of the farmer, pesticide applicator and the bee keeper to understand the problem of each other to face the challenge judiciously for mutual benefit. The stake holder must know the followings

Causes of bee poisoning

- Insecticide application directly on blooming crops/weeds.
- Drift of insecticides on to blooming crops/ weeds.
- Insecticides dust gathered with pollen and taken back to bee colony.
- Drinking of contaminated water by bees.

Symptoms of bee poisoning

- Larger than normal numbers of dead bees in front of the hives.
- Sudden increase in the aggressiveness of the bees.
- Regurgitation of nectar from the honey stomach.
- Confusion and fighting at hive entrance.
- Immobility or slow movement of the bees.
- Dead brood in the comb or in front of the hive.
- Death of queen in case of severe insecticide poisoning

In order to protect the crops with pesticide with least hazardous effect on honey bees the following steps should be taken up by different stake holders viz pesticide applicators, farmers and bee keeper .

Pesticide applicator

- Use insecticides less toxic to bees as none of the insecticide is fully safe to bees.
- Go for ground application which is less hazardous than aerial application.
- Apply insecticides in late evening.
- Apply insecticide(s) on crop prior to blooming.
- Do not apply insecticides in cool weather.
- Choose least hazardous insecticides formulations. (E.C. formulations are less hazardous).

- Apply insecticides with repellant.
- Keep contact with bee keeper and inform him about his chemical protection schedule.

Farmer

- Use mechanical, cultural and other non chemical methods of insect pest management practices.
- For perennial insect pest preventive method for early season control should be applied where ever possible.
- Understand the bee keeper's problem with insecticide poisoning of bees.
- Understand own problem with insecticide poisoning oh the honey bees.

Bee keeper

- Establish a local bee keeper-farmer-applicator communication network prior to the spray season.
- Post his name, full address and phone number in large letter in each apiary.
- Choose apiary sites away from the intensively sprayed areas.
- Understand the farmer's pest control problem.
- If spraying is inevitable move the hives from the treated area or restrict movement of bees to outside for a day or two.

Table 9: Relative hazards of conventional pesticides to honey bees.

Highly toxic pesticides	Moderately toxic pesticides	Less toxic pesticides
Carbaryl 50% WP	Malathion 50EC	Endosulfan 35 EC*
Carbophenothion 20 EC	Methyl demeton	Menazon 70 DP
Cypermethrin 10 EC	Monocrotophos 40 EC	Phosalone 35 EC
Decamethrin 20 EC	Trichlorfan 50 EC	Pyrethrins,
Dichlorvos 100 EC	Diazinin 20 EC	Rotenone,
Dimethoate 30 EC	Ethyl parathion 46%	Ryania,
DDVP 100 EC	Fenitrothion 100 EC	Sulfur,
Monocrotophos 36 WSC	Fenthion 100 EC	Tetradifon,
Oxydemeton-methyl 25 EC	Metacid 50 EC	Toxaphene,
Parathion	Methyl Parathion 50 EC	Trichlorfon
Phosphamidon 100 EC	Dithane M 45 75 WP	Zineb,
Phorate	Foltaf 80 WP	Ziram
Permethrin 25 EC	Difolitan 50 WP	
Quinalphos 25 EC	Hexacap 50 WP	
Sumithion 50 EC	Bavistin 50 WP	
Thiometon 25 EC		

*Banned

Climate change, decline in pollinators and honey bees

Honey bees are very thermo sensitive and their normal activities is much influenced by the temperature (Table-7).Climate change and change in weather conditions not only affect the honey bees or the pollinators but also affect crop productivity due to its adverse effects on pollinators other than plant phenology. The report of IPCC (International Panel on Climate Change) states that the most plausible effect of climate change on plant-pollinator interaction can be expected due to the result from increase in temperature. Estimates from IPCC indicate that average global surface temperature will further increase by between 1.1 ^{0}C and 6.4 ^{0}C during the 21st century and this increase will be greatest at high latitudes. (IPCC 2007). In such a case the pollinators may either adapt to changed environment, or immigrate to new areas or get extinct. Climate change has also altered the phenological response of plants. In the last two decades, the US and many countries in Europe have been experiencing a decline in both domesticated and wild populations of honeybees. The phenomenon of Colony Collapse Disorder(CCD) got its name in late 2006 when a drastic rise in the number of disappearances of Western honeybee colonies in North America and Belgium (Europe) was witnessed. No reason has yet been discovered about the cause of CCD but most scientists agree that it is probably due to a mix of biotic and abiotic factors. Pathogens such as *Varroa destructor* mites, *Nosema apis* & Israel Acute Paralysis Virus are thought to be the major cause of CCD, as they weaken the bees by disrupting their immune system. Pesticide misuse is another major factor causing the population decline of honeybee populations. Pesticides especially neonicotinoids have also been reported affecting the bee colonies. Many broad-spectrum pesticides which are harmful to fauna other than just agricultural pests have been banned in many countries but continue to be in use in the tropical and subtropical countries. This decline in bee populations with severe consequences for crop yields and economy has been propounded as the 'pollinator crisis'. In India, though the CCD situation has not been experienced but poor performance of honey bees and decline in their population in nature due to loss of habitat owing to urbanization, large scale deforestation, agricultural intensification, monoculture practices, misuse of pesticides, reduction in nesting area and food resources for the pollinators, environmental pollution has been widely realized. In 2007 assessment of the scientific data on the issue, the UN Environment Programme stated: "*Any loss in biodiversity is a matter of public concern, but losses of pollinating insects may be particularly troublesome because of the potential effects on plant reproduction and hence on food supply security.*" Due to loss of entomophily many of our crops experience inadequate pollination resulting in reduced yields, occasional crop failures and poor crop quality. The United Nations says that the world's bees are facing multiple threats and unless something is done to halt their decline, there could be serious long-term consequences for food supplies. Therefore, it is very important to conserve biodiversity for the sake of our sustenance, poverty alleviation, improving health & prosperity of present and future generations and to deal with climate change. Under such situation apiculture in IFS is a mean to support conservation of bees *vis a vis* biodiversity and safe guard our environment.

Conclusion

Agriculture is the backbone of the livelihood security system of million of people in the country and will always remain the most important sector at both national and global level as food is one of the prime requirements for any living organism. Sustainable Agriculture is another concern of the present day scenario where in Integrated Farming System holds a special position. Apiculture is one of the eco friendly approaches for the development of Sustainable agricultural environment. Integrated Farming System has gained a lot of importance in the recent years because it has the ability to evolve suitable strategy for augmenting the income of the farmers. Another main reason for adoption of Integrated Farming System is the need for vertical expansion of land due to decrease in the level of agricultural land area. The Integrating farming in present context relies upon raising various crop and other allied enterprises in an unit area in a coordinated framework in as compactable manner as possible with optimum utilization of all available inputs, reducing the use of off-farm inputs enhancing the use of on-farm inputs, as well as effective waste recycling with an objective for obtaining enhanced crop production and productivity keeping in view the socioeconomic acceptability, environmental safety and sustainability of the farming system. Integrated farming system are often less risky, if managed efficiently and they benefit from synergisms among enterprises. Inclusion of apiculture in IFS is unique in this context. Scenario where other natural resources are already stressed, protecting pollinators would prove beneficial. Wild pollinators can neither inhabit nor meet the high pollinator demand of intensively managed agricultural systems. Such systems depend heavily on commercial beekeeping where honey bee colonies are either purchased or rented by farmers through organized markets. In India, agricultural practices and current policies indicate increasing dependence on commercial beekeeping. Though pollinator crisis is not yet being experienced in India, it is important to consider the possible effects of such a crisis and the imperative it poses on adopting pollinator-friendly policies to avoid possible catastrophes in one end and achieving sustainability in food production on the other.

References

Abrol,D.P. 1997.Bees and Bee keeping in India. Kalyani Publishers,Ludhiana,719 p

Atwal, A.S.(ed.). Insect pollinators of crops. Punjab Agricultural University Press, Ludhiana, 116 p.

Gatoria, G.S, Chhuneja,P.K,Butter, N.S and Singh Jaspal 2007.Advances in the management of Apis mellifera Linnaeus, ICAR sponsored winter school for apiculture scientists from SAUs & ICAR institutes (November 06 to 26 2007) Department of Entomology, Punjab Agricultural University, Ludhiana.

Lakra, R.K. 2006, Honey Production: Tremendous scope for expansion: the Hindu: Survey of Indian Agriculture. pp 216-218.

Satapathy,C.R.and Jena B.C.2009. Bee keeping an avenue for women empowerment *Bee World*, 2nd issue ,July – September,2009.Published by National Bee Board,New Delhi

Satapathy,C.R.2009 .Bee flora and Pollination ,Chapter (Block -2) in Introduction to beekeeping published by IGNOU,New Delhi.pp.1-24.

Satapathy,C.R., Patnaik H.P. and Mohapatra,L.N.2010.Beekeeping for improving livelihood. NAIP-Sustainable Rural Livelihood and Food security to rainfed farmers of Odisha,Technical Bulletin No 5.

Saraswat, B.L. 2011.Beekeeping as a fifth and most important input for overall sustainable development of agriculture. *Souvenir* Biennial group meeting of AICRP on Honey bees and Pollinators held at OUAT, Bhubaneswar from 11th to 13th February, and 2011.pp. 63-68.

Satapathy, C.R.and Padhi, J.2012. Achievements of AICRP on Honeybees and Pollinators, Department of Entomology, OUAT, Bhubaneswar, p. 24.

Thakur, R.K. 2011.Role of bumble bees in crop pollination. *Souvenir* Biennial group meeting of AICRP on Honey bees and Pollinators held at OUAT, Bhubaneswar from 11th to 13th February, 2011. pp. 69-74.

Verma L. R. and Satapathy, C.R. 2012. Honey bee management pollination for sustainable agriculture.Lead paper presented and published in *Souvenir* on the occasion of *National Symposium on Eco-friendly Approaches to Pest Management for Sustainable Agriculture* held at OUAT, Bhubaneswar during 24 – 25th November, 2012. pp 64-69.

World Commission on Environment and Development,1987. Our Common Future, also known as the Brundtland Report.Oxford University Press. pp 383.

Integrated Farming System Practices: Challenges & Opportunities, pp 383-415

New India Publishing Agency, New Delhi, India

15

Post Harvest Management and Value Addition in Sustainable Agriculture

S.K. Dash

The stages through which the commodity passes from harvest till consumption are considered as post harvest stages. Post harvest management is one important component of agriculture and is the key to make more food and nutrition available to people. In integrated farming system, varieties of foods are produced. However, if they are not stored properly and/or processed, there may be both quantitative and qualitative losses. The produce may be attacked by different extraneous organisms including microorganisms, birds, rodents, etc. These spoilage agents may impair the quality of food, degrade the nutritional value and even make the food unfit for consumption. Besides, value addition of the food produce is another area where the overall profitability of the farming system can be improved. Thus proper management of post harvest operations and value addition are very important.

Importance of post harvest management and value addition

It has been estimated that about 10% of the food grains and about 30-40% of the fruits and vegetables are lost every year in a country like India due to improper post harvest management and lack of value addition. New avenues to

---

Department of Agricultural Processing and Food Engineering, College of Agricultural Engineering and Technology, Orissa University of Agriculture & Technology, Bhubaneswar - 751003, Odisha, India increase food production are being explored to meet the demand of growing population of the country. However, prevention of post harvest losses is one basic step that can be taken up to make more food available for the people. In addition, proper post harvest management and food processing can help in the following aspects.

1. Maintenance of the quality of food commodities till consumption
2. Prevention of seasonal market gluts and better use of surplus production, regulation of prices and orderly marketing
3. Availability of food in offseason.
4. Value addition of the commodity and increase in the variety of foods available for consumption
5. Generation of additional employment and income, particularly in rural area

In general, the crops are classified depending on the moisture content and storability as follows.

- *Durables*. The normal storage life (also known as shelf life) of the crops are few months to one or two years. The moisture content is in the range of 8-20%. The food grains as rice, wheat, corn, millets, pulses, oilseeds come under this category.
- *Semi-perishables*. The normal shelf lives of the crops are few weeks to few months. The examples of such commodities are potato, yam, onion, garlic, etc.
- *Perishables*. The normal shelf lives of the crops are few days only. The moisture content is more than 70%. Examples are most fruits and vegetables, meat, fish, milk, etc.

Accordingly, the post harvest management will also differ for the different groups. For instance, the food grains have as such a shelf life of several months and slight delay in the post harvest operations may not cause visible damage to the commodity. But such delay is not permissible for fruits and vegetables. Even there are some fruits which can be stored for 7-10 days in room conditions, whereas some spoil even in a day or two.

Food spoilage agents

The major agents causing the spoilage of food in post harvest stages are the microorganisms, insects, rodents, birds, etc. who consume the food or infect the food to a level that it becomes inedible. However, these agents can attack and spoil the commodity only if the environment surrounding the food is conducive to them. A good example is that the dried grains can be stored for longer period than the wet grains. The temperature of storage is also important and it is a common fact that the shelf life (storage life) of fruits and vegetables can be increased in refrigerators. In addition, the chemical and biochemical factors, e.g. the respiration of the commodity (the fresh fruits and vegetables are living bodies and they respire and the rate of respiration directly affects their storage life), other biochemical reactions occurring in the food also affect the storage life and spoilage. Thus, once the potential spoilage agents are identified, the protection of food becomes easier.

The major losses to the grains are due to birds, rodents and insects. They consume the stored produce and damage the crops by infestation (insect infestation) and by their excreta. The fruits and vegetables are mostly spoiled by microorganisms and by drying and shriveling. Thus, post harvest operations and management are to be planned considering the nature of commodity and the type of end use.

General post harvest operations

The basic post harvest operations are aimed at three broad aspects.

1. Conversion of raw materials to edible form
2. Processing and value addition
3. Storage and preservation

As the different types of commodities differ in their physical, chemical and textural characteristics, their post harvest operations also differ. Their end forms and uses also differ, which require the application of different types of food processing and value addition methods.

The processing operations can also be broadly classified as primary, secondary and tertiary processing. When agricultural commodities are processed to improve their quality without changing the form by such operations as cleaning, grading and drying, it is known as primary processing. Primary processing can also include the operations to convert the inedible raw material to edible form like conversion of paddy to rice, or wheat to flour, etc. However, secondary processing basically aims at value addition. The conversion of rice to rice flour and puffed rice, dhal to besan, etc. are such operations. Similarly preparation of extruded products from rice flour and dhal powder, corn flour, etc. are examples of tertiary processing operations.

Milling and value addition of food grains

As discussed earlier, the post harvest operations differ for different types of commodities. However, there are certain unit operations which are carried out for all types of grains (though the methods and machines may vary) to add value to them. The unit operations are cleaning, sorting and grading, drying and conditioning. For example, cleaning and drying are essential before storage of any grain. Sorted and graded (and also packed) grains command a better price in the market. These unit operations are briefly explained below.

Cleaning

The basic objective of cleaning is to improve the grade of the commodity, it helps to protect the milling machines and save in storage space. Further, the extraneous materials present in grain are always at higher moisture content than the average moisture content of the grain and help growth of insects and microorganisms. So the shelf life of the grain is improved when these extraneous materials are removed (Dash *et. al.,* 2012).

The most common equipment used for cleaning is the screen or a set of screens. The screens are of different types and can be changed to meet the varied needs of different types of grains and for different varieties. Some important points to be remembered during cleaning or for the selection of cleaning equipment are as follows.

- Two screens can be assembled one above the other to separate the large size impurities (on upper screen), clean grain (on the lower screen) and the small impurities below the lower screen.
- The screens can be operated by electricity or diesel operated motors. Small size screen assemblies can be suspended by hangers and can be operated by hand.
- In commercial cleaners, an air blower (or aspirator) is also fitted with the screens to blow away the lighter materials like chaff, straw, etc.
- In some large grain processing units, some advanced types of grain cleaners/ graders are used such as the magnetic separators (which can separate iron nails, nuts, bolts and other magnetic materials) and specific gravity separators or destoners (to separate stones of same size as the grains, which are not separated by screens).

Drying

The dried grains store better than wet grains, primarily because the microorganisms can not easily attack the dried grains. Depending on the type of grain, there is a safe moisture level for storage. However, the grains are usually harvested at a higher moisture content than that. For example, the paddy grain is harvested at a moisture content of 20-24% and its safe storage moisture content is about 12-13.5%. For most cereals and pulse grains like wheat, maize, pigeon pea, etc. the safe storage moisture level is lower than 10% and that for oilseeds is lower than 8%. If the grain is stored for future use as seed, then the safe storage moisture content will be about 1-2% lower than the moisture content that is used for storage as food. The grain should be dried as quickly as possible to reduce losses.

When the grain is kept in the field/drying yard for some days after harvest it looses some moisture, and it is known as natural drying. The drying process is very slow. The drying rate is increased if grain is dried under sun in thin layers (say 1-3 cm thick layers) and the process is sun drying. The sun drying is advantageous in that no extra energy is required for drying and there is no drying cost. But it is dependent on weather conditions and there are losses due to birds, rodents, etc. But if the grain is not properly stirred (mixed) during drying, there is uneven drying and cracks in the grains. This is important when whole grains are required for our consumption such as rice. The cracks can not be seen in the naked eye, but during milling they cause breakage of the grain. Tempering of the grain in between drying process (giving a rest for about 2-3 hours during the drying process by gathering the grain and covering that under gunny bag) reduces breakage during milling.

For large capacity drying and for a good control on the drying process, mechanical dryers can be used. Both batch and continuous types of mechanical dryers of different capacities are available. These are generally operated by electricity. The quality of grain obtained by mechanical drying is better than sun dried materials. Now a day, solar dryers are also available for the purpose. They give a better quality product than open sun drying and can be of benefit during bad weather.

Milling of rice

Rice is an important crop in the integrated farming system and the provision to mill rice in the farm itself can add income to the enterprise. Earlier, the rice was milled mostly by manual hand pounders, chakkis or leg pounders; now most of the rice is milled in hullers and modern rice mills.

The hullers (Fig. 1) are popular rice milling machines and are very well suited for small amounts of paddy. However, the hullers have two major disadvantages, as follows.

1. As a huge pressure is applied on the grain for taking out the husk and bran layers in a single operation, there is a lot of breakage of rice. The breakage becomes very high for raw rice and for long grain varieties of rice.

2. In the huller, the husk and bran come out through a single outlet as a mixed powder form, which is used (and can be used only) as cattle feed. But the husk and bran, if obtained separately, can have many beneficial uses. In particular, the rice bran is a good source of oil (100 kg rice bran can give about 20-25 kg oil) and the oil can be used both for cooking and other purposes. The husk is a very good fuel and is commonly used in the parboiling and drying operations in rice mills. It has also many agricultural and industrial uses.

In the modern rice mill, the dehusking and bran removal are performed in two stages so that the husk and bran are obtained separately. The equipment used are known as sheller (rubber roll sheller) and polisher, respectively. In addition, to improve the performance of the rubber roll sheller and polisher and also to get good quality rice and bran, some other machines as cleaners, husk aspirators, etc. are also fitted in a modern rice mill. The modern rice mills are available in different capacities. Single pass modern rice mills with metallic polishers are also available with capacities as low as 500 kg paddy per hour (Fig. 2).

The modern rice mill gives higher output of rice. But the husk and bran are obtained separately, which can not be used as cattle feed. Besides, proper facility for use of husk and bran are also not available in all areas. These are some reasons why the hullers are still popular in many areas. However, considering the extra yield of rice and potential benefits from the by-products, the modern rice mills can give higher income to compensate the loss caused by loosing the cattle feed (Dash *et. al.*, 2012)

Fig. 1: A huller

Fig. 2: A single pass modern rice mill

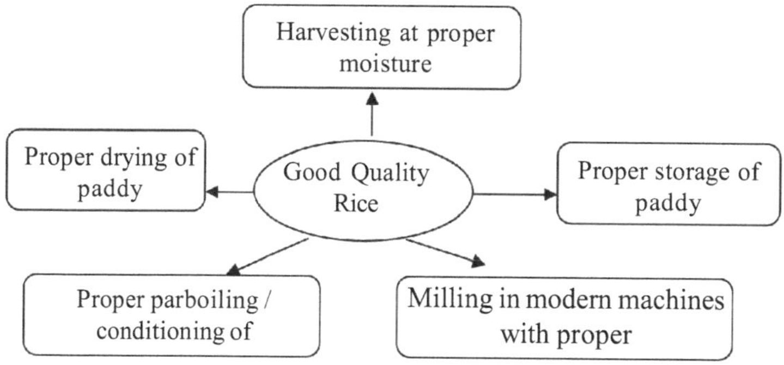

Fig. 3: The five finger rules for getting higher yield and good quality rice (Adapted from Dash, et al., 2012)

It is not only the milling machine, but also the pre and post-harvest operations, that affect the output from rice milling. A modern rice mill alone can not give the best recovery of rice. In fact, the series of pre- and post-harvest operations also significantly affect the output of rice from a mill. The five finger rules (related to post harvest activities) for getting the best output from rice milling can be as shown in Fig. 3.

Conditioning/ parboiling of rice

Parboiling is a hydrothermal treatment that improves the hardness of rice and nutritional quality. As the hardness is increased, the breakage during milling is reduced. Parboiled rice also stores better as it is less attacked by the insects due to increased hardness. The nutritional quality is also improved because the steam and water applied during conditioning carry most of vitamins and minerals from outer layers to inner part of the grain and distribute them there. Subsequently the loss of these nutrients during milling and cooking is reduced. Of course, many people do not like parboiled rice due to the slight change in colour and as it takes more time to cook.

There are 3 basic steps in parboiling, namely, soaking, steaming and drying. The steaming should be uniform and different designs of steaming tanks are available. The CFTRI (Central Food Technological Research Institute, Mysore, India) design is a popular model, where the treatment is almost uniform and the quality of rice is highly acceptable. Small parboiling tanks for 1 bag of paddy per batch have also been developed by the CRRI, Cuttack, which give uniform conditioning and are suitable for small enterprises/ household uses. Normally it takes 4-6 hours per batch for the parboiling process (Sahay and Singh, 1994).

Value added products from rice

Preparation of rice products as flaked rice, puffed rice, popped rice, rice powder, rice cakes, extruded products, etc. from rice can add to the income of the farming system. These types of products can be called secondary rice products.

Dhal milling

Different pulses like pigeon pea (arhar), blackgram (urad), lentil (masoor), greengram (moong) can be grown in the integrated farming systems. The common products from pulses are polished whole pulse, splitted dal, roasted pulses, fried products, food mixes (sattu), *badi, papad* and ready-to-eat extruded products.

The basic dhal milling process involves separation of the outer husk layer and splitting of dhal. However, to reduce the losses and breakage during these operations proper conditioning of dhal is very important. The conditioning is carried out after cleaning the grains and is done by first pitting or scratching the grains, mixing with oil (200-600 g/q) and heaping for few hours, then sun drying for 2-4 days with 5-7 % water application in between and then tempering. As the husk in the pulses like red gram, green gram, black gram, etc. are held more tightly with the kernels as compared to pulses like bengal gram, lentil, peas, the conditioning takes more time for the former group.

The basic unit operations in dal milling after conditioning are dehusking, husk separation, splitting, grading and polishing. Small scale dhal mills of about 50-200 kg/hour capacities are available which can meet the need of small farming systems (Fig. 4 and 5). A Small Cleaner cum grader used in commercial dal mills in shown in Fig. 6.

Fig. 4: CIAE dhal mill

Fig. 5: PKV mini dhal mill

Fig. 6: A cleaner cum grader used in the dhal mill

Pulse based products

Bengal gram and peas are consumed in roasted form. The unit operations involved in roasting are cleaning, grading, moistening, roasting, puffing, splitting, husk separation, grading and packaging. Fried pulses can be obtained after soaking and frying. Besan making consists of size reduction (by pulveriser/ attrition mill), flour sifting followed by packaging. It is used for preparation of pulse based sweet dishes and snacks. Badi and papad making units can also be integrated into the system.

Milling of oilseeds

The farming systems having a section to grow oilseed crops can add to their income by incorporating the provision of processing of oilseeds and selling oil and other value added products rather than selling the raw oilseeds.

Oilseeds contain on an average of 20-45% oil and hence are mostly used for oil extraction. However, some oilseeds are consumed in kernel form. Groundnut, sesamum and sunflower have great potential for food uses. Kernels of groundnut are used after roasting and salting as snack food. Kernels of white variety of sesame are preferred for confectionery.

Table 1: Different components of some oil seeds

Oil seed	Husk/Hull	Kernel	Oil
Groundnut	25	75	33
Mustard	18-20	80-82	33-41
Sesamum	14-18	82-86	40-49
Sunflower	30-40	60-70	37-42

Oil is obtained from the oilseeds normally by pressing (or mechanical expression) or by solvent extraction. The commercial solvent extraction plants involve very huge capital investment and hence are not suggested for small enterprises.

Whatever may be the method of extraction, all oilseeds require some form of conditioning prior to extraction. The method of conditioning and extracting the oil greatly depends on the type of oilseed. The different unit operations that are carried out in oil mill include preparation of raw material by drying, cleaning, dehulling (as in groundnut), grinding or flaking, heating or conditioning and then pressing (or solvent extraction) to extract the oil. The oil is then clarified and filtered. The oil cake obtained after pressing is very nutritious and can be used for animal feed or as an ingredient in other food products.

During pressing in the expeller a lot of frictional heat is generated which tends to increase the temperature of oil. Care should be taken that undue heating of oil to a very high temperature is avoided. Improper processing may cause dark colour and poor protein quality of oil.

Extracted oil, if properly processed, has a shelf life of up to 6 months. During conditioning and processing, the raw material (oil seeds or nuts) is heated, which inhibits the activity of enzymes and micro-organisms that may cause rancidity. For improving the storage life of oil, it may also be slightly heated after extraction to remove as much water as possible; it also improves the quality of the oil. Oils should be stored in a cool, dry place away from direct sunlight and heated to prevent chemical changes that can lead to rancidity.

The equipment needed to set up a small or medium scale oil extraction enterprise falls into three main categories:

a) Pre-extraction equipment: Dehullers, Seed/ kernel crackers, Roasters

b) Extraction equipment: Manual presses, Ghanis, Screw press/oil Expellers

c) Equipment for primary refining of the oil: Filters, Settling tanks.

The oil expeller and filter used in commercial oil mills are shown in Fig. 7 and 8. As mentioned before, the solvent extraction plants require a huge investment and may not be recommended for small enterprises, and hence are not mentioned here.

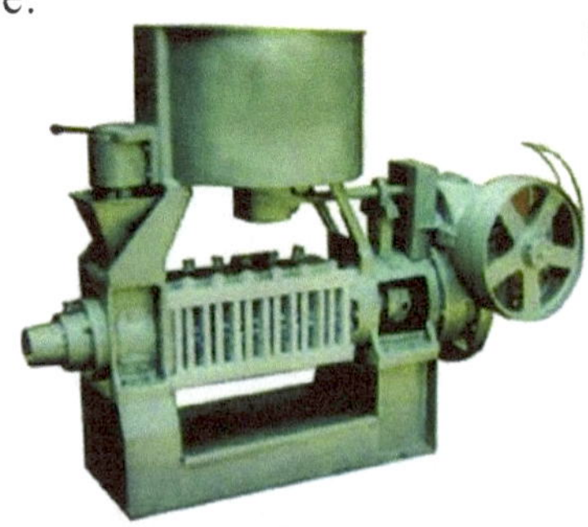

Fig. 7: Oil expeller

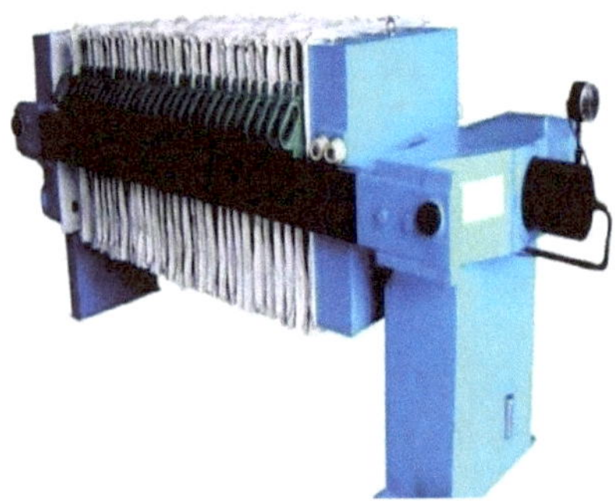

Fig. 8: Oil filter

The oilcake is mostly used for poultry and cattle feed. It can also be modified for human consumption to the form of flour, protein concentrate and textured vegetable proteins. In screw press extraction, pressure is kept as low as possible to avoid high frictional temperature and resultant damage to both oil and cake. Cereal based extruded snack foods, food mixes (sattu), sweets like laddus, etc. can also be prepared from oilseeds.

Some other pulse-oil seed based processing enterprises are preparation of extruded products, weaning foods, snack foods like mixture and sweets, roasted/puffed products, feed from the by-products of dhal and oil mills, etc. A seed processing unit can also be incorporated, which can not only meet the seed requirements of the farm for the next season, but also add to the overall income of the farm.

Processing and value addition of maize

Maize is considered a promising option for diversifying agriculture in upland areas of India. Some important aspects to be considered to minimise post harvest losses in maize are as follows.

- Harvesting the kernels with 25 to 30 percent moisture content.
- Drying the cobs immediately before threshing/shelling. Use of proper machines for threshing and winnowing to reduce the losses.
- Cleaning the grains to remove the admixtures to improve the grade of product and protect the milling machines
- Shelling the maize with proper machine to avoid physical damage to kernels that normally happens in traditional beating methods. A damaged kernel is more susceptible to pests and moulds and cause damage to the germs.
- Drying the kernels up to safe moisture content of 12% before storage.
- Grading to classify the product to different grades to obtain better price.
- Use of proper storage method and equipment/ device and good packages to protect the grain.

Milling of maize for transforming into food and industrial products is done by two methods, viz. dry milling and wet milling. In the dry milling method, the maize is ground in a hammer mill. This meal is suitable for many rural applications and often considered inferior by the trade, and does not command good price. So in improved processes, the germ and hull are removed from maize before milling. The process involves first adding some moisture to the maize grain (to make it to about 21% moisture content) and then tempering the grain for some time. Thereafter the bran and germ are separated using a plate and roller mill and a degerminator. The meal is then shifted by a vibrating screen and subsequently after drying the endosperm, the milling is carried out by different types of size reduction machines as hammer mill, roller mill, attrition mill, etc..

The other method, known as wet milling, is used in the industries for obtaining the endosperm and germ separately so as to extract starch from the grain and oil from corn germ.

Different value added products such as flour, semolina, grits, corn starch, glucose, gluten meal, popped corn, corn flake, extruded product, bakery product, corn oil and feed can also be prepared from maize.

The preparation of corn flakes involves making the corn kernels into flaking size 'grits'. The corn grits are cooked in steam pressure cookers, at temperatures exceeding 100°C. This cooking process lasts for an hour that softens the hard grits. During cooking, the steam increases the water content in the batch to about 30-35%. The grits are then dried hot-air driers and are pressed/ squeezed using rollers. The flakes are then tumble toasted in huge cylindrical ovens. The air in the ovens is heated by 600°C gas flames. The drum is slightly inclined and the flakes while moving from one end to the other end of the drum due to inclination and rotation are exposed to the hot air for a short period. The flakes can then be coated with chocolate and sprayed with flavours, minerals, etc. Malted barley is also added to enhance the flavour of the corn flakes. The corn flakes are packed in air tight bags. The common equipment used in processing and value addition of maize are shown in Fig. 9-12.

Fig. 9: Maize degermer/dehuller

Fig. 10: Hammer mill

Fig. 11: Roller mill for flaking

Fig. 12: Twin screw extruder

Preparation of popcorn is also another interesting enterprise which can be done in a small scale. Properly conditioned corn can be popped in a suitable device by suddenly exposing the corn grains to high temperature and/ or pressure.

Processing and value addition of ragi

Finger millet (commonly known as ragi) is an important food crop, which can be fitted into our normal cropping system. It has a lot of nutritional and therapeutical advantages and hence, it is becoming very popular these days. Properly processed and value added ragi products have a good market demand.

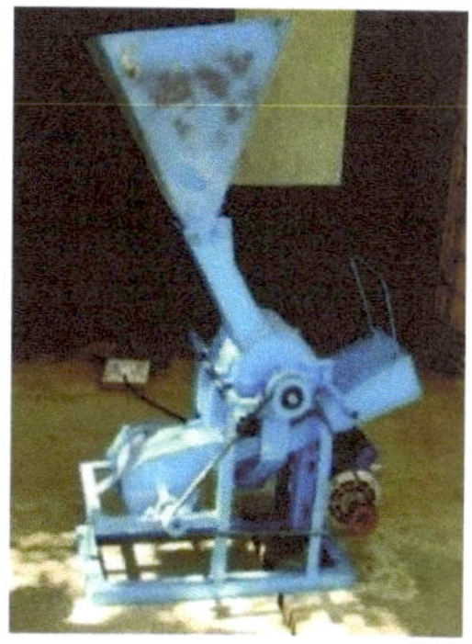

Fig. 13: Vivek thresher for pearling of ragi

The post harvest handling processes of ragi are threshing and pearling (removal of the upper thin and smooth coating called glumes of ragi millets). The present practice of threshing is to beat the panicles manually over a hard surface. Pearling of ragi is conventionally done by hand and foot pounding. The removal of grains is usually done by pearling followed by winnowing or sieving. However, pearlers that are specifically suitable for threshing and pearling of ragi are now available. The Vivek thresher developed by VPKAS, Almora for ragi has a threshing and pearling capacity of 36 and 44 kg/h, respectively (Fig. 13). The Pearled negi grains obtained after threshing are shown in Fig. 14.

The common hammer mills and attrition mills are used for converting the ragi to powder form. Ragi flour can be substituted for cereals and pulses and utilised in products like baked, roasted, steamed, fried, boiled and fermented products at a lower cost with higher nutritional value.

Fig. 14: Pearled grains from the ragi thresher

Ragi malt. The ragi has high malting power and with a type of starch that is more resistant to hydrolysis, it can be used for producing a variety of nutritionally designed food for infants to old. Malted grain is used to make weaning food, instant mixes, pharmaceutical products, malted shakes, confections and flavored beverages and some baked goods. Malted ragi flour is called 'ragi malt' and used in the preparation of milk beverages.

For this, clean ragi grain is soaked overnight in adequate amount of water at room temperature (say about 25-30°C). The water is drained in the morning and the grain is spread on aluminium tray (with a bed thickness of less than 0.8 cm), covered with moistened muslin cloth and germinated for 48 h. During germination, the millets are gently mixed in order to aerate and prevent from matting; sprayed with water and covered with wetted muslin cloth twice a day. The germinated millet (green malt) is dried in a mechanical dryer at 50±2°C for 10 to 15 h (to the final moisture content of 9±1%) and the rootlets are removed by rubbing and winnowing. The malted grain can be either used directly or further milled in attrition or hammer mill to semolina or flour. The malted ragi flour can be used along with germinated green gram flour to formulate a high calorie dense weaning food having excellent nutritional qualities.

Ragi can also be used for preparation of baked products and multigrain flour. Ragi cakes, cookies are the examples. The cakes supplemented with 50% malted ragi flour has excellent sensory and nutritional properties. The other grains included in multigrain foods include corn, oats, soybean, barley, buckwheat, wheat, flax and different millets.

Extruded products

A food extruder is a versatile machine capable of performing operations like mixing, kneading, shearing, heating (cooking), shaping and forming, etc. Extrusion cooking combines the heating of food products with the act of extrusion to create a cooked and shaped food product. The typical food products that can be made by extrusion technology include pasta, ready-to-eat (RTE) cereals, snack products, pet foods, confectionary foods, precooked and modified starches, beverage bases, texturized proteins : meat analog, fish paste (surimi), etc.

Pasta production, e.g. macaroni and spaghetti relies mainly on the process of cold extrusion. Low temperatures and pressure are needed to keep the pasta from cooking. However, high temperatures and pressures are necessary to produce the snacks. The starch has to be gelatinised so that the product expands when it moves from high pressure (within the extruder) to a low pressure (outside the extruder) and its expanded shape is retained. Different spices and condiments are used to improve the flavour and taste of products. Nevertheless it is a promising area to add value to the raw materials prepared in a farming enterprise.

Table 2. Value added products / processed products from different food grains

Sl. No	Crop	Product range
1.	Rice	Puffed rice, popped rice, rice flour, dosa and idli mix
2.	Wheat	Flakes, puffed wheat, bread, cakes, extruded products, spaghetti, vermicelli, biscuits, malt, etc.
3.	Maize	Puffed corn, flakes, corn oil, starch, corn flour, semolina, corn grits,
4.	Ragi	malt, flour, weaning foods
5.	Pulses	Dhal, weaning food, ready to serve snack foods, papad, besan, ladoos, chikkis, etc.
6.	Oil seeds	Edible oil, Oil cake for cattle feed, chikkies, shell for fuel

Storage of food grains

Proper storage of food grains is very important to reduce losses and to maintain their quality. The crop has to be protected from the different spoilage agents as microorganisms, birds, rodents, insects, etc. The temperature and moisture content of the grain and storage environment greatly affect the storability of the grain during storage. High relative humidity and high moisture contents and moderate temperatures (say 25-37°C) of stored produce are favourable for the development of pest organisms. Thus, the quality of the stored produce can be maintained through necessary steps like drying of the produce, good storage hygiene, cooling the grain by controlled ventilation and pest control.

Different types of traditional storage devices made up of locally available materials as straw, wood, mud, etc. are used for storage of food grains. However, they can not protect the grain from the storage pests and from the influence of environment. Modern storage devices like the metal bins and silos and well managed godowns are recommended for keeping the grain for longer period.

The Pusa bin is a suitable structure which can be used for storage of grains in high humidity areas. It has a LDPE (Low density polythene sheet) covering the grain in the brick chamber. The bin can be constructed adjacent to the walls of a room and can store even up to 10 MTs grain. The cover and plinth (CAP) device can be used for short term storage of grains during the harvest season.

The following should be kept in mind during storage of food grains.

- A clean, dry and cool store together with clean and dry grain are the first and most important steps for successful storage of grain.
- Before storage, the grains should be cleaned and graded. Impurities in the grain including straw, weed seeds and dirt, etc. not only decrease the value of food-grains, but also accelerate the deterioration during storage.

- No food-grain with moisture content higher than the safe acceptable level should be kept in storage.
- The grains should be spread over plastic sheets or cemented floor while drying, otherwise it will pick up moisture from the ground. The grains should be kept cool and dry between the time of harvest and storage.
- Preventive and curative measures are to be taken as and when necessary.

Post harvest management of fruits and vegetables

A wide variety of horticultural crops under the categories of fruits, vegetables, spices, flowers, plantation crops and medicinal plants are incorporated in to the integrated farming system. It is worthwhile to note that most of these are perishable commodities and have a shorter shelf life as compared to the food grains. Thus, in most of the cases these commodities are sold out immediately after harvest. However, proper post harvest management, storage and value addition of these commodities at farm level can substantially improve the cost-benefit ratio of the farming enterprise. The other benefits of proper post harvest management have been discussed earlier.

There are two broad types of activities in post harvest management of horticultural crops. First, the commodities have to be stored in such a way that they remain fresh-like for longer period without much loss in nutritive value, and the second, the form of the commodity may be changed such as the preparation of pickles, jams, jellies, etc. Beyond the basic objectives of storage and preservation, preparation of a variety of products as potato chips, canned foods, dehydrated mushroom, etc. constitute an important area of food processing/ post harvest management.

Important considerations for value addition and processing of horticultural crops

It is very important to note that all varieties of fruits/ vegetables are not suitable for consumption in fresh or raw form, nor all varieties are suitable for processing. Thus, the availability of sufficient quantity of processable variety of the commodity must be assured before deciding on the processing and value addition aspects. Similarly the stage of harvesting, i.e. the maturity of the crop at the time of harvest is very important. For most of the processing operations as canning, drying, etc. the fruits should be harvested at firm ripe stage. Late harvested fruits can be used for preparation of juices/ syrups, etc. The fruits should be harvested with utmost care. Improper harvesting methods may cause external damage as well as tissue damage, which not only reduce the cost of the product, but also tend to accelerate the spoilage. The shelf life of the commodity is also affected by many pre and post harvest factors as the part of the crop selected, condition of food at harvest, temperature of harvest, storage, distribution and retail display, etc.

The basic post harvest operations for fruits and vegetables to be sold in fresh form are pre-cooling, cleaning/washing, sorting and grading, transportation and storage (Verma and Joshi, 2000). Different types of equipment are available for each of these unit operations to make the work easier and to do more work in less time (Fig. 15-16). Precooling helps to remove the field heat and extends the shelf life of commodity. Cleaning and washing, in addition to removing the dirt and undesirable materials, also reduce the microbial load and thus offer extended shelf life. Sorting and grading of the commodity help in removing the damaged as well as inferior quality crops and classifying the crop into different grades. Before the processing operations also, it is very important to sort the commodities on the basis of different properties as they directly affect the processing operations and quality of final products. Preliminary sorting can be done in the field itself to remove the damaged ones, and the final sorting is done in the processing/ packing centres.

The time of storage as well as transportation for the commodities to be sold in fresh from should be as less as possible. The temperature should also be low. Refrigerated vans can help maintain the quality of the produce during transportation. If such facilities are not available, then natural air cooling, evaporative cooling or ice cooling, etc. can be adopted to maintain a low temperature during transit. Handling of the commodities with utmost care to avoid any type of impact and compression on the commodity is desirable. The use of plastic crates than gunny bags can help reduce the compression load for many commodities during storage and transportation.

Fig. 15: Vegetable Washer

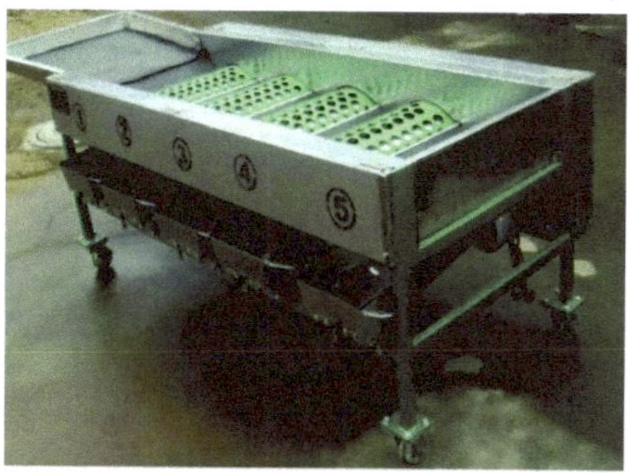

Fig. 16: Size sorter with drums

Storage and packaging of fruits and vegetables

During storage, the fruits and vegetables have the natural tendency to deteriorate as they are living bodies. Many factors affect the process of deterioration, out of which the breakdown of produce by different chemical and biochemical factors, respiration of the commodity, enzyme action, activity of microorganisms, insects, etc. are important. Controlling the temperature and relative humidity of the storage environment can greatly reduce the damage by these biochemical processes and extraneous factors.

Refrigerated storage

Refrigerated storage is also known as chilled or cold storage, in which normally a temperature of -1° to 8°C and about 80-95% relative humidity (RH) are maintained. Cold storage helps in reducing the rate of biochemical and microbiological changes, and hence extends the shelf life of fresh as well as processed foods. Chilling is often used in combination with other methods as fermentation or pasteurisation to improve the shelf life of mildly processed foods (Srivastava and Kumar, 1994).

The recommended storage temperatures and relative humidities vary with the type of commodity. Storage at temperatures above these recommended values shortens the life and affects the nutritional value. Similarly, some fruits and vegetables undergo some undesirable changes when the temperature is reduced below a specific level, which is known as chilling injury. The examples are the black colouration of banana or lemon when stored in the refrigerator. The browning of inside tissues and external surface, failure to ripen, blemished skin are some common symptoms of chilling injury. Chilling injury occurs to banana at less than 12-13°C, lemons at less than 14°C, mangoes at less than 10-13°C and melons, pineapples and tomatoes at less than 7-10°C.

The recommended temperature, relative humidity and expected storage lives of some fruits and vegetables are given in Table 3 (Thompson, 1996).

Table 3: Optimum storage conditions for some common fruits and vegetables

Food	Temperature, °C	Relative humidity (%)	Storage life (days)
Banana	11-15.5	85-95	7-10
Bean (snap)	7	90-95	7-10
Brocccoli	0	95	10-14
Carrot	0	98-100	28-42
Cherry	-1	90-95	14-20
Cucumber	10-15	90-95	10-14
Egg plant (Brinjal)	7-10	90-95	7-10
Lemon	10-14	85-90	30-180
Lime	9-10	85-90	40-140
Lettuce	0-1	95-100	14-20
Mushroom	0	90	3-4

Contd.

Food	Temperature, °C	Relative humidity (%)	Storage life (days)
Potato	3-10	90-95	150-240
Spinach	0	95	10-14
Strawberry	-0.5-0	90-95	5-7
Tomato	4-10	85-90	4-7
Watermelon	4-10	80-90	14-20

Cold stores can be constructed of different capacities with matching refrigeration equipment. Small size cold rooms are also available to meet the requirements of small producers/ traders and retail stores (Fig. 17). Cold rooms can also be self-constructed, purchased as prefabricated units (new or used). However, the advice of experts should be taken in this regard.

Potatoes for processing should not be stored at very low temperatures as that will help production of sugars which darken when heated during processing. When loading commodities, particularly potatoes into bulk storage, even distribution of the produce is important for proper ventilation.

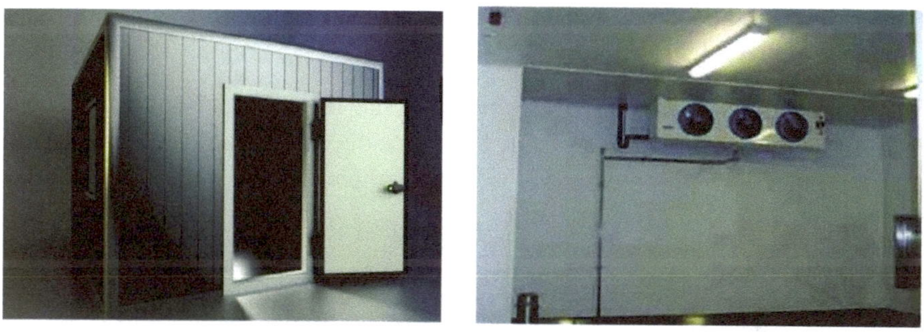

Fig. 17: Small cold room (a) outer view (b) inner view

Ventilated storage

For bulk storage of onion or garlic, ventilated storage systems are beneficial. In such systems, the air can easily flow into all parts of the storage chamber and the heat of respiration is taken out. If produce is stored in cartons or bins, stacks must allow free movement of air. Rows of containers should be stacked parallel to the direction of air flow and be spaced six to seven inches apart. An adequate air supply must be provided at the bottom of each row and containers should also allow air flow into them.

Low cost ventilated onion storage structures can be made of bamboo/ wire mesh, etc. (Fig. 18). It is also important to allow some curing of the crop before handling and storage. Curing helps to dry out the external layers and prevents further water loss during storage.

Fig. 18: Low cost onion storage structures
Source: www. kvkdelhi.org; www.msamb.com

Some important points to be remembered for keeping of fresh commodities in cold stores are as follows.

1. Only good quality produce should be stored. They should be sorted out prior to keeping the produce in storage.
2. The stored produce should also be inspected at regular intervals and the spoilt and infected produce, if any, should be immediately removed.
3. The containers should allow air flow into them. They should be strong enough to withstand stacking. The space between two rows of containers should allow proper ventilation through them.
4. Reusable containers and sacks should be disinfected in chlorinated or boiling water before reuse.
5. The materials which require dry conditions for storage should be placed on sacks or cartons on the floor. Placing materials on cartons also helps to reduce the chance of fungal infection, while also improving ventilation and/or sanitation in the store room.
6. Commodities stored together should be capable of tolerating the same temperature, relative humidity and level of ethylene in the storage environment. As different types of commodities require different storage conditions, multi chamber cold stores should be preferred in farm situations.
7. There should be regular inspection of the produce and suitable action be taken immediately if any undesirable symptom is noticed.

Evaporatively cooled storage

The cold storage facility requires high capital investment and huge amount of power for operation. Continuous power is also not available in most rural places, particularly during summer months, when the cold storage facility is required most. Besides, it is not suitable for rural or on-farm storage where most of farmers would like to store the commodities for only a few days to accumulate sufficient quantities before carrying them to the markets situated far off and in urban areas. Most importantly, the cold store facility is not available in all places. Therefore, a low cost alternative to cold stores can be adopted which is known as evaporatively cooled storage.

A common form of evaporatively cooled storage structure, which is also known as the '*Zero energy cool chamber*', is made up of side walls in two layers of brick with a gap of 7.5 cm in between (Fig. 19). The annular gap is filled with river bed sand. The floor of the structure is made of a single layer of brick spread over 5 cm soil layer on the ground. This is done to prevent moisture seepage through walls and accumulation of water on the floor of the structure. The top cover of the structure is made of 2.5 cm thick layer of wood-wool pads, suitably sandwiched between two layers of welded wire mesh/ bamboo splits. There is no provision for mechanical ventilation. The side walls and the top cover are kept completely wet during the period of storage.

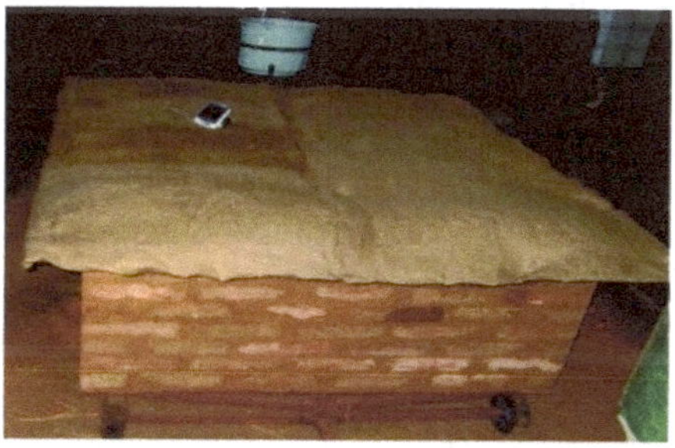

Fig. 19: Evaporative cool chamber for perishables

It has been observed that the cool chamber is capable of reducing the inside temperature by up to 15-16°C under dry climatic conditions. However, as the principle is based on evaporative cooling, the system would not work properly in high humidity areas such as in places near the sea or river, as well as in rainy season.

Portable evaporatively cooled storage structures have also been developed with a metallic frame made up of square iron bars, flats and angles(Fig. 20). The side walls can be made up of sponge/ aspen fibre pad sandwiched between plastic nets as support. The structure also performs satisfactorily as regards to cooling of commodities. The structure is light weight and can be conveniently shifted from one place to another and can be used as temporary storage structures for short term storage or during transportation in trucks/ rail cars. They can be installed within a short time at any place to meet immediate demands.

Fig. 20: Portable evaporatively cooled storage chamber developed at OUAT

Another simple method for storing small quantities of produce is to use any available container, and create a cool environment for storage by burying the container using insulating materials and soil. For example, a wooden barrel can be used as storage container and straw for insulation.

Frozen storage

Frozen storage is another type of low temperature storage, where a temperature of less than -10°C is maintained. The frozen storages can increase the life of fruits and vegetables for a much longer period than cold stores. Frozen peas, frozen mushroom are some examples of such products available in the market.

Controlled atmosphere (CA) and modified atmosphere (MA) storage

The controlled atmosphere (CA) and modified atmosphere (MA) are some potential devices to increase the shelf life of fruits and vegetables and other respiring commodities. For many fruits, both respiration and production of ethylene gas (produced naturally by many commodities) increase after harvest, which increase the ripening and accelerate spoilage of the fruit. These activities require oxygen and by controlling the oxygen and carbon dioxide in the atmosphere, the shelf life of the fruit can be increased and the quality be maintained for longer period. The MA packaging (MAP) can be conveniently applied for highly respiring commodities. In MAP, the low level of oxygen and high level of carbon dioxide are generated by respiration of the commodity and the oxygen in the package do not go to very low level (less than 2% is harmful) and the carbon dioxide do not reach more than 6-8 per cent levels as the selected packaging material has certain permissible level of gas permeability. Therefore, proper selection of the packaging film is very important and any film can not be used for any type of commodity.

Modified atmosphere packaging with plastic is advantageous for fruits and vegetables as the packaging materials are easily available and skill required is low. As the rate of spoilage is slowed down, it allows handling of fruits at little higher temperatures, which prevents chilling injury and gives uniform ripening. It is very beneficial for chilling sensitive fruits such as mango, banana, papaya, tomato, etc. and most vegetables. As it reduces respiration rates, decreases ethylene production and reduces sensitivity to ethylene action, there is delayed senescence (deterioration) as indicated by retention of chlorophyll, textural quality and sensory quality of fruits.

If modified atmosphere is desired quickly (as it is required for highly perishable commodities), some air from the package can be taken out and a recommended gas mixture is injected into the package. Oxygen or ethylene absorbers or carbon dioxide emitters can also be used to get the effect quickly.

The real benefit is obtained if both low temperature and modified atmosphere are used together (with recommended temperature and as well the gas concentrations).

Ripening of fruits

Controlled ripening of fruits is another important post harvest operation, particularly for fruits like mango and banana. The ripening process of fruits can normally take place in the plant. However, provision of ripening chambers in the farm can allow the fruits to be harvested prior to full ripening. It will be helpful for scheduling the harvesting operation and reducing the damage during transportation. Ripening chambers can also give even and controlled ripening. These are specifically designed rooms in which ripening agents (usually the ethylene gas) are used to speed up the ripening process. The ripening process is controlled by controlling the temperature, relative humidity, and the ethylene and carbon dioxide concentrations of the storage environment. Ripening of mango can also be induced by treating with ethral, which is an ethylene releasing compound. Normally mango is treated with ethral (1ml/ lit of water) at 52°C for 3 minutes. It not only induces early fruit ripening, but also reduces the development of post harvest diseases to a minimum in mango.

Calcium carbide is also used for ripening fruit artificially in many places. Calcium carbide reacts with water to produce acetylene, which acts as an artificial ripening agent. However, industrial-grade calcium carbide may contain traces of arsenic and phosphorus which are dangerous to human health. Hence the use of carbide gas for ripening of fruits is prohibited in India.

Food packaging

A properly graded and packaged food increases consumer acceptance and commands higher price in the market. Different types of packaging materials can be used for different types of commodities and processed products.

The main types of packaging materials used in food are the paper and paper boards, many types of plastic films, boxes and containers, foam, textile, glass, etc. Flexible film packaging materials are used for grains, fruits and vegetables. The PET bottles and HDPE bottles are used for fruit drinks, ready to serve beverages, ketchup, yoghurt, milk and juices, etc. Glass bottles are also used for different types of drinks, coffee, etc. Textile bags are used for grains or for fruits and vegetable commodities for bulk transportation. Papers and cartons are used for secondary protection of primary film packages and for protection of fruits and vegetables during transit.

The package should be able to protect the food from impact, shock, light, oxygen, etc. depending on the product requirements. Each type of package material has its own advantages and limitations depending on the nature of commodity and consumer needs. Now a day, many types of laminated films and combination of materials are used for different types of packaging applications. It is not only important to use proper type of package, but also to use proper mode even during storage, transportation, loading and unloading.

Value addition of fruits and vegetables

Value addition of fruits and vegetables can substantially add to the income of farm enterprises. Varieties of products can be prepared from each type of fruits, vegetables, spices, etc. The common preservation and value addition methods of fruits and vegetables have been shown in Fig. 21.

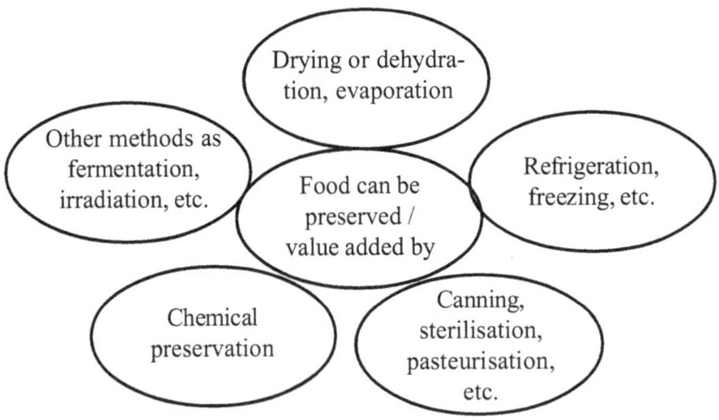

Fig. 21: General methods of preservation and value addition of fruits and vegetables

Drying and dehydration

Drying and dehydration are the methods of reducing moisture from the food to such a low level that the microorganisms can not grow and the biochemical reactions are inhibited. Dried products, if packed properly, have a very good shelf life. They also enjoy good demand in big cities and other countries. The good examples are dried chilli, peas, spices, aonla, potato chips, etc. Now a day, even dried vegetables like cauliflower and cabbage, dried fruits like pineapples, apples, etc. are available. Actually, it is one of the most common methods of value addition of fruits and vegetables. As the dried product can be stored for longer period, surplus crop in the harvest season can be dried and stored to avoid distress sale.

Drying of the fruits and vegetables can be carried out under sun or in mechanical dryers. The drying under sun is dependent on weather conditions. Besides, there is no control on the drying parameters. Hence, sun drying gives inferior quality product. Further, drying outside often contaminates the product and there is loss due to birds, rodents, etc. Dried product often becomes dark in colour.

Fig. 22: Large solar tunnel dryers

The mechanical dryers give comparatively better quality product as the drying temperature and other conditions can be controlled. But the mechanical dryers are costly and usually require electrical power. Therefore, mechanical drying is suggested only in case the product is to be sent to the urban markets and high for value customers, as it can fetch a better price there.

The mechanical dryers can be either batch type and continuous type, and depending on the volume of the commodity handled and the manpower available, the type and size of the mechanical dryer has to be selected. As a compromise between the mechanical drying and sun drying, drying in green house type dryers or low tunnel dryers can be considered (Fig. 22).

The general method of preparation of dried fruits and vegetables (or spices, etc.) is shown in (Fig. 23).

The sulfuring (treating fruits with sulfite solution or placing the fruits in sulfur fumes chamber) is carried out to prevent enzymatic activity and subsequent browning of the product during drying. However, sulfite-sensitive people should not eat food treated with sulphur. For vegetables, instead of sulfuring, blanching is done. Blanching is defined as the method of heat treatment that helps to inactivate enzymes. Usually the temperature is 70-100°C and time maintained is 2-5 minutes for hot water blanching. Individual quick blanching (IQB) methods use steam for a time of less than 1 minute. Vegetables such as onion, green pepper and garlic are not blanched because it reduces their flavour/ pungency. Blanching is often combined with peeling and/ or cleaning.

Syruping (addition of sugar solution) is done particularly for sour fruits so that they become sweet while loosing some moisture to the sugar solution. Usually the fruit slices are kept in sugar solution for 4-6 hours and then dried. The product is also known as osmo-dried fruits. Salt solution is used for vegetables in place of syrup for getting osmo-dried vegetables.

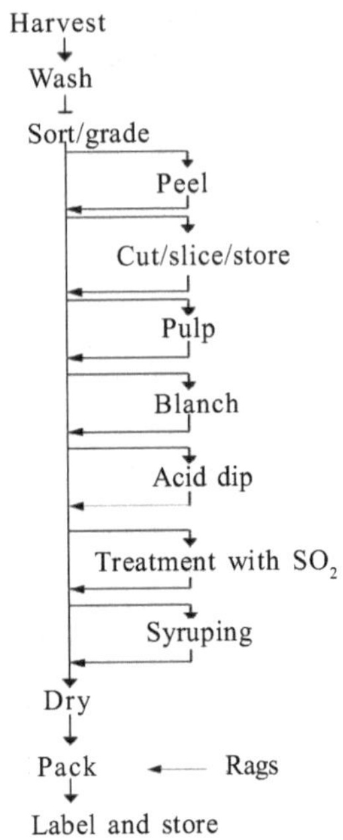

Fig. 23: Process flow chart for preparation of dried fruits and vegetables

Canning

The method involves killing of microorganisms originally present in the food by heating at a high temperature (usually more than 100°C, i.e. more than the boiling temperature of water) for a specific time. Canned peas, mushrooms, pineapples, etc. are available in the market. Canned products have good demand in international market. The method of canning requires high pressure sterilizers (also known as retorts) to sterilize the contents and other ancillary facilities. The sliced commodities are taken in a can (or a glass bottle/ container) along with sugar or salt solution and exposed to very high temperature. The specifically designed canning equipment, exhaust boxes, can seamers, bottle filling and sealing machines, etc. are available.

Most of the spice pastes are pasteurised at high temperatures (usually between 85-100°C) to increase their shelf life.

Preservation by using chemicals

Some chemicals are used along with the recommended methods to prepare products like jam, jelly, marmalade, squash, fruit syrup, pickles, etc. The process followed for each commodity and each type of product will be different and the amount of chemical additives will also be different. There are certain chemicals which are permitted by law as food additives and there are also maximum permissible limits for most of the chemicals, which should be strictly followed (Srivastava and Kumar 1994). The Basic unit operations in the proparation of Jam and Jally, Marmalade and fruit juice are shown in Figure 24-25.

Food acidification is one important way of preventing deterioration by creating an un-favourable medium for development of micro-organisms. This acidification can be obtained by two ways: natural acidification and artificial acidification. Fermentation is a way of natural acidification. The acetic acid, citric acid,

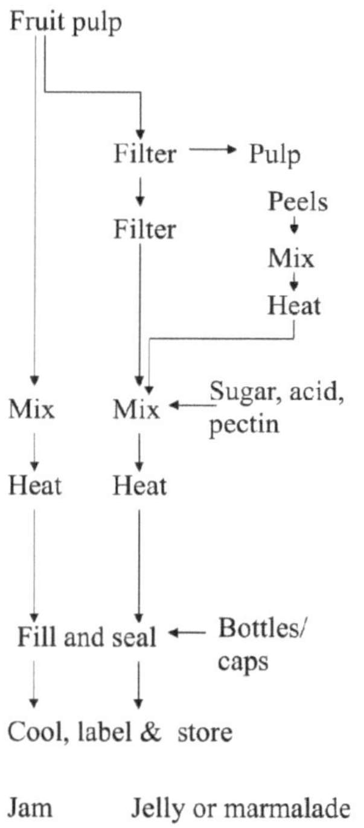

Fig. 24: Flow chart for preparation of jam, jelly and marmalade

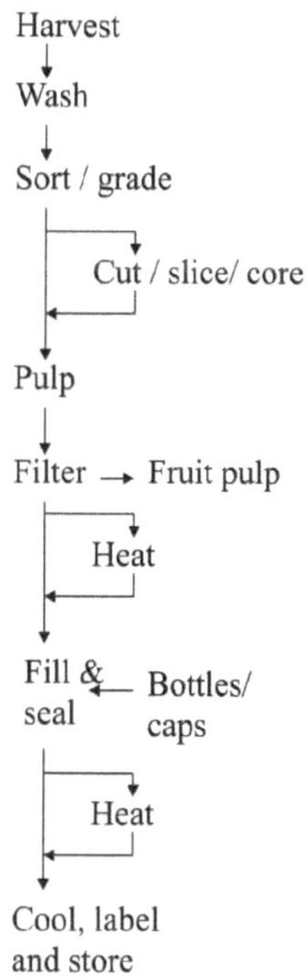

Fig. 25: Process chart for fruit juice preparation

vinegar, potassium metabisulphite are the examples of preservatives which increase the storability of fruits and vegetables. Even the common salt and sugar are also good preservatives. Sugar is used in the preparation of jam and jelly; salt is used in preparation of pickles, etc. The sulphur dioxide, carbon dioxide and chlorine are some gaseous preservatives. Proper dosage of the chemical food preservatives is important, and anything extra becomes harmful or unacceptable.

The preparation of jams, jellies etc. use both evaporation of moisture (also known as concentration of liquid foods) and addition of some preservatives.

The frozen peas, frozen corn, minimally processed chilled fruits and vegetables are the examples of value addition by low temperature. In addition to the above methods, fruits and vegetables are also preserved by fermentation, irradiation, etc.

Further, most of the fruits, vegetables and spices are very good sources of nutraceuticals and medicinal compounds, and are being used for extraction of nutraceuticals and other active ingredients.

Processing of fruit juices

The processing operations for fruit juice/ pulp are shown in Figs. 24-25. Different types of pressing machines as basket press and hydraulic press, pulping machines, homogenisers, etc. are available for preparation and processing of fruit juices.

Minimally processed fruits and vegetables

Now a day in the retail stores, fresh fruits and vegetables with minimum of processing are also available and a section of people have special preference for that. Though minimal processing has a broad definition, but the normal operations that are carried on fruits and vegetables are cleaning and trimming, peeling, slicing, packaging and storage under refrigeration. The minimally processed food contribute to convenience of the consumers and particularly helpful for the working women.

Processing and value addition of egg

Proper storage and distribution of egg is very important. Packaging is an important component in delivering quality eggs to buyers. Packaging protects the eggs from micro-organisms, such as bacteria, natural predators, loss of moisture, tainting, temperatures that cause deterioration; and possible crushing while being handled, stored or transported. Eggs also need to breathe, hence the packaging material must allow for the entrance of oxygen. The material used must be clean and odourless so as to prevent possible contamination and tainting.

There are many types of egg packages, and the materials are also different. The common package is the filler tray made up of wood pulp. The fillers are placed in boxes or cases for transportation. The cases are usually made of card board or wood. Now a day plastic fillers are also used, which can be reused and are washable. The fillers can be covered with plastic coverings and be used as packages for final sale to the buyer. Egg packages made up of paper board holding 2-12 eggs are also available.

The ideal temperature for storage of eggs is usually between 10° and 13°C. The storage room must be kept at a constant temperature and humidity must be checked so that the loss of water due to evaporation is minimum. The storage room must be free from products and materials which create off-flavour. For the long term, eggs are best stored alone. Eggs should be stored so that they are allowed to breathe. There should be proper air circulation in the storage room.

In cold storages the temperature between -1.5° and 0°C is recommended. At a temperature of -2.5°C eggs freeze. The relative humidity should be between 80 and 85 percent at a cold storage temperature of -1° C. At cold storage temperatures of about 10° C the relative humidity should be between 75 and 80 percent. The most important factors in successful cold storage are proper selection and packaging of eggs, proper temperature, humidity and air circulation and periodic testing for quality. Only good quality eggs must be stored, and hence, it is best to candle all eggs before storage. The average storage life for eggs is between six and seven months.

The processing of eggs involves breaking and pasteurizing the eggs. The pasteurised egg can be packaged in liquid, frozen or dried form. Dried whole eggs provide a convenient product and is often used in food processing industries. Dried eggs can be reconstituted with water to make pancakes, scrambled eggs, etc.

Processing of fish

Fish in its natural form can be sold in live form or in chilled/ frozen form. When live fish are transported they need oxygen, and the carbon dioxide and ammonia that result from respiration must not be allowed to build up. Most fish transported live for a long distance are placed in water supersaturated with oxygen.

Refrigeration and freezing are used to keep the fish fresh for longer period. In refrigeration, the temperature is dropped to about 0°C; in freezing the temperature is dropped below -18°C. Usually the chilling of fish is done by distributing ice (usually crushed ice or slurry-ice) uniformly around the fish. Proper maintenance of cold chain is very important if the material is taken to a long distance. Freezing is intended for long-term preservation, but it is relatively expensive in terms of equipment and operating costs.

Different types of fish products can also be prepared in the farm to increase income and employment. Some of these are as follows.

Cooked fish. Cooking can be done in boiling water or oil. Cooked fish products are usually for immediate consumption and require no sophisticated packaging. The shelf-life can be extended for a few days by using refrigerated storage and the product should be covered to prevent recontamination.

Cured fish products

Curing involves the techniques of drying, dry salting/brining (soaking in salt solution) or smoking. These techniques reduce the water content in the flesh of the fish, and thereby prevent the growth of spoilage microorganisms. Cured products have a long shelf life.

Dried fish. In order to prevent spoilage, the moisture content of fish needs to be reduced to 25 per cent or less. The percentage will depend on the oiliness of the fish and whether it has been salted. Traditionally, whole small fish or split large fish are dried under sun. However, solar or mechanical dryers provide a better control on drying conditions and give better quality product. Further they also prevent attack by insects or vermin and reduces contamination by sand and dirt. Mechanical dryers can work even in bad weather conditions. However, it is only advantageous to use such dryers if there is a market for a higher-quality product or if the fish would otherwise be lost.

Salted fish. Increase in salt content in fish reduces water activity and improves shelf life. Thus, salting is a common method in fish preservation. However, there are some bacteria who are salt loving and hence to inhibit their activity drying of fish is required. Thus when salt is applied to fish two processes occur. First the salt enters the fish and increases salt concentration and second water comes out due to osmosis. The traditional method of application of salt involves rubbing salt into the flesh of the fish or keeping alternate layers of fish and salt (recommended levels of salt usage are 30-40 per cent of the prepared weight of the fish). A better method is dipping in salt solution, in which the fish is immersed in a pre-prepared solution of salt (36 per cent salt). In this method the salt concentration can be more easily controlled, and salt penetration is more uniform. After brining the fish is dried to remove the surface moisture.

Smoked fish. The preservative effect of the smoking process is due to drying and the deposition in the fish flesh of the natural chemicals of wood smoke. Fish can be smoked in a variety of ways, but as a general principle, the longer it is smoked, the longer its shelf-life will be. Smoking can be categorized as *cold smoking*, in which the temperature is not high enough to cook the fish (usually less than 35°C) and h*ot smoking*, in which the temperature is high enough to cook fish. Hot smoking is often the preferred method because the process requires less control than cold processing and the shelf-life of the hot-smoked product is longer, because the fish is smoked until dry.

Proper packaging of cured fish products is important to prevent moisture absorption by the fish and recontamination by insects and micro-organisms.

Fermented products

The use of fermentation as a low-cost method of fish preservation is commonly practiced all over the world. The process increases the acidity of the fish and therefore prevents the growth of spoilage and food-poisoning bacteria. There are many different types of fermented products and their nature depends largely on the extent of fermentation which has been allowed to take place. They can be categorized as: fish which retains its original texture, pastes, liquids/sauces. The fermented products are usually packed in glass bottles, plastic or even earthenware pots. The containers should be air-tight to develop and maintain the airless conditions required for good fermentation and storage.

Waste produced during fish processing operations can be solid or liquid. The solid wastes which include skin, viscera, fish heads and carcasses (fish bones), can be recycled in fish meal plants or it can be treated as municipal waste. The liquid wastes include blood water and brine from drained storage tanks, and water discharges from washing and cleaning. This waste may need holding temporarily, and should be disposed of after suitable treatments without damage to the environment (Chooksey and Basu, 2003).

Processing and value addition of milk

Milk is an important product in the farming system. The milk produced in the farm can be sent to the nearby dairy plant as quickly as possible. The milk should be collected under hygienic condition and then chilled to less than 5°C in bulk coolers. The milk then has to be transported to the dairy plants in bulk milk pick up vehicles and the temperature during transit is kept below 5°C.

If sufficient quantity of milk is available, then milk can be pasteurised and value added in the farm itself. For 100–500 litre quantities of milk, milk pasteurization with a plate pasteurizer is not recommended and batch pasteurisation is preferred. Besides, the milk can also be used for preparation of products as yoghurt, cheese, ghee, butter, butter milk, etc. For each of the product standard methods and equipment of different capacities are available.

Processing of honey

The simplest processing is to remove the honeycomb from hives and sell or consume it as "cut-comb" honey. The process involves collecting pieces of sealed and undamaged honey comb, cutting them into uniform sized pieces and packing them carefully in bags or cartons. Because the honeycomb is unopened, it is readily seen to be pure, and it has a finer flavour than honey that is exposed to air or processed further. However, the honeycomb is easily damaged by handling and transport and a lot of care is required during packing, stacking and transportation.

Strained honey. This is the honey that is processed to a minimal extent. It is prepared by removing the wax cappings of the honeycomb using a long sharp knife. The knives have to be previously heated in warm water. After breaking the honeycombs to pieces, the honey is strained to remove wax and other debris. The first straining by a coarse strainer removes the large particles and subsequent straining through cotton or muslin cloths help remove the finer particles. The clear honey is collected in a clean, dry container. When most of the honey has drained (the time may be over many hours depending on the temperature) the combs are squeezed inside a cloth bag to take out the residual honey. The wax is collected and formed into a block by melting it gently in a warm water bath or solar wax extractor. The strained honey is then stored in large containers or in small retail glass jars or plastic bags. This beeswax, which is obtained as a by-product is used as wax polish or for candle-making.

Packaged honey. The wax cappings are removed from the honeycombs as for strained honey. Electrically heated honey knives or "planes" are also available for larger scales of production. When extracting honey from top-bar frames, the frame is placed over a dish, and the thin layer of wax capping is cut from the bottom to the top of the frame. The honey falls down into the dish below. The frame is then turned and the capping on the other side is removed to take out the remaining honey. The honey that is stuck to the wax cappings is strained using cloth bags. The frame is then placed in a honey extractor. Both manually or electrically operated honey extractors are available which extract the honey by spinning the frames at high speed. They can be either "tangential" or "radial" type machines.

In a tangential machine, the frames lie against the barrel of a drum and the outer side of the frame empties when the drum is spinning. The frames are then turned so that the other face of the honeycomb faces outwards, and the machine is spun until this side is empty. The machines give more complete extraction than the radial machine and this design is more compact and cheaper than radial types. In a radial machine, the frames sit between rings, arranged like the spokes of a wheel and honey is extracted from both sides simultaneously. Radial machines can hold more frames than a tangential machine (e.g. a 20-frame radial extractor compared to an 8-frame tangential machine).

The honey is collected in a pan, and filtered through a nylon or stainless steel filter unit. The unit has sequentially arranged filters to have coarse filtration initially and fine filtration towards the end. The clear honey is collected and packed in moisture proof packages so that the atmospheric moisture should not enter inside. The glass or plastic containers with suitable air tight caps can serve for the purpose.

References

Chooksey MK & Basu S. 2003. *Practical Manual on Fish Processing and Quality Control*. CIFE, Kochi.

Dash, SK, Bebartta, J P, Kar A. 2012. Rice Processing and allied activities. Kalyani Publishers, New Delhi.

Sahay KM & Singh KK. 1994. *Unit Operation of Agricultural Processing*. Vikas Publ. House.

Srivastava RP & Kumar S. 1994. *Fruit and Vegetable Preservation. Principles and Practices*. International Book Distr.

Thompson AK. 1996. *Post Harvest Technology of Fruits and Vegetables*. Blackwell.

Verma LR & Joshi VK. 2000. *Post Harvest Technology of Fruits and Vegetables*. Vols. I-II. Indus Publ.

Integrated Farming System Practices: Challenges & Opportunities, pp 417-437

New India Publishing Agency, New Delhi, India

16

Mechanization for Small Farmers in Integrated Farming Systems

S. K. Swain

Mechanization of agriculture is essential for optimal utilization of vital inputs for crop production system such as soil, water, seed, fertilizer, pesticide and labour force leading to higher productivity and reduced cost of production for greater profitability. Thus with the support of mechanization, the agrarian sector as an enterprise needs to become sustainable and economically competitive which coincides with the objectives of the Integrated farming System concept. Mechanization also imparts capacity to the farmers to carry out farm operations with dignity, comfort and freedom from drudgery, making the farming agreeable vocation for educated rural youth as well. It helps the farmers to achieve timeliness and precisely meter and apply costly input for better efficacy and achieve higher productivity with reduced application of inputs. Small and marginal farmers now make use of high capacity agricultural machines on custom hire basis to enjoy the economic benefits. As such, the broad subject of agriculture can be understood as production agriculture, which takes care the field operations in the crop production system for higher productivity and post production agriculture, better known as post harvest technology which looks for reducing the loss of produce and value addition for economic gains. The IFS outlines the need of sustainability for farmers with limited resources of land, water and economy while mechanization aims at maximizing the profitability in agricultural

College of Agricultural Engineering and Technology, Orissa University of Agriculture & Technology, Bhubaneswar-751003, Odisha, India system with a commercial attitude. Since the availability of labour force during peak hours of need and hike in wage very often creates hindrance in production and post production agriculture, mechanization can be the right answer to harness the benefits of IFS approach.

Commercialization is essential for agricultural development, which, amongst other things, entails mechanization of agriculture to reduce the cost of production and to increase the yield of crops. In some instances, agricultural commercialization may take place even without mechanization. However, due to the ever increasing agricultural labor scarcity in developing countries, an extensive scale agricultural commercialization may not be possible without mechanization.

The various farm operations involved in crop production system are land leveling, tillage for seed bed preparation, sowing and planting, weeding and interculture, spraying, harvesting and threshing. The mechanization of all these operation has been appreciated by the farmers which are discussed in the following paragraphs.

Land leveling

Land leveling is basically a vital operation through which the land is made suitable prior to start the crop production system. Conventionally, the tractor operated leveler is used where the operator manages the leveling of the land by visual observation; thus the leveling is never very accurate. Precision laser land levelling system will improve crop establishment and care, reduce the amount of effort required to manage the crop, and will increase both grain quality and yields. Performance evaluation of the laser land leveler in farmers' field in Sonepur district of Odisha state indicated that the saving in irrigation water was within 31-35% along with increase in yield within 18-26 %. Furthermore, the weeding cost was drastically reduced by 70 % saving considerable labour, cost and time. In spite of higher initial investment of Rs 3.5 lakhs for the equipment, the cost of leveling was calculated as Rs3150/- per ha as compared to Rs3250/- per ha in case of conventional leveling because of higher output of the laser leveler as 7.5 hrs/ha in comparison to 10 hrs/ha in conventional method. The saving in time along with saving in fuel, weeding cost and water requirement along with increase in yield and higher fertilizer use efficiency achieved by the laser leveler have been appreciated by the farmers.

Tillage for seed bed preparation

The first and foremost field operation in crop production system is tillage for seed bed preparation which creates a favourable condition prior to sowing of a seed or planting a seedling. The animal drawn indigenous ploughs and tractor operated cultivators are conventionally used by farmers for this job where the furrow section is triangular and depth of tilling is also limited. Furthermore, during ploughing operation uniform tillage is never achieved which results in non-uniform growth of plants and reduced yield. For primary tillage operation, mould board plough is recommended in general for all types of land where a trapezoidal furrow section is formed during ploughing with complete inversion of furrow slice. Of course, disc plough is recommended for grassy or trashy lands as mould board plough does not perform adequately under such condition. The standard disc ploughs are used for primary tillage operation under such conditions where mould board ploughs usually does not work satisfactorily. The productivity of crop also improves with higher depth of cut, achieved by these ploughs although the farmers have to incur additional cost towards this operation. The capacity of a tractor operated two bottom mould board plough or disc plough is around 0.02 ha/h and the cost of such equipments ranges from Rs40, 000/- to Rs60, 000/-. Similarly bullock drawn mould board ploughs have been developed for small and marginal farmers by which the quality of tillage can be enhanced comparing to the traditional indigenous plough (Pande, 2010).

The secondary tillage operation is taken up at least four to seven days after primary tillage to create the proper soil condition i.e. seed bed for sowing seeds or planting seedlings. The tractor drawn cultivator, disc harrow and levellers are the common secondary tillage implements while animal drawn disc harrows are hardly being used by farmers. The cost of the tractor operated cultivator is around Rs 30,000/- and its actual field capacity is 0.2 ha/h. In most parts of the country, this implement is being used as primary and secondary tillage implement since the cost of operation is reasonably low and the coverage is quick enough for land preparation. However, the depth of operation is limited to 5-7.5 cm only and the field is never prepared good enough to take up sowing or transplanting subsequently.

The power tiller and tractor operated rotavators have of late being appreciated by the farmers for seed bed preparation both for dry and wet land condition because of the better quality of tilth in lesser time. The power tiller operated rotavator has got limitations of less depth of operation and speed apart from the fact that the operator has

Fig. 1: Tractor Rotavator

to walk behind the power tiller. Incidentally, the tractor operated rotavators not only provide better tilth with higher depth of operation but also the quality of land preparation is just too good and quick to facilitate sowing or planting operation by seed drill or transplanter with ease (Fig. 1). The fuel consumption of tractor with rotavator is only 4.5-5.5 l/h even though the weight and size of rotavator is considerably higher. The cost of the rotavator is within the range of Rs 1.00 to 1.5 lakhs approximately depending upon the no of blades mounted corresponding to the width of operation. The capacity of tractor operated rotavator is around 0.2 ha/h and it is considered by far the best secondary tillage implement for quality seed bed preparation (Swain, 2013).

Puddling is the secondary tillage operation under wet land condition which facilitates transplanting of paddy seedlings. The small and marginal farmers generally use indigenous plough twice or thrice for puddling operation before leveling and transplanting subsequently. The improved bullock drawn puddler or Rotary blade puddler has the capacity of 2.0 acres/day and the cost of this implement is around Rs 4, 500/-. For puddling operation the tractor or power tiller operated rotavators are preferred due to better quality of work within less time and cost. During secondary tillage operation under wet land condition, single cage wheels are fitted with the real wheels of tractor or sometimes double cage wheels replaces the rear wheels in order to increase the traction. In case of power tiller the pneumatic tyres are replaced with cage wheels before working with rotavator and the adjustable gauge wheel is replaced with an adjustable float.

Fig. 2: Tractor operated Post Hole Digger

For planting of fruit saplings, such as banana, coconut, Pomegranate etc. farmers generally dig holes with spade and pic axe which involves huge labour, time and cost. Post hole diggers operated by engines, power tiller and tractor have been developed and are used by famers (Fig.2). The capacity of such diggers is around 60 pits/h. The shape of the pit is cylindrical with diameter varying from 15 cm to 60 cm and depth varying from 30 cm to 90 cm depending upon the power source and type of plants. The tractor operated Post Hole Digger is mounted on the rear side of tractor through 3-pt linkage system. It consists of cardan shaft, mast, gear box, auger and hole opener. The Digger consists of a stabilizer system, which keeps the auger perpendicular to the ground while in operation and control swinging of augur while moving from one point to another point. This equipment is suitable for 540 RPM PTO with minimum tractor hp required being 35 hp. The depth of operation depends upon the soil type and field condition.

Sowing and planting

Generally, for higher production and productivity, desired seed rate and optimum plant population are essential for which seed drills have been developed. A seed drill is used for sowing small seeds where row to row distance is maintained while a planter is used for sowing small to bold seeds where row to row as well as hill to hill distance is maintained. A seed-cum fertilizer drill is basically a seed drill with additional provision for placement of fertilizer along with seed simultaneously at desired depth and rate. A manually drawn seed drill is used to sow small to bold seeds in two or three rows. It has got a handle attached to the frame to be pulled by one person while a pair of ground wheels drives the metering mechanism inside the seed box for delivery and placement of seed at desired depth and rate in the furrow. The Animal drawn Seed-cum-Fertilizer Drill has got the similar arrangement as that of a manually operated drill excepting the hitching facility, provision for three to five rows and a cut off device for stopping the delivery of seeds during turns in the field (Fig. 4). In simplest form, a funnel with a plastic pipe is often attached with a country plough to drop the seeds manually. A bullock drawn single row multi crop seed cum fertilizer drill has been developed where a seed and fertilizer box is attached to a conventional narrow plough (Fig. 3). While the plough is used to open the furrow, the metering mechanism gets the drive from a drive wheel attached to the seed and fertilizer box. The inclined plate in the metering mechanism is changed according to the type and size of seeds.

Fig. 3: Bullock drawn single row seed-cum-fertilizer drill

Fig. 4: Bullock drawn five row seed-cum-fertilizer drill

The power tiller and tractor operated seed-cum-fertilizer drill has got provision for sowing seeds and fertilizer in 5, 7, 9 or 11 rows depending upon the power available in tractor (Fig. 5 and 6). It has got a hitching arrangement matching to the three point linkage system of tractor. It has separate boxes for seed and fertilizer, mounted on the main frame having separate metering mechanism to deliver requisite quantity of seed and fertilizer in the furrow. Furthermore, the land needs to be ploughed twice and properly levelled before operat-

Fig. 5: Power tiller operated seed-cum-fertilizer drill

Fig. 6: Tractor operated seed-cum-fertilizer drill

ing the seed cum fertilizer drill. This device saves time, labour and seeds which results in reduction in cost of sowing. Of course, the intercultural operations such as weeding and hoeing etc. can be carried out effectively through mechanical method with lower labour, time and cost apart from reduction in drudgery. The cost of a tractor operated seed cum fertilizer drill is around R55, 000/ - and its actual field capacity is 0.28 ha/h.

Under conservation agriculture, the use of tractor drawn zero till drill has drawn attention among the scientists and farmers where minimum soil disturbance is essentially required for conserving the soil structure (Fig. 7). Moreover, the cost of land preparation (primery and secondary tillage) is eliminated to make the cropping system highly cost effective while the productivity is ensured due to better germination by proper utilization of residual moisture. The output of the tractor operated zero till drill is 0.2 ha/h and its cost is approximately Rs 65,000/-.

Fig. 7: Tractor operated Zero till Drill

Fig. 8: Sugarcane Set Cutter Planter

For line planting of specific crops like potato, sugarcane, tractor operated sugarcane set cutter planter and potato planter have been developed which are operated under thoroughly prepared upland condition. The multipurpose sugarcane set cutter planter comprises of a set of furrow openers, set hopper, cutting unit, fertilizer box, fungicide container, compaction roller and power transmission unit (Fig. 8). The main frame of the planter holds two way mould board shaped furrow openers for opening the furrows. The spacing between furrow openers is adjustable. A slanting chute at an angle of 55^0 to the horizontal is provided for conveying of the sugarcane to the cutting unit. The operator has to put the cane in the chute which slips down by gravity towards the cutting unit. The lengths of the sets remain uniform with the forward tractor speed within 3 kmph.

In conventional method, potato tubers are manually placed in a furrow opened by a hand drawn ridger or bullock drawn ridger which consumes huge labour, time and cost. The tractor operated potato planter has two seed hoppers, two seed picking conveyor belts with high density rubber cups, three ridger bodies, gauge wheel, a lugged drive wheel, chain and sprocket drives and two operator seats (Fig. 9). The potato seeds are placed in the hopper. The seed picking conveyor, which is run by the drive from the lugged ground wheel through chain and sprockets, picks potato tubers from the hopper and delivers them on to the field. Then the ridger bodies places soil over the potato tubers and form a ridge. The operator can pick potato seeds from the hoppers and place them on the conveyor in case of miss by the conveyor belt. The equipment is operated by a 35 hp tractor and can plant potato tubers with a spacing of 60-70 cm. The depth of planting is adjusted by the gauge wheel. The spacing between the rubber cups on the conveyor belt decides the plant to plant distance which varies within 15-20 cm (Fig. 10).

Conventional method of manual transplanting using root washed seedlings is the most labour and time consuming operation in paddy cultivation system which involves huge drudgery. Furthermore, the shortage of labour during peak hours of need followed by the wage hike very often delays the transplanting operation compelling the farmers to transplant old age seedlings which causes significant reduction in yield. The self-propelled rice transplanter (Riding type)can reduce the drudgery, labour, time and cost involvement to a great extent (Fig. 11). This transplanter works with mat type seedlings for transplanting instead of the conventional root wash seedlings which are used in manual transplanting. Therefore, the technique of raising the mat type seedling is quite important for mechanical transplanting. The transplanter is operated by a 4.5 hp diesel engine and transplant eight rows at a time.

Fig. 9: Tractor operated Semi-automatic Potato Planter

Fig. 10: Tractor operated Automatic Potato Planter

Fig. 11: Riding type (8 row)Transplanter

Fig. 12: Riding type (6 row) Transplanter

Fig. 13: 4 row Walk behind Paddy Transplanter

The field needs to be puddled and levelled properly before operating transplanter. It takes around 2.5 to 3.0 hours to cover one acre as compared to around15 man days required in conventional manual random transplanting method. The row to row spacing is about 23.8 cm and the distance between hills can be varied with settings of 14 cm, 16 cm and 18 cm. The cost of this transplanter is around Rs1, 90, 000/- and the field capacity of the transplanter is around 0.16 ha/h. The transplanter saves huge labour, cost and time apart from reduction in drudgery and facilitates the use of mechanical weeder later for intercultural operation. Of late, a four row walk behind self-propelled paddy transplanter with rotary type transplanting mechanism has been developed which has been appreciated by the farmers (Fig. 13). The capacity of this transplanter is around Rs 3.50 lakhs and the output is found to be nearly same as that of riding type transplanter. Recently, one riding type four wheel drive six/eight row self-propelled paddy transplanter has been introduced where the output is 0.13 ha/h (Fig. 12). The costs of these transplanters are around Rs16.00 lakhs and are getting popular among farmers day by day as there are service providers available who make these transplanters available for the farmers through custom hiring at a much cheaper cost compared to conventional manual method of transplanting.

The paddy transplanting operation under wet land condition is certainly labour consuming although transplanters have been made available. After puddling and leveling operation, sometimes farmers go for broadcasting of pre-germinated paddy seeds. Of late, a manually operated eight row drum seeder has been developed, where the operator pulls it to cover eight rows with row to row spacing of 20 cm (Fig. 14). Under this situation, the field is left with some patches of minor depressions with water logged condition due to improper leveling and some patches of major depressions created due to foot marks of the operator who pulls the drum seeder. Thus some seeds get buried resulting in improper plant population. For this purpose bullock drawn and power operated drum seeders have been developed which makes raised beds over which the pre-germinated paddy seeds are allowed to fall (Fig. 15 and 16). The capacity of these drum seeders are around 0.2 ha/h. Under controlled conditions this method can reduce huge labour, time and cost as compared to transplanting without sacrificing the yield (Swain et al., 2013).

Fig. 14: Manual (8 row) drum seeder

Fig. 15: Bullock drawn drum seeder

Fig. 16: Power operated drum seeder

Weeding and interculture

The weeders can only be used when the plants are sown or transplanted in line with specified spacing. Operation of weeders further facilitates the plants to develop better root growth for higher water and nutrient uptake leading to better yield. The upland weeder has a wheel on the front to which a frame containing five nos hook type weeding elements fitted (Fig. 17). The weeder is moved forward and backward by a handle for weeding in inter row space within 2 -3 weeks after date of sowing under upland condition. The capacity of this wheel finger weeder is 0.016 ha/h and its cost is around Rs 1300/-. This weeder is suitable for crops where row to row spacing is kept more than 20 cm. The power weeder has been developed to cover more area with least effort. This weeder is operated by one 6.5 hp two/four stroke petrol/diesel engine and one operator controls the machine. The power weeder can operate in line sown or line planted wide spaced crops where row to row spacing is more than 45 cm (Fig. 18). Thus, it is very much suitable for vegetable and fruit crops. The cost of this type weeder is around Rs 65,000/- and its capacity is 0.07 ha/h. The wet

Fig. 17: Wheel finger weeder

Fig. 18: Power weeder

Fig. 19: Cono Weeder

Fig. 20: Mandava weeder

land weeders basically has got a float made of either plastic or GI sheet, fitted at the front of the weeder, which helps the forward movement by sliding. The weeding elements in these weeders are normally rotary type fitted around a shaft and the shaft gets drive from the movement of the float.

The cono weeder is a manually operated push-pull type weeder used in line transplanted paddy fields under medium and low land condition having output of 0.014 ha/h (Fig. 19). This is suitable for sandy-loam soil condition and is used once within two weeks and later again within three to four weeks after transplanting. The mandava weeder is also used for line transplanted paddy (Fig. 19). This weeder has bean made lighter with one set of weeding hooks fitted around a rotating shaft being used for weeding purpose. This is suitable for all type of soils (Nayak et al., 2014).

The power weeder for low land paddy has got a float made of plastic, placed centrally on which a gear box with two rotating weeding "L" type elements is mounted (Fig. 21). The gear box gets the drive from a 1.75 hp petrol engine mounted on a frame, fitted onto a pair of handles. The operator walks behind the power weeder to guide it while it covers four rows of line transplanted paddy with row to row spacing of 25-30 cm. The cost of this weeder is Rs 35,000/- and the output of this weeder is 0.1 ha/hr (10 hrs/ha) and it consumes around 700 ml of petrol per hour.

Fig. 21: Power weeder for low land paddy

Fig. 22: Brush Cutter

Brush Cutters are available for cutting grasses, bushes and other unwanted plants either in orchards or areas nearby to the crop (Fig. 22). It is a light weight engine operated equipment used for cutting grass, hedge, small bushes and also crops like paddy, wheat etc. with minor modifications in the cutting arrangement. This equipment is operated by two stroke or four stroke petrol engine. The equipment has got three types of cutting arrangements i.e. a circular saw for cutting thick stems, a two blade cutter for small bushes and a rotor with two threads for cutting grasses. The engine is carried by the operator on his back by specially arranged adjustable straps to facilitate movement while cutting. The operator has to wear special protection gears such as glass, apron and helmet for safety while operating the equipment.

Spraying and dusting

Sprayers and dusters are machines to apply fluid chemicals in the form of droplets and dust respectively. A hydraulic knapsack sprayer is a manually operated sprayer which works under hydraulic pressure (Fig. 24). Its tank capacity is up to 15 litres, with provision for mechanical agitation of the spray liquid. The worker uses his left hand to operate the lever handle of the sprayer as the lever maintains constant pressure and pumps the spray liquid in to the lance held by his right hand. The output is usually 0.4 ha per man day. But it is possible to obtain higher output by replacing the lance with a 2-3 nozzle boom. During operation it is usually possible to deliver the spray at almost constant pressure. The hydraulic knapsack sprayer can spray at 3 to 4 kg /cm^2 pressure to prevent a possible spray drift. This type of sprayer is a low cost, easy to maintain, and a small holding farmer sprayer. Its performance particularly satisfactory for spot treatment, band application of herbicides and blanket application if it is provided with boom.

Fig. 23: Ultra Low Volume Sprayer

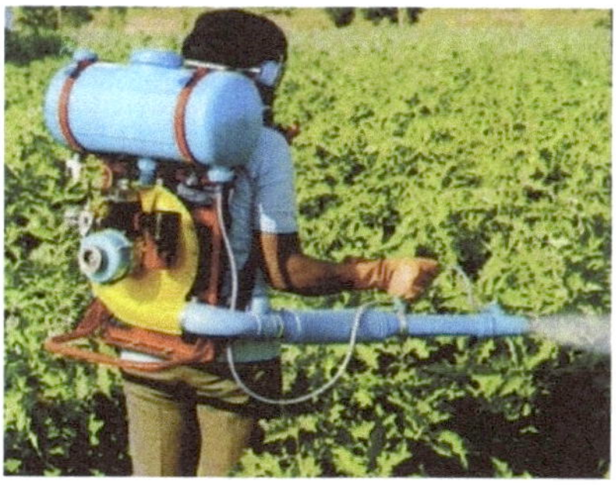

Fig. 24: Motorized Knapsack Mist Blower cum Duster

A motorized pneumatic sprayer is a low volume sprayer suitable for spraying concentrated spray liquids (Fig. 23). A blast of air acts as a carrier of herbicide concentrates in these sprayers which are called blowers. The air is forced-through the spraying jet of the delivery hose of the blower and a nozzle tube ejects the spray liquid. The air blast atomizes the spray liquid into fine droplets. Thus, in these sprayers air acts as the carrier. Faster the air is pumped into the spraying jet, more vigorous is the atomization. The equipment is fitted with petrol engine of about 1.2 H.P.

Fig. 25: Tractor operated Aero Blast Sprayer

The tractor mounted aero blast sprayer is suitable for orchards having trees up to 10 m height (Fig. 25). The sprayer consists of a plastic tank of 300 litre capacity, triplex pump and centrifugal blower of 1 m^3/ s capacity. The delivery of the chemical and air in two spouts is controlled by a set of flow rate control and bypass valves. The sprayer has provision to change the orientation of spouts with the help of ratchets and pawls to achieve maximum coverage of plant canopy. Tractor power is used through PTO shaft for operation of the blower and pump. The effective working width of the machine is 10 m and average field capacity is 0.87 ha/h.

Harvesting

Farmers normally give lot of importance for harvesting operation to be completed in time after putting their best effort to grow the crop. Sickle is the most common harvesting tool for harvesting field crops and cutting vegetations; but involves more labour, time and cost apart from drudgery. Power operated mowers and reapers are available for easy and quicker harvesting.

Fig. 26: Power Reaper

Fig. 27: Reaper Binder

The self-propelled power reaper has got either a diesel engine or a petrol start kerosene run engine as the prime mover (Fig. 26). Normally, it has got the cutter bar of 1.2 m width while the height of cut is adjustable from 10-30 cm. The reaper has got a set of crop dividers which guide the crop to the cutter bar. The crop after being cut is carried through a conveying unit by a set of star wheels to the right side of the reaper. The reapers are suitable when the crop stands erect in the field and the field condition is nearly dry. However, the reaper can still harvest when the crop stands at an angle up to 60^0. The fuel consumption is around 1.3 l/h. The machine takes about 5-6 hours to harvest one hectare crop field in comparison to 30 man days per hectare in conventional manual harvesting by local sickle. Apart from the saving in labour, cost, time and reduction in drudgery, the machine saves the crop from unexpected bad weather. The approximate cost of the reaper is Rs 95,000/-. Reaper binders are also available for the farmers which cuts and binds the harvested crop simultaneously (Fig. 27) . There is a seat for the operator in reaper binder and the binding threads are made of polythene generally. The cost of reaper binder is around Rs 3,00,000/-.

Harvesting of groundnut by manual hand picking is a highly labour and time consuming activity and the farmers mostly apply flood irrigation in their fields prior to harvesting to facilitate uprooting which often creates problem of germination if harvesting is not accomplished quickly. Harvesting of groundnut crop mechanically is an operation in which whole crop are removed from the soil without detaching or damaging the pods entangling the crop, by loosening the soil with a tool. The tractor operated groundnut digger-shaker has got a blade which penetrates in to the soil below the pod remaining zone (Fig. 28). As the tractor moves forward, the groundnut plants along with pods and soil clods move backward riding over the picker conveyor assembly where the soil clods are separated from the plants. The groundnut plants with pods fall on the back of the digger. This machine requires around 5 hours to harvest one ha area as compared to around 20 to 25 man days/ ha in conventional manual hand uprooting method. The average depth of cut by the blade of the machine has been observed to be 8 to 10 cm and the average harvesting efficiency was found to be 94.18%. Moreover, the primary field preparation for the subsequent crop could be achieved simultaneously with digging operation which would help a farmer to utilize the residual moisture for better germination and higher yield

Fig. 28: Tractor operated Groundnut digger

Fig. 29: Tractor operated Potato Digger

of subsequent crop. The performance can be more effective provided the ground nut field is maintained nearly weed free. Similarly, potato digger elevators are available for digging and windrowing of the potatoes where the picker conveyor consists of roller chain of iron bars (Fig. 29). These equipments save time, labour cost and the crop from various types of losses.

Threshing

Threshing operation carries huge significance for the farmers as they need to complete this operation in time to allow them to go for the subsequent crop. The hold-on type threshers are commonly used for threshing of paddy by the farmers of Odisha where the rotation of the drum is achieved by a pedal operated lever connected to the rotating drum through an eccentric mechanism and a pair of gears for a desired speed of rotation. The rotation can also be achieved directly by an electric motor of desired power with a 'V' belt joining the motor pulley and a pulley on the threshing cylinder shaft having desired diameters. The threshing of paddy grains is effected by striking action against the wire loops fixed on to the cylinder in staggered manner. This the power operated paddy thresher can be used for threshing groundnut by changing the threshing slats fitted with straight pegs. Generally the cleaning operation of the threshed material is done separately by a hand/pedal/power operated winnower to get the clean grain.

In case of feed-in type threshers the threshing and winnowing operation are accomplished simultaneously to get the clean grain (Fig. 30). The entire crop or

Fig. 30: Tractor operated Feed-in type Thresher

Fig. 31: Power operated Maize Sheller

Fig. 32: Groundnut Thresher

Fig. 33: Power tiller operated Feed-in type Thresher

harvested plants are fed in to the threshing cylinder through a feeding chute. The clean grains are separated subsequently from chaff by passing through a set of sieves. The performance of a thresher primarily depends on correct setting of the cylinder speed, concave clearance, blower speed etc. These type of threshers can be used for green gram, groundnut etc (Fig. 32). by changing the threshing elements on threshing cylinder and the sieves with desirable apertures. Power tiller operated axial flow thresher are also available where the entire paddy crop in field for threshing (Fig. 33). The output of these threshers are 5-6q/ha. Power operated maize shelters are used to increase the shelling capacity up to 8q/ha where the whole maize cubs are fed and clean maize grains are collected (Fig. 31).

Combine harvester

Combine harvester is a machine by which the harvesting, threshing and winnowing of paddy, wheat and other crops are accomplished simultaneously in the field. The combine harvesters are either wheel type which works generally under dry land condition or track type which works under comparative moist /wet condition. The combine harvester can be of tractor on top type where the tractor is dissembled and mounted on top of combine and is assembled again to work as a tractor when the job is over. It can be self propelled type where the engine is solely used for combine operation throughout the year. Generally, the combines are operated by tractor with 55.0 hp or above or engines of similar power range (Fig. 34).

Fig. 34: Track type Combine

The capacity of a combine is around 0.4 ha/h which reduces lots of labour, time, cost and drudgery. Furthermore, the field is made available early for subsequent crop to be grown under residual moisture condition. The cost of combines ranges from Rs 18.00 lakhs to Rs 23.00 lakhs. Combine harvesters are generally available to the farmers on custom hiring basis and the hiring charge is around Rs2000/- per hour (Fig. 35).

Fig. 35: Wheel type Combine

The agriculture and the allied sectors such as dairy, poultry, mushroom, fishery which are the constituent partners for uplifting the

Fig. 36: Chaff Cutter

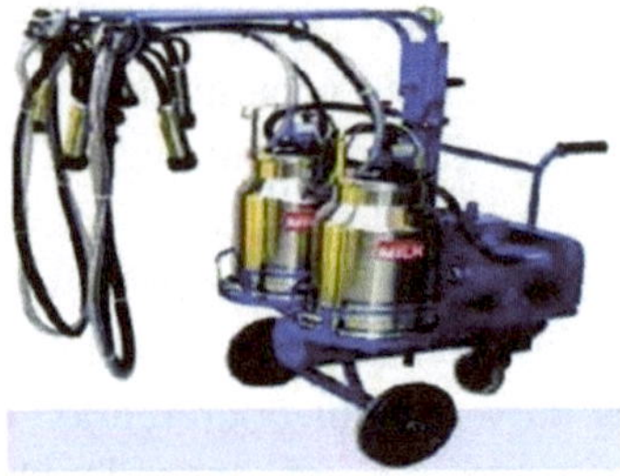

Fig. 37: Milking pail

socio-economic status of the farmer under Integrated Farming system concept are now looking for intervention of improved machinery for the similar reasons. The actual development of mechanization at global level in various fields has left Indian scenario way behind; nevertheless the farmers are realizing the need of mechanization in these allied sectors and venturing this new generation technology. The dairy sector is an integral component in IFS where some machinery and equipments are essentially required such as the chaff cutter machine, milking pails, milk cans and minor implements. On farms maintaining more than 20 milch animals, machine milking may be economical and more convenient as compared to hand milking. Installation of fans and mist less cooling devices in animal sheds for protection against heat stress is also essential if one wishes to keep high yielding crossbred cows. Dairy farms with 50 or more milch animals may also require a milk cooler, electricity generator set and a utility vehicle for the procurement of farm supplies and marketing of produce besides a tractor with implements for the cultivation of fodder crops and their harvesting, transportation chaffing, processing etc.

Milk is extracted from the cow's udder by flexible rubber sheaths known as liners or inflations that are surrounded by a rigid air chamber. A pulsating flow of ambient air and vacuum is applied to the inflation's air chamber during the milking process. When ambient air is allowed to enter the chamber, the vacuum inside the inflation causes the inflation to collapse around the cow's teat, squeezing the milk out of teat in a similar fashion as a baby calf's mouth massaging the teat. When the vacuum is reapplied in the chamber the flexible rubber inflation relaxes and opens up, preparing for the next squeezing cycle. The extracted milk passes through a strainer and plate heat exchangers before entering the tank, where it can be stored safely for a few days at approximately 42 °F (6 °C). At pre-arranged times, a milk truck arrives and pumps the milk from the tank for transport to a dairy factory where it will be pasteurized and processed into many products. Dairy farming is an important source of subsidiary income to small/marginal farmers and agricultural laborers. The manure from animals provides a good source of organic matter for improving soil fertility and crop yields. The gober gas from the dung is used as fuel for domestic purposes and also for running engines for drawing water from well. The surplus fodder and agricultural by-products are gainfully utilized for feeding the animals. Almost all draught power for farm operations and transportation is supplied by bullocks. Since agriculture is mostly seasonal, there is a possibility of finding employment throughout the year for many persons through dairy farming. The main beneficiaries of dairy programmes are small/marginal farmers and landless labourers.

Poultry farming is a very lucrative aspect of agriculture under livestock farming because of the higher return against investment. To achieve maximum result in poultry farming business, there are certain equipments such as feeders, brooders, incubator, chick box fly try etc. essentially required. Feeders are equipments used for feeding poultry birds. The food is deposited in the feeder and the birds feed from it. The amount of feeders provided for a poultry farm should be according to amount of birds available. It is important that to keep the feeders clean to ensure the safety of the birds (Fig. 38). The Chick Feeder is designed to helps new born chicks to consume feed easily from day old. It reduces feed wastage and thus increases the feed conversion ratio. The quantity of feed level in the pan can be regulated with the help of the adjustment slots provided in the feeder bucket. The chick grill restrict the chicks getting inside the feeder thus control feed wastage, Chick feeder extension (funnel) is also available for easy pouring of feed.

Fig. 38: Feeder for poultry birds

It is essential that the temperature of the poultry farm be regulated especially during cold weather. The heater or brooder is an equipment used in regulating and increasing the temperature of the poultry farm. These helps to keep the birds warm when the weather is cold. This is an instrument used in hatching eggs. Egg hatchery with an incubator can be described as a means of hatching of eggs in an unnatural way. These means can be employed when there are many eggs to be hatched. The chick box is an equipment where the poultry birds are kept for egg laying. It has a roll away egg tray attached to it so that when eggs are layed, they roll away and the birds will not trample on the eggs. This particular equipment help in preventing egg damage. Fly trap is an equipment used in controlling the number of flies around a poultry farm. Poultry Plucker Rubber Finger is equipment applied to chicken dressing machine. These rubber fingers are fixed to the bottom and side plate of the dressing machine in order to produce many dressed chicken in a short period. Egg Tray is an equipments used in setting the eggs. Just like the name, it is a tray-like equipment where the eggs are placed for sampling. Poultry incubator controller is an equipment used for controlling the incubator and timer counter. It displays the temperature and humidity condition of the incubator. The ventilation fan is an equipment used for ensuring maximum ventilation in the poultry farm. It is also used in reducing the temperature of the poultry farm during a hot weather. Laying nest is another equipment that help the birds for laying of eggs. One of the advantages of this equipment is that it increases the egg productivity of the poultry birds. Egg Scale is an equipment used in weighing the weight of the eggs. It helps the poultry farmer know the eggs that are fertile enough for hatchery because it is assumed that an under weight egg does not have what it take to form a chick. Egg washer is an equipment that makes use of a powder called the egg washing powder. Water is added into the egg washer and then the egg washing powder is added. It is used for washing the eggs before delivery. Neat water is required for growth and digestion in poultry birds just like in humans. Therefore, the drinkers are used for supplying water to the birds. It must be ensured that the drinkers are washed regularly to avoid disease (Fig. 39, a,b&c).

Cages and Coops are the poultry equipments used for keeping poultry birds, suitable for small scale poultry farming. Dressing Machine is an equipment used for feathering birds after slaughter. The use of a dressing machine makes chicken dressing easier, clean and hygienic. Lastly, the use of protective clothing for humans is very necessary. Special protective clothing like hair caps, disposable sleeves, boots and coverall are required to avoid transfer or contamination from the birds to man or from man to the birds. Also, it is important to ensure that visitors disinfect their hands before touching the birds.

Fig. 39a: Drinker for chicks

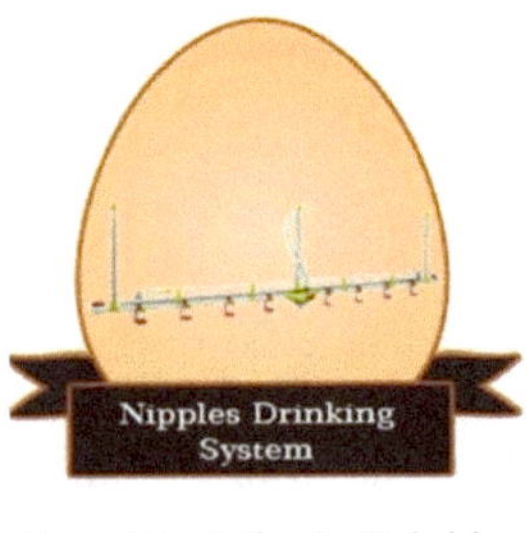

Fig. 39b: Nipple Drinking system for chicks

Fig. 39c: Brooder for chicks

Fig. 40: Automatic Feeding

Fig. 41: Paddle wheel aerator

Fishery is another important component of IFS where the farmer gets much higher return as compared to crop production system with less involvement of labour, cost and time. But, the farmers hardly take any measure to get better return from this sector in our state. They should go for mechanical automatic feeder for feeding in the pond in stead of traditional bag feeding method by which the feed cost is reduced by 28% with saving in labour by 21% (Fig. 40). Aeration is another important aspect which is done naturally by ducks or by means of adding more water and flashing the water (Fig. 41). By putting a paddle wheel aerator in action, the survival rate can increase by 18% and the increase in production as well as productivity can be remarkably improved.

The farmers generally go for conventional method of drying fish in sun which can be modernized with use of solar fish dryer for production of quality dried fish to gain better consumer preference (Fig. 42 and 43). The productivity also increases which gives the farmer better return.

Fig. 42: Convetional drying of fish

Fig. 43: Solar fish dryer

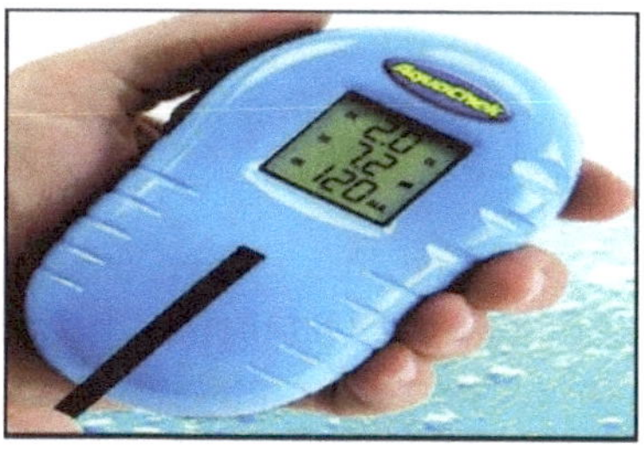

Fig. 44: Aqua check meter

It is very important to check the quality of water in the tank being used for fishery for maintaining the health of fishes as well as to ascertain if any remedial measure is required. A device called aqua check meter can be used which is easily accessible for measurement of water quality parameters as compared to the conventional chemical method of water analysis (Fig. 44). Most importantly the farmer should go for digging the pond with mechanical dozers within very short time and at competitive price as compared to manual method of digging a pond. Considering the availability of labour, wage hike and their efficiency, the mechanical method of digging a pond is certainly preferable.

The role of mushroom cultivation for economic development of a farm family has been proved to be highly remarkable since this enterprise is taken up by the family members quite effectively while the return is assured and just remarkable against the cost involvement. The use of improved machinery in

Fig. 45a: Boiling kettel

Fig. 45b: Autoclave

Fig. 45c: Laminar flow

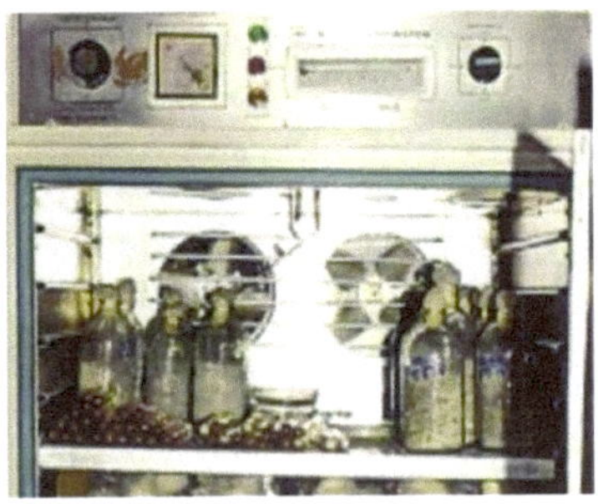

Fig. 45d: BOD incubator

this enterprise certainly can boost the quality and quantity of mushroom with reduction in energy expenditure.

The common equipments required in a spawn laboratory are grain cleaner, boiling pans, pH meter, Autoclave, BOD incubator and Laminar flow cabinet (Fig. 45a,b,c &d). The power operated grain cleaners are used to maintain the quality of grain A grain boiler is used for boiling the grains while a bag filling machine can be used for filling the grains in the bags quite quickly and smoothly.

A pH meter is normally kept to check pH of the medium. Autoclave is used for sterilization of spawn medium and oven for sterilization of glassware. If boiler is available steam operated autoclaves can be used for better efficiency. A small clinical autoclave can also be kept for sterilization of culture medium. BOD incubator is needed to incubate cultures and master cultures. Laminar flow cabinets (normally 4 ft. horizontal) are needed for isolation and multiplication of cultures and spawn inoculation. The use of machineries will reduce the labour requirement by 50%, power requirement by 60 % and labour efficiency is increased. The contamination of spawn is also reduced.

Mechanization for the major components of Integrated Farming System namely crop production, dairy, fishery, poultry and mushroom will certainly enhance the net profit of the farmer with reduction in labour, cost, time and drudgery. The quality aspect can still be taken care of with mechanization which will make the Integrated Farming System approach more lucrative and realistic for adoption at grass root level.

References

Nayak, B.R., Swain, S.K. and Mandal, P. 2014. Practical manual on weed in management in Horticultural Crops. College of Horticulture Chiplima, Odisha.

Pandey, M.M. 2010. Agricultural Engineering in India. Present status and future perspective. Vol. 34(3)www.India Journal.org.

Swain, S.K. 2013. Practical manual on Farm implements and Machinery. College of Agriculture, Chiplima, Odisha.

Swain, S.K., Nayak, B.R., Khande, C.M. and Dash, A.K. 2013. Effect of different establishment methods on yield and economics rice (*oryze sativa* L.). *Madras Agricultural Journal.* 100(7-9):747-750.

http://www.gourmetmushroom.com

http://www.milkwell.com

http://www.facebook.com/grwelagrovet/pest

Integrated Farming System Practices: Challenges & Opportunities, pp 439-454

New India Publishing Agency, New Delhi, India

17

Integrated Farming Systems: Transformation of Climate Smart Agriculture in South Asia

A.K. Sahoo and *V.P. Singh

Introduction

It is obvious beyond doubt that the climate is changing. Climate has also been changing in the past, but its rate of change has been unprecedented in recent past years. It is true that no one could stop the climate from changing, but could do a lot for mitigating some of its effects and formulate adaptation mechanisms to it. Agriculture, especially the small holder producers have been at the receiving end of the climate change effects, resulting into not only in the fluctuating and unstable productivity, loss of natural resources and consequent livelihoods, but also to their survival.

Climate change and food insecurity are two most pressing challenges human beings are facing today and has emerged as the biggest twin challenge of this century. There are evidences that the increase in greenhouse gases caused by human activities is responsible for global warming and consequent climate change (FAO, 2007). Under the six emissions scenarios the global average surface temperatures are expected to increase by 1.1°C - 6.4°C this century [IPCC,

Directorate of Research, Orissa University of Agriculture & Technology, Bhubaneswar-751003, Odisha, India
*World Agroforestry Centre (ICRAF) South –Asia, NASC Complex, Veda Prakash Shastri Marg, Pusa, New Delhi-110012, India

2007]. Climate change has its impact on almost everything on the earth and so on the food production. The agriculture plays a very important role in the economy of most of the nations of the developing world. For India, agriculture is the base of nation's economy, largely rainfed and dominated by smallholders (Singh 2008 and 2010). As per reports from a number of governments, non-government and intergovernmental organizations, the countries and agriculture especially practiced by smallholders are particularly vulnerable to climatic changes and they have least capacity to adapt to these changes. In most of the developing countries of the world, the agriculture is not merely a matter of cultural practice but it is the source of livelihoods of millions of smallholders and their dependents. Changing climate as predicted will not only affect the agricultural productivity but also heavily hit the lives and livelihoods of the smallholder families and ultimately lead to food insecurity to the concerned nation and the world (Singh et al. 2012). Business-as-usual scenarios of population growth and food consumption patterns indicate that agricultural production will need to increase by 70 percent by 2050 to meet global demand for food. South-Asian countries made tremendous progress in food production in 1960s, 1970s and 1980s. with an impressive growth of food production, India, Pakistan, Bangladesh and Nepal transformed themselves from countries with a chronic food deficit to countries that were almost self sufficient in early 1990s. Except Afghanistan, all these countries had exported some quantities of food grain in late 1990s. The dynamic growth in the agriculture sector has, however, recently been lost. Productivity of major food grains has slowed and has been decline for some crops (Kumar et al. 2008) with food production failing to keep pace with population growth (G. Rasul and A. Schild 2009). As a result, South-Asian countries are now finding it difficult to meet their population most basic food and nutritional needs and remain vulnerable to food insecurity (Tab.1).

Impact of climate change on food productivity

The impacts of climate change will reduce productivity and lead to greater instability in production in the agricultural sector (crop and livestock production, fisheries and forestry) in communities that already have high levels of food insecurity and environmental degradation and limited options for coping with adverse weather conditions (FAO,2009). The agriculture sector is not only among the most vulnerable sectors to the impacts of climate change; it is also directly responsible for the 14 percent of global greenhouse gas emissions. In addition, the sector is a key driver of deforestation and land degradation, which account for an additional 17 percent of emissions [FAO, 2010a]. The agricultural sector can be an important part of the solution to climate change by capturing synergies that exist among activities to develop more productive food systems and improve natural resource management (Sahoo, 2014a). Sustainable utilization of natural

Table 1: Key features of agriculture and food security in South- Asia.

Country	Population		Agricultural Land									
	2007 millions	Pop. Growth rate (%)	Total land area (m ha) in 2003	Agrl. Land % of land area	Arable lands (% of Agrl. land)	Per capita land (ha) 2003-05	Irrigated crop land (%)2002	Cereal production (m tones*)	Cereal production (Kg per capita 2003-05)	% of under nourished pop.	% GDP agriculture (2004)	% of Pop. in agriculture
Afghanistan	28.1	1.36	65.2	-	20.8	-	29.6	0.12	-	35	36	23.36
Bangladesh	142	1.9	14.4	69.6	88.2	0.1	54.5	25.95(11.2)	285	30	20	28.53
Bhutan	0.71	2.2	4.4	-	26.9	0.08	24.2	0.10(12.4)	-	-	26	50.47
India	1095	1.6	32.7	60.9	88.6	0.3	33.6	196.84(4.5)	219	21	20	25.45
Nepal	27	2.1	14.718	34.8	55.8	0.1	34.5	4.6(8.5)	288	17	39	42.93
Pakistan	156	2.0	79.6	35.2	78.6	0.3	80.5	25.0(8.4)	203	20	22	17.24
Maldives	0.30	2.5	0.30	-	28.6	-	-	-	-	-	8	10
Sri Lanka	21	0.9	6.561	36.4	38.9	0.2	33.3	-	155	22	15	19.17

*Figures in parenthesis shows variation in domestic cereal production

Source: World Development Report 2008; (Allauddin and Qulggin 2008)

resources will require management and governance practices based on ecosystem approaches that involve multi-stakeholder and multi-sectored coordination and cooperation. This is a crucial element for the transformation to climate resilient agriculture. More productive and resilient agriculture is built on the sound management of natural resources, including land, water, soil and biodiversity (Nanda and Garanayak 2010). Conservation agriculture, agroforestry, improved livestock and water management, integrated pest management and ecosystem approaches to fisheries and aquaculture can all make important contributions in this area. According to IPCC 4th assessment report, agriculture is currently responsible for about one third of the World's GHG emissions and this share is projected to grow, especially in developing countries. To support food security and boost incomes, agricultural systems in developing countries will be under pressure to increase productivity sustainably and strengthen the resilience of agricultural landscapes. Strategies exist to sequester carbon and reduce greenhouse gas emission reductions in the agricultural sector. Many of these strategies also improve food security, foster rural development and help communities adapt to climate change. However, tradeoffs may have to be made when seeking to reach different development goals, such as climate change mitigation and adaptation, sustainable agricultural production and poverty reduction.

Climate smart agriculture- FAO new initiative

Recently, FAO has developed a new tool box "Climate Smart Agriculture" looking to global climate change scenario and existing land-use policies, which is defined as "Agriculture that sustainably increases productivity, resilience (adaptation), reduce / remove GHGs (mitigation) and enhance achievement of national food security and development goals" is now the only panacea to achieve food security and future challenges on climate change [FAO, 2010b and 2013]. Climate Smart Agriculture promotes agricultural best practices, particularly integrated crop management, conservation agriculture, agroforestry, improved seeds and fertilizer management practices, improved water and energy management practices as well as supporting increased investment in agriculture sector. It encourages the use of all available and applicable climate change solutions in pragmatic and impact focused manner .While resilience is a key, Climate Smart Agriculture is broader and calls for more innovations and pro-activeness on climate smart practices towards agriculture resilience which change in the way the farming to adapt and mitigate while sustainably increasing productivity. Further, Climate Smart Agriculture practices propose the transformation of agricultural policies and agricultural systems to increase food productivity and enhance food security while preserving the environment and ensuring resilience to changing climate. It has different meanings depending upon the scale at which it is being applied. For example, at the local scale, it may provide opportunities for higher production through improved management techniques such as more targeted use of fertilizers. At the national scale it could mean providing a framework that incentives sustainable management practices. And at the global scale it could equate to setting rules for the global trade of carbon. It is not clear how actions at one scale may affect the others. For smallholder farmers in developing countries, the opportunities for greater food security and increased income together with greater resilience will be more important to adopt climate-smart agriculture than mitigation opportunities.

In the face of climate change, agriculture must be 'carbon neutral', farming activities would take out of the atmosphere a quantity of greenhouse gas equal to or greater than the amount they emit (Verchot 2007). At the very least, these activities should aim to meet future food requirements without further increases in emissions. Some practices with high potential for synergies over a wide range of circumstances can, however, be identified, such as avoiding bare fallow; incorporating crop residues; diversifying crop rotations to incorporate food producing cover crops and legumes; improving fodder quality and production; expanding low energy irrigation; expanding agroforestry; and adopting soil and water conservation techniques which contribute significantly for resilient agriculture and food production (Rabindranath et.al 2008;Lasco & Pulhin,2009; Verchot,2009 and Sahoo, 2014 b.).

Due to intensive agriculture several ills have also appeared in Indian agriculture, such as declining factor productivity, degradation of natural resource base, environmental degradation including ground water depletion & contamination and declining farm profitability & productivity (Gill *et al.*, 2009). To tackle such problems, farming systems approaches to research has been widely recognized, where whole farm is viewed as a system and interactions among the various components are taken into consideration (Nanda et.al.2007). Efficient integration of crops with animals (cow, buffalo, pig, goat, sheep, fish), birds (poultry, pigeon, duck), multipurpose trees and agro-forestry systems and other enterprises (bio-gas, apiary, mushroom, etc.) clearly showed the best advantages over conventional system of cropping under irrigated and rainfed conditions as well as in tribal areas (Jayanthi *et al.*, 2001). Integrated Farming System (IFS), which focused around a few selected inter-dependant, inter-related and often inter-linking production systems seems to be the viable alternative to ensure food, nutritional and economic-security, particularly to the large chunk of small and marginal farmers of the state (84%), which operate under complex-diverse-risk prone environment. Adoption of Integrated Farming System provides high opportunities of productivity enhancement, employment, income generation, nutritional security and a climate resilient agricultural production system by diversifying and integrating different components of farming viz., crop, horticulture, livestock and fisheries. The systems based on multiple recycling of carbon, energy and nutrients would also help minimize environmental loading with pollutants. Mixed farming involving crop-livestock integrations has become a way of life and means to livelihood in rural areas of the state, where livestock and livelihoods are intimately related. About 85% of livestock in the state are owned by the landless, marginal and small farmers and 80% of rural households depend on livestock from where they draw 30% of their annual income for sustenance. About 70% of total poultry are of local backyard breeds, which were again owned by the down trodden ones. The livestock component acts as a stabilizing factor in the system, thus needs strengthening of the crop and livestock linkage to enhance the economic viability and sustainability of the farming systems. Further, Livestock and livelihoods are intimately related and the ownership of livestock is more egalitarian than that of land (Swaminathan, 2010). Among various Integrated Farming Systems, pond based system-encompassing fish-rice-duck/ poultry-vegetables and fish-cow/pig-duck/poultry-vegetables are feasible particularly, in rainfed lowlands. In this system, marginal land/wet lands are brought into productive use, animal wastes are used to fertilize fish pond/refuge and crop lands, while the lands in return produce crops which serve as food for animal, fish and men and the pond provides food for fish, duck and men. As system's dykes are comparatively wide, housing units for cow/ pig/duck/poultry along with cultivation of economic varieties of plants such as vegetables and fruit plants are carried out, using pond water for life saving irrigation at different growth stages. Rice-fish-duck culture has great potential in this ecosystem. Ducks are eco-friendly who used to destroy harmful weeds, snails and insects from rice fields and improve soil fertility. Duck droppings serve as feed for fishes. Thus fish and duck together would ensure high profit due to higher fish yield, duck eggs, duck meat and enhanced rice yield. Profit of margin will be enhanced much more than any individual system. Besides, this can provide gainful employment to family members and improve their nutritional status. The dykes to be constructed for preventing escape of fish from the integrated system can be used for growing vegetables and other short gestation fruit trees like papaya, banana and drumstick to make the system more economically viable. Vegetables such as cucurbits, radish, brinjal, okra, leafy vegetables during *kharif* season and tomato, french bean, radish, bitter gourd, cucumber, cauliflower, cabbage, brinjal, pumpkin and leafy vegetables (coriander, amaranths and Indian spinach) can be grown during winter season. Vegetables such as snake gourd, bitter gourd, ridge gourd, bottle gourd or ash gourd can be grown throughout the year on raised platforms and okra, poi, cowpea, bitter gourd, ridge gourd and pumpkin in the field during summer season. Therefore, rice-fish-duck farming should be encouraged particularly in the rainfed lowland areas. Sunken and raised system need to be popularized in the low land waterlogged areas to enhance water productivity.

In the rainfed upland ecosystem, crop diversification, management of soil and water on the basis of watershed, creating water-harvesting structure in 10% area towards lower reach in sloppy land for life saving irrigation, growing suitable crops, livestock and adopting ancillary avenues of income appear important. The income from cultivation of cereals, pulses, oilseeds, spices, fiber crops, forestry species, cash crops, vegetables, fruits and flowers along with other enterprises like dairy, poultry, food processing, mushroom, honey bee and combination of all these enterprises or integrated farming systems utilizing the by-products of the main crop and engagement of family labour can contribute to better livelihood security. So, farming system aims for enhanced productivity, profitability, sustainability, climate resilient agricultural production, balanced food and nutrition, clean environment, recycling of resources, optimum utilization of farm resources like labour, animal power, machinery, finance & minimum dependence on external inputs, adoption of improved technologies, regulation of farm income, creation of employment opportunities, high input-out put ratio, solving food, fodder, feed, fuel & fertilizer crisis, encouraging afforestation, promoting agro-based industries, and ultimately improving standard of living of the farming community (Gill *et al.*, 2009).

Climate smart agriculture: integrated farming system practice and CDM (clean development management) opportunities

The small holders, who do sequester carbon and also reduce emissions, individually as well as collectively through different farming system practices, and qualify for the minimum tradable carbon amount are usually left isolated from this opportunity, mainly because they do not have the capacity to comply with the rigid and expensive requirements of the carbon market schemes, especially the CDM protocol. Moreover, in most cases, they are also not organized in groups, which further preclude their participation in such ventures. This fact is of enormous importance to the Indian small holders, who comprise the majority of farmers, collectively contribute to a substantial load of carbon sequestration, and yet are unable to benefit from such opportunities, and are left wandering alone. By way of investment from developed countries, the CDM projects could also lead to a large positive impact on programs which are aimed at forest conservation and regeneration, through afforestation / reforestation, reclamation of degraded lands, integrated farming system, conservation agriculture practices and socio-economic development of rural communities, in addition to the global environmental benefits. Now this is being looked upon as a major benefit with market instruments favoring the active involvement of industry in the mechanism. The potential for CDM related projects is quite high in India (UNFCCC 2008a and 2008b). Although it is not mandatory for Indian industries to meet any sort of targets under Kyoto protocol, they prefer to favor this phenomenon as it strengthens their claim for using the wastelands for this purpose in the country. India has the biggest scope for generating Certified Emission Reductions (CERs). Few carbon projects under regulated market mechanism (CDM) and voluntary carbon market mechanism has already been initiated in India under regulatory framework of UNFCCC (Table-2, Table 3 and Table.4). However, the country currently has the largest number of CDM projects, but China is largest in terms of carbon tonnage. The Government of India has set up a National Designated Authority (NDA) to deal with the CDM projects and promote climate finance benefits in the country.

The global carbon market has been linearly expanding in recent years. From a meager of about 8 billion Euros in 2005, it has grown to about 92 billion Euros in 2008. Through this opportunity, a number of agencies, corporate houses and some large NGOs have benefited from it. The paid up benefits so far have been for reducing emissions, carbon sequestration, or both. Few such benefits have also accrued in India, but mostly for reducing the emissions. However, only large companies, which could qualify for the minimum tradable amount of carbon, afford to properly write the application, prove the additionality in emission reduction, and/or carbon sequestration and get their efforts/ projects validated, mostly by external agencies at usurious costs, sometimes running to between 100 and 200 thousand dollars per case, have been able to cash on this opportunity.

Integrated farming systems for evergreen agriculture

Throughout the world, agriculture is faced with an immense challenge: how to increase yields to feed a growing population from depleted natural resources such as soil, water and biodiversity etc. in the face of climate change. So there is an urgent need to increase biomass production in farming systems with richer sources of organic nutrients to complement whatever amounts of inorganic fertilizers a smallholder farmer can afford to apply. Many of the farmers of Zambia, Malawi, Niger and Burkina Faso are meeting this challenge through the adoption of evergreen agriculture by integrating agroforestry and conservation farming (www.worldagroforestry.org/evergreen_agriculture). Evergreen agriculture, where trees are intercropped in annual food crops systems, sustains a green cover on the land throughout the year. Its bolsters nutrient supply through nitrogen fixation and nutrient cycling, which is urgently needed in rainfed situations; increases direct production of food, fodder, fuel, fibre and income from products produced by the trees; improves micro climate and soil water relations enabling greater adaptation to climate change; increases

Table 2: Mitigation Carbon Projects related to Afforestation/Reforestation for regulated markets (CDM)

Sl. No	Title of CDM project	Activity planned	Total Emission Reductions of the project (t CER)	Duration of the project	Annual emission reductions (Tons/year)	Starting Year
1.	Small Scale Cooperative Afforestation CDM Pilot Project Activity on Private Lands Affected by Shifting Sand Dunes in Sirsa, Haryana	Planting high timber value trees in degraded & barren lands to improve the livelihoods of local people	231920	20	11596	2008
2.	Reforestation of severely degraded landmass in Khammam District of Andhra Pradesh, India under ITC Social Forestry Project	Planting high timber value trees in degraded & barren lands to improve the livelihoods of local people	1733753	30	57792	2001
3.	The International Small Group and Tree Planting Program (TIST), Tamil Nadu, India	Planting high timber value trees in degraded & barren lands to improve the livelihoods of local people	107810	30	3594	2004
4.	India: Himachal Pradesh Reforestation Project – Improving Livelihoods and Watersheds	Planting high timber value trees in degraded & barren lands to improve the livelihoods of local people	828016	20	41401	2006
5.	Improving Rural Livelihoods Through Carbon Sequestration By Adopting Environment Friendly Technology based Agroforestry Practices- Orissa & Andhra Pradesh	Planting high timber value trees in degraded & barren lands to improve the livelihoods of local people	146888	30	4896	2004
6.	Bagepalli CDM Reforestation Programme-Karnataka	Planting high timber value trees in degraded & barren lands to improve the livelihoods of local people	2116531	20	105827	2008
7.	Agro-forestry Interventions in Koraput district of Orissa	Planting high timber value trees in degraded & barren lands to improve the livelihoods of local people	219895	30	7330	2009

Contd.

Sl. No	Title of CDM project	Activity planned	Total Emission Reductions of the project (t CER)	Duration of the project	Annual emission reductions (Tons/year)	Starting Year
8.	Rehabilitation of Degraded Wastelands at Deramandi in Southern District of National Capital Territory of Delhi through Reforestation	Planting high timber value trees in degraded & barren lands to improve the livelihoods of local people	451743	30	15058	2008
9.	Araku Valley Livelihood Project- Andhra Pradesh	Planting high timber value trees in degraded & barren lands to improve the livelihoods of local people	1641701	20	820851	2010
10.	Arranging CERs through CDM over Mango/Cashew/Indian Gooseberry plantations- Odisha	Planting high timber value trees in degraded & barren lands to improve the livelihoods of local people	9268390	30	308946	2014

Table 3: Mitigation Carbon Projects related to Agriculture

Sl. No	Title of Carbon project	Activity planned	Total Emission Reductions of the project (t CER)	Duration of the project	Annual emission reductions (Tons/year)	Starting Year
1.	Improvement in Energy Efficiency through Micro-Irrigation Systems (MIS) in cultivation of Banana Crop in Jalgaon, Dhule Nadurbar and Nashik districts, Maharastra State, India-CDM	Replace flood method of irrigation with drip method of irrigation	71568	7	10224	2011
2.	System of Rice Intensification (SRI) program for reduction of methane emissions and water consumption in rice fields of IndiaVCS-001-Bihar	To utilize water efficiently and reduce the Methane emissions from the paddy fields	567756	10	56776	2014

Table 4: Mitigation Carbon Projects related to Afforestation/Reforestation under Voluntary Carbon Market (VCM)

Sl. No	Title of CDM project	Activity planned	Total Emission Reductions of the project (t CER)	Duration of the project	Annual emission reductions (Tons/year)	Starting Year
1.	Indian Farm Forestry Development Co-operative Limited (IFFDC) Watershed Forestation Project-Chhattisgarh	Planting high timber value trees in degraded & barren lands to improve the livelihoods of local people	320467	64	5007	2002
2.	TIST Program in India CCB-001 Tamil Nadu	Planting high timber value trees in degraded & barren lands to improve the livelihoods of local people	107810	30	3594	2004
3.	Araku Valley Livelihood Project Andhra Pradesh -	Planting economic value plants in degraded lands to raise the livelihoods of local people	1413768	20	69,997	2010
4.	Community-based reforestation project on degraded lands in Uttar Pradesh, India by Indian Farm Forestry Development Co-operative Limited- Uttar Pradesh	Planting economic value plants in degarded lands to raise the livelihoods of local people	169554	30	5652	2008
5.	Plantation Project on wastelands by Sun Plant Agro Limited- Haryana	Planting economic value plants in degraded lands to raise the livelihoods of local people	56379	33	1708	2006
6.	Reforestation of degraded land in Chhattisgarh, India-Chhattisgarh	Planting economic value plants in degraded lands to raise the livelihoods of local people	320467	64	5007	2002
7.	TIST Program in India, VCS 001-Tamil Nadu	Planting economic value plants in degraded lands to raise the livelihoods of local people	331410	30	11047	2004

Table 5: Transformation towards Climate Smart Village in the eve of Food Security and Climate Change

Key Interventions for Climate Smart Village

Sl. No	Weather Smart	Water Smart	Carbon Smart	Nutrient Smart	Energy Smart	Knowledge Smart
1.	Seasonal Weather Forecast	Aquifer recharge	Agroforestry	Site Specific Nutrient Management	Fuel efficient Engines	Farmer-Farmer Learning
2.	ICT based Agro advisory	Rain water harvesting	Conservation Agriculture	Precision Fertilizer	Biofuels	Farmer Network on Adaptation of Technology
3.	Climate analogues	Alternate Drying and Wetting	Zero tillageor Minimum tillage	Catch cropping / Legumes	Solar Solutions	Seed and Fodder Banks
4.		Laser land leveling	Change in land-use towards evergreen agriculture		Energy efficient system use	Market Information
5.		Community Management of Water	Livestock Management		Residue Management	Off-Farm Risk Management
6.		On farm Water Management				Homestead Agroforestry / Kitchen Garden

carbon accumulation in food crop farming; enhances biodiversity in annual crop / agroforestry system; and reduces deforestation due to enhanced potential in rainfed agriculture. Evergreen agriculture is primarily based on five important principles, such as minimizing soil disturbance, maintaining land/ soil cover, practicing crop rotation, good agronomic management practices, and intercropping fertilizer and high value trees. The evergreen agriculture now experienced by its wide application in Africa, where diversity and complexity is a common feature of agricultural systems. The most promising results are coming from the integration of fertilizer trees into cropping systems. These trees improve soil fertility by drawing nitrogen from the air and transferring it to the soil through their roots and leaf litter. Scientists have been evaluating various species of fertilizer trees for many years, including *Sesbania, Gliricidia and Tephrosia.* Currently, *Faidherbia albida* is showing particular promise as a possible corner stone of Evergreen Agriculture in the future. Now, promising results have also been observed from decades of research conducted in India and Bangladesh. A new initiative in Asia is immerged by developing a broad alliance of governments, research institutions and local and international development partners committed to spearheading the Evergreen Agriculture Programme in South Asia via the Partnership of National Agriculture Research and Extension Systems (NARES), thus the establishment of the South Asian Network on Evergreen Agriculture (SANEA). Bangladesh, India and Sri Lanka have pledged providing resources for the network and similar support is forth coming from other countries. After the emergence of Climate Smart Agriculture by UN FAO, it is now need to transform the whole agriculture practice by developing smart villages to ensure food security without further disturbing to the environment (Table-5). Further, underpin new framework by simulating biophysical, climatic, social, economic, innovation, human capital and policy buffers for MRV (Monitoring, Reporting and Verification) of CSA (Climate Smart Agriculture) growth system for effective management and tenural security of the program.

Conclusion

Transformation, now need in global agriculture in the context of climate change and food security (Sahoo et al.2014c). In agriculture based countries, where agriculture is critical for economic transforming smallholder systems, is not only important for food security but also for poverty reduction, as well as for aggregate growth and structural change. In developing countries, increasing productivity to achieve food security is clearly a priority, which is projected to entail a significant increase in emissions from the agricultural sector in developing countries. It is the time to promote Integrated Farming System through Climate Smart Agriculture practices for a food secure world through the provision of science-based efforts that support sustainable agriculture and enhanced livelihoods while adapting to climate change and conserving natural resources and environmental services with focus on developing countries. Further, it will be a part of the solution by contributing to climate change mitigation, through carbon conservation, sequestration and substitution, and establishing ecologically designed agricultural systems that can provide a buffer against extreme events. So, Climate Smart Agriculture is important both for climate change mitigation as well as adaptation through reducing vulnerability, diversifying income sources, improving livelihoods and building capacity of smallholders to adapt to climate change. Further, strong local institutions are in need for identifying, coordinating and recognizing informal rights and strengthening customary systems while scaling up Climate Smart Agriculture.

References

FAO. 2007. Adaptation to climate change in agriculture, forestry and fisheries: Perspective, framework and priorities, Interdepartmental Working Group on Climate Change, Food and Agriculture Organization (FAO) of the United Nations, Rome.

FAO. 2009. Climate Change Mitigation Finance for Smallholder Agriculture: A guide book to harvesting soil carbon sequestration benefits. FAO, Rome.

FAO. 2010a. Carbon finance possibilities for Agriculture, Forestry and Other Land Use Projects in a Smallholder Context. FAO. Rome.

FAO, 2010b. "Climate Smart Agriculture" - Policies, Practices and Financing for Food Security, Adaptation and Mitigation. Food and Agriculture Organization of the United Nations , Rome.

FAO, 2013. "Climate Smart Agriculture"– Source Book, Food and Agriculture Organization of the United Nations, Rome.

Gill, M.S., Singh, J.P. and Gangwar, K.S. 2009. Integrated farming system and agricultural sustainability. *Indian Journal of Agronomy* 54 (2):128-139

IPCC. Climate Change Synthesis Report, 2007. Bernstein L, Bosch P, Canziani O, Chen Z, Christ R, Davidson O, Hare W, Huq S, Karoly D, Kattsov V, Kundzewicz Z, Liu J, Lohmann U, Manning M, Matsuno T, Menne B, Metz B, Mirza M, Nicholls N, Nurse L, Pachauri R, Palutikof J, Parry M, Qin D, Ravindranath N, Reisinger A, Ren J, Riahi K, Rosenzweig C, Rusticucci M, Schneider S, Sokona Y, Solomon S, Stott P, Stouffer R, Sugiyama T, Swart R, Tirpak D, Vogel C and Yohe, G. Cambridge University Press, Cambridge

Jayanthi, C., Rangasamy, A., Mythili, S., Balusamy, M., Chinnusamy, C. and Sankaran, N. 2001. Sustainable productivity and profitability to integrated farming systems in low land farms. In: *Extended Summaries*, pp. 79-81. (Ed:A.K.Singh, B.Ganwar, Pankaj and P.S.Pandey), *National Symposium on Farming System Research on New Millennium,* PDCSR, Modipuram.

Kumar, P; Mittal, S and Hossain, M .2008 Agricultural Growth Accounting and Total Factor Productivity in South-Asia: A Review and Policy Implications. Agricultural Economics Research Review, 21:145-172.

Lasco R, Pulhin F. 2009. Agroforestry for Climate Change Adaptation and Mitigation. An academic presentation for the College of Forestry and Natural Resources (CFNR), University of the Philippines Los Banos (UPLB), Los Banos, Laguna, Philippines.

Nanda, S.S. and Garnayak, L.M. 2010.Integrated farming for livelihood security in coastal ecosystem. Proceeding of National Symposium on Resource Management Approaches towards Livelihood Security held at University of agricultural Sciences, Bengalore, Karnataka. Dec.2-4, 160-167pp.

Nanda, S.S. 2007. Farming systems for different agro-ecosystems. *Presentation in the Seminar on Road Map for Agriculture Development in Orissa* held during 6-7 November OUAT, Bhubaneswar, Odisha, India.

Rasul, G and Schild, A. 2009. Food Security and Agricultural Sustainability in South-Asia. Proceedings of National Seminar on Food Security and Sustainability in India, Nov.07-08 at New Delhi, India, 32-41 pp.

Ravindranath, N.H.; Chaturvedi, R.K. and Murthy, I.K.. 2008. Forest conservation, afforestation, and reforestation in India; Implications for forest carbon stocks. Current Science, 95: 216-222.

Sahoo, A.K. 2011. Climate Mitigation and Carbon Finance: Global Initiatives and Challenges. Book Published by New India Publishing Agency (ISBN 978-93-81450-02-4), New Delhi, India.

Sahoo, A K. 2014a. Climate Smart Agriculture: Transformation towards Green Economy and Global Power. International Journal of Tropical Agriculture. Vol.32 (1-2) January-June 2014 pp.317-324.

Sahoo, A K 2014b. Technological Challenges on Climate Smart Agriculture: A Global Issue. Proceedings of Annual Group Meet of All India Coordinated Research Project on Agroforestry on "Agroforestry Options for Climate Resilient Farming" held on 26th-28th July, 2014 at Bhubaneswar pp.89-91.

Sahoo, A.K., Nanda, S.S., Nayak.D and Singh,V.P.,2014c. Climate Smart Agriculture: Global transformation towards resilience and food security. In Extended Summaries of Odisha Chapter, Indian Society of Agronomy & Orissa University of Agriculture and Technology, Bhubaneswar Pub. pp.18-20.

Shames, S. and Scherr, S.J. 2010. Institutional Models for Carbon Finance to mobilize Sustainable Agricultural Development in Africa. US AID, USA

Singh, V.P. (2008). Agroforestry for food, livelihoods and environmental security. Paper presented during the India-Africa Summit, 10-12 Nov. New Delhi, India.

Singh, V.P. 2010. Addressing the Challenges of Climate Change for Securing Sustainable Livelihoods in India. IFAD Investment Strategy Paper for India. In: COSOP Document for India, IFAD, Rome. 27 pages.

Singh,V.P.; Nayak, D. and Bohra, B. 2012. Agroforestry Science for Achieving an Evergreen Agriculture. In the Proceedings of the National Seminar on Agroforestry - an Evergreen Agriculture for Food Security and Environment Resilience, 02-04 February 2012, Navsari Agricultural University, Navsari (Gujarat) 396450, India.

Swaminathan, M. S. 2010. Enhancing the disaster resilience of agriculture. *The Hindu Survey of Indian Agriculture*: 7-10pp.

UNFCCC. 2008a. Challenges and opportunities for mitigation in the agricultural sector Technical paper.

UNFCCC. 2008b. Ad Hoc Working Group on Long-Term Co-operative Action under the convention. Fourth Session Poznan 1-10 December 2008.Ideas and submissions from intergovernmental organizations. Ideas and proposals on the elements contained in paragraph 1 of the Bali Action Plan. Paper No1 FAO and IFAD. Financing climate change adaptation and mitigation in the agricultural and forestry sectors.

Verchot, L.2007. Design and best practices for agroforestry and community forestry carbon projects: A work plan for ICRAF in the CPR Alliance, ICRAF, Nairobi, Kenya.

www.worldagroforestry.org/evergreen-agriculture

Integrated Farming System Practices: Challenges & Opportunities, pp 455-470

New India Publishing Agency, New Delhi, India

18

Conservation Agriculture for Smallholder Farming Systems

P.K. Roul, A. Pradhan and K. N. Mishra

Introduction

The challenge of increased food production to meet food security needs still persists 50 years after the start of the Green Revolution as population growth continues to increase in many developing countries. There appears to be no alternative but to increase agricultural productivity and the associated individual factor productivities (such as water, labor, nutrients, energy and capital) to meet the food demand by 2050 and to alleviate poverty and hunger. Thus, needs technological initiatives for further agricultural intensification. But already happening negative effect of crop production intensification on essential natural resources such as soil, water, biodiversity, and associated ecosystem services has forced agriculture stakeholders to search for novel initiatives that focus on sustainable production optimization rather than just intensification. Furthermore increased food supply is a necessity though not sufficient for eliminating hunger and poverty. What is important is who produces the food, has access to the technology and knowledge to produce it, and has the purchasing power to acquire it. Of the developing world's 5.5 billion people, 3 billion live in rural areas – nearly half of humanity. Of these rural inhabitants, an estimated 2.5 billion are in households involved in agriculture, and 1.5 billion are in smallholder households (World Bank, 2008 and Herren, 2010). Poor farmers need low-cost and readily available technologies and practices to increase local food production and to

Orissa University of Agriculture & Technology, Bhubaneswar-751003, Odisha, India raise their income. At the same time, land and water degradation is increasingly posing a threat to food security and the livelihoods of rural people who generally live on degradation-prone lands. There are several other key issues while relating the situation to Indian context.

- Even though India's agricultural growth has been sufficient to move the country from severe food crises of the 1960s to aggregate food surpluses today, most of the increase in agricultural output over the years has taken place under irrigated conditions. On the other hand, the opportunities for continued expansion of irrigated area are limited, as a result of which there is increased emphasis on rainfed, or unirrigated agriculture to help meet the rising demand for food projected over the next several decades.
- Rainfed agriculture accounts for about two-thirds of total cropped area and nearly half of the total value of agricultural output. Nearly half of all food grains are grown under rainfed conditions, and hundreds of millions of poor rural people depend on rainfed agriculture as the primary source of their livelihoods but there is involvement of high risk at both environment and socio-economic level.
- In India, rains are received within the short period of the rainy season. It rains for short periods of time. But it rains very ferociously. The high velocity raindrops break up the surface of the soil and carry the fragments away making them unproductive and degraded. Again the current cultivable land which is 43% of total geographic area, is shrinking because of rapid urbanization and use of agricultural land for other purposes.
- India is home to 22 percent of the world's poor and a bulk of it is represented by agricultural wage earners, small and marginal farmers (Ansari and Akhtar, 2012). Of these, 75 percent live in rural areas having total dependence on agriculture and generally produce much of what they eat. Small land holdings and their low productivity along with uncertainties in occurrence of rain, hit the small farmers and laborers worst further making their situation critical.

All these call for identification of opportunities for stimulating agricultural growth and reducing poverty and environmental degradation in rainfed areas. A set of soil-crop-nutrient-water-landscape system management practices known as conservation agriculture (CA) has the potential to address all of these issues.

Conservation agriculture (CA)

CA is a concept for resource-saving agricultural crop production that strives to achieve acceptable profits together with high and sustained production levels while concurrently conserving the environment. CA is based on enhancing natural biological processes above and below the ground. Interventions such as mechanical soil tillage are reduced to an absolute minimum, and the use of external inputs such as agrochemicals and nutrients of mineral or organic origin are applied at an optimum level and in a way and quantity that does not interfere with, or disrupt, the biological processes. CA is characterized by three principles which are linked to each other, namely: 1.Continuous minimum mechanical soil disturbance. 2. Permanent organic soil cover. 3. Diversified crop rotations in the case of annual crops or plant associations in case of perennial crops.

The first key principle in CA is practicing minimum mechanical soil disturbance which is essential to maintaining minerals within the soil, stopping erosion, and preventing water loss from occurring within the soil. In the past agriculture has looked at soil tillage as a main process in the introduction of new crops to an area. It was believed that tilling the soil would increase fertility within the soil through mineralization that takes place in the soil. Also tilling of soil can cause severe erosion and crusting which will lead to a decrease in soil fertility. Today tillage is seen as a way as destroying organic matter that can be provided within the soil cover. No till farming has caught on as a process that can save soils organic levels for a longer period and still allow the soil to be productive for longer periods (FAO 2007). Also with the process of tilling requires the time and labor for producing that crop. When no-till practices are followed, the producer sees a reduction in production cost for a certain crop. Tillage of the ground required more money due to fuel for tractors or feed for the animals pulling the plough. The producer sees a reduction in labor because he or she does not have to be in the fields as long as a conventional farmer.

The second key principle in CA is maintaining a permanent organic soil cover, which is much like the first principle in dealing with protecting the soil. This principle can allow for growth of organisms within the soil structure. This growth will break down the mulch that is left on the soil surface. The breaking down of this mulch will produce a high organic matter level which will act as a fertilizer for the soil surface. If the practices of CA were being done for many years and enough organic matter was being built up at the surface, then a layer of mulch would start to form. This layer helps prevent soil erosion from taking place and ruining the soils profile, also when soils are covered under a layer of organic cover, the ground is protected so that it is not directly impacted by rainfall (Hobbs *et al.* 2007). This type of ground cover also helps keep the temperature and moisture levels of the soil at a higher level rather than if it was tilled every year (FAO 2007).

The third principle is the practice of crop rotation with more than two species. According to an article published in the *Physiological Transactions of the Royal Society* called "The role of conservation agriculture and sustainable agriculture," crop rotation can be used best as a disease control against other preferred crops (Hobbs *et al.* 2007). This process will not allow pests such as insects and weeds to be set into a rotation with specific crops. Rotational crops will act as a natural insecticide and herbicide against specific crops. Not allowing insects or weeds to establish a pattern will help to eliminate problems with yield reduction and infestations within fields (FAO 2007). Crop rotation can also help build up a soils infrastructure. Establishing crops in a rotation allows for an extensive buildup of rooting zones which will allow for better water infiltration (Hobbs et al. 2007).

Benefits of conservation agriculture

In the field of CA there are many benefits that both the producer and conservationist can obtain. On the side of the conservationist, CA can be seen as beneficial because there is an effort to conserve what people use every day. Since agriculture is one of the most destructive forces against biodiversity, CA can change the way humans produce food and energy. These benefits include less erosion possibilities, better water conservation, improvement in air quality due to less emission being produced, and a chance for larger biodiversity in a given area. On the side of the producer and/or farmer, CA can eventually do all that is done in conventional agriculture, and it can conserve better than conventional agriculture. CA according to Theodor Friedrich, who is a specialist in CA, believes "Farmers like it because it gives them a means of conserving, improving and making more efficient use of their natural resources" (FAO, 2006). Producers will find that the benefits of CA will come later rather than sooner. Since CA takes time to build up enough organic matter and have soils become their own fertilizer, the process does not start to work over night. But if producers make it through the first few years of production, results will start to become more satisfactory.

CA is shown to have even higher yields and higher outputs than conventional agriculture once CA has been establish over long periods. Also, a producer has the benefit of knowing that the soil in which his crops are grown is a renewable resource. According to New Standard Encyclopedia, soils are a renewable resource, which means that whatever is taken out of the soil can be put back over time (New Standard, 1992). This could be very beneficial to a producer who is practicing CA and is looking to keep soils at a productive level for an extended time. The farmer and/or producer can use this same land in another way when crops have been harvested. The introduction of grazing livestock to a field that once held crops can be beneficial for the producer and also the field itself. Livestock can be used as a natural fertilizer for a producer's field which will then be beneficial for the producer the next year when crops are planted once again. The practice of grazing livestock in a CA helps the farmer who raises crops on that field and the farmer who raises the livestock that graze off that field. Livestock produce compost or manures which are a great help in producing soil fertility (Pawley W.H., 1963). With the practices of CA and grazing livestock on a field for many years can allow for better yields in the following years as long a practices are continued to be followed.

Smallholder farming situations of tribal farmers in Odisha

India being the second largest tribal dominated area after Africa has as many as 427 tribal communities which are largely concentrated in central and North-East India region. Among them, Odisha is one such tribal dominated state with the largest number of tribal communities (*i.e.* 62). The total tribal population of the state is 8.15 million and of 30 administrative districts, 6 districts are declared as fully tribal districts, Keonjhar being one of them. It is one of the most backward areas of the state having tribal people more than 56% of total population and a literacy rate as low as 24%. The district consists of tribal households with more than 90% of the population depending on agriculture of farm size less than 1 ha. Upland agriculture in this area is predominantly rainfed having low crop yields as the lands suffer from poor soil fertility and productivity, poor moisture retention, susceptibility to water erosion and other external pressures of development and environmental conflicts of climate change (SMARTS, 2009). The predominant upland crop in this tribal region is maize (*Zea mays* L.), grown during the rainy season (June to September), followed by mustard (*Brassica campestris* L.), an oilseed grown in the post-rainy season (October to January). All the crop produce are primarily used for household consumption. After the crop harvest, fields are generally left fallow for the remainder of the year but are subject to unmanaged grazing by livestock. All the crop residues are generally used for fuel purpose and there is no legume cultivation in the fields since past 40-50 years. Again continuous monocropped maize is not sustainable as it degrades the natural base and is labor intensive (Mloza-Banda, 2005; Ito et al., 2007). Furthermore, seasonal rainfall distribution dominates the climate related crop production constraints in rainfed agriculture (Harris, 1996). In growing seasons of good rainfall distribution crop yields are limited by low soil fertility and poor agronomic practices. Conventional practices include cross-plowing to prepare the soil for sowing and low inputs of organic matter and nutrients, primarily as uncomposted farmyard manure, which is insufficient to meet crop demands and replace nutrients lost in harvested yields (Wang *et al.,* 2009). Deep tillage along with high intensity rainfall and no residue cover makes the land more degraded through soil erosion. There has been increasing realization that the solution to maintenance and improvement of soil fertility cannot be solely through use of inorganic fertilizers (Ngwira *et al.,* 2012). A shift towards more sustainable cropping system such as conservation agriculture may help in reversing soil degradation and improving crop production as well as socio-economic condition of the tribal farmers in the area. Conservation agriculture enables farmers to improve their resource base using their own resources. First, farmers should keep the soil covered as far as possible - either by leaving the crop residue after the harvest or by letting something else grows on the soil. Two; tillage of fields should be kept to a minimum and three; they should rotate crops spatially and temporally.

CA for smallholder tribal farming situations

There is no such one-size-fit-all ready-to-use blue print recipe of conservation agriculture for smallholder farms. The actual crop and soil management practice is site specific and depends on climate, cropping history, land type, culture and socio-economic condition of the farmers, the combination of all of which determines a set of treatments known as conservation agriculture production systems (CAPS). First with respect to tillage, conventional way of tribal farmers is to go for 2-3 times plowing before sowing which needs to be shifted to minimum tillage. Again for determination of cropping system, it was known from baseline survey that Keonjhar is considered as one of maize patch area of the state as tribal farmers generally grow maize rather than rice (*Oryza sativa* L.) which is a most common and traditional practice in all other area. As maize is the important *kharif* (rainy season) crop and mostly gown under rainfed upland conditions, it is associated with high risk due to inadequate and erratic rainfall. Under such conditions, intercropping with legumes has been reported beneficial to single cropping system (Jaganathan *et al.,* 1974; Kalra and Gangwar, 1980). Several benefits are derived from intercropping which include stability of production, insurance against crop failure, better resource use and income, and employment generation (Rao and Willey, 1980; Koshta and Karanjkar, 1986). On the basis of crop growth morphology and duration, a variety of crop combinations consisting of cereals and legumes can be proposed such as maize and cowpea for intercropping. In the areas of either low (<600 mm) or high (>600 mm) annual rainfall, cereals and legumes of varying maturity are used to ensure efficient utilization of above and underground resources (Ofori and Stern, 1987). Maize seems to dominate as the cereal component and can be combined with different legumes like pigeon pea, mung bean and cowpea. Among all legume options, cowpea is to be chosen due to availability of variety and high market price in the study area. As for third principle of residue cover, mustard and horsegram (*Macrotyloma uniflorum*) will be grown as post rainy season cover crop but after harvest, the residues are to be returned back to the fields for cover mulch rather than using them as fuel.

Field experiments on CA

The objective of the study was to quantify the effects of maize-based CAPS on productivity and certain basic soil characteristics. The soils in the study area (20°502 55.382 2 N, 85°342 30.612 2 E at 499 m above the MSL), which developed from colluvial-alluvial deposits, are classified as *Fluventic Haplustepts*. The local climate is hot and sub humid, with a mean annual rainfall of 1473 mm. A 3-year field experiment was conducted in the Research Farm of Orissa University of Agriculture & Technology located at Keonjhar in order to investigate the effects of CA techniques on maize-based agriculture and soil quality. Each year, during the 1st seasons (June-October) of the cropping cycle, four treatments (T) were applied. T_1 consisted of performing conventional tillage and planting maize (*Zea mays*) only. T_2 consisted of performing conventional tillage and planting maize intercropped with cowpea (*Vigna unguiculata*).T_3 consisted of performing minimum tillage and planting maize only. T_4consisted of performing minimum tillage and planting maize intercropped with cowpea. The size of each plot during this season was 40m$^{2.}$ All treatments were applied in triplicate using randomized block design. During the 2nd season (November-January) of the cropping cycle, the residual effects of the first four treatments (main plot) and the direct effects of applying cover crop treatments (sub plot) were tested using a split-plot design. The following treatments were applied: NCC, no cover crop (fallow); T, Mustard (*Brassica campestris*) as a cover crop; and H,Horsegram (*Macrotyloma uniflorum*) as a cover crop. The size of each sub-plot during this season was kept 12 m^2 by dividing each main plot into three sub-plots. Conventional tillage (CT) involves tilling the field three times by moldboard plowing to a depth of 20–25 cm, to which no vegetative material was added, where as in case of the minimum tillage treatment (MT), one shallow plowing is done to a depth of 10 cm and the chopped vegetative material of the main crops (maize and cowpea) and cover crops (horsegram and mustard) was added. Soil samples (from 0–5 cm, 5–10 cm, and 10–20 cm below the surface) collected at the beginning of the study (2011), and at the end of the 3rd cropping cycle (2014) were analyzed to determine their organic carbon content (Page *et al.*, 1982), water-stable aggregates (Kemper and Rosenau, 1986), and microbial biomass carbon content (Vance *et al.*, 1987). The maize equivalent yield was determined according to the current market price of the crops.

Experimental findings on CA

The elevation of soil organic carbon due to residue retention and minimum soil disturbance is related to the physical protection of C macro-aggregates in MT systems. The SOC content (Table 1) of plots receiving MT treatment increased significantly (27.9% in 0-5cm and 15.2%in 5-10cm soil depth) from the initial proportions of 6.82 and 6.44 g kg^{-1} in the top two soil layers, 0–5 and 5–10 cm below the surface, respectively. A reduction in SOC (9.8% and 15.7% in the top two layers, 0–5 cm and 5–10 cm, respectively) was observed in plots receiving CT treatments. The SOC pool was 13.0% and 2.6% higher in plots with cover crops than in plots receiving NCC treatment (6.88 and 6.06 g kg^{-1}in the top soil layers, 0–5 cm and 5–10 cm, respectively). Neither CT nor MT treatment had any significant effect on the SOC of the deepest soil layer, 10–20 cm. The combined effect of MT and residue retention on the accumulation of SOC is greater than that of either MT or residue retention alone. In plots receiving MT treatment, the organic matter is more rapidly incorporated into the top few centimeters of the soil than in plots receiving CT, where it is redistributed up to a depth of 20 cm. (Kushwaha*et al.*, 2001). The considerable buildup of SOC in the top two soil layers of the plots receiving MT treatments in combination with a cover crop is attributable to a greater residue input (5.5 to 6.5 t ha^{-1}of biomass), as well as lower rates of biological oxidation due to decreased tillage-induced soil inversion (Hel *et al.*,2009). The lower physical impact of the MT system increases aggregate stability, resulting in lower aggregate turn over; therefore, MT systems provide greater physical protection from decomposition to aggregate SOC. This results in the retention of larger amounts of SOC. The higher rate of macro-aggregate degradation induced by the intense soil disturbance of the CT system exposes SOC formerly incorporated in the top soil layers to microbial decomposition (Zotalleri *et al.*, 2007).

Table 1: Effect of tillage and cropping systems on soil organic carbon and microbial biomass carbon

Treatments	SOC (g kg^{-1})			MBC (μg C g^{-1})		
	0-5cm	5-10cm	10-20cm	0-5cm	5-10cm	10-20cm
Initial	6.82	6.44	4.72	82.3	78.6	74.7
Main plot						
CT-M	5.78	5.21	4.67	86.9	83.4	78.0
CT-M+C	6.51	5.64	4.80	101.2	91.2	80.9
MT-M	8.36	7.09	4.84	162.3	124.9	72.5
MT-M+C	9.08	7.74	5.08	191.7	136.9	76.5
LSD (0.05)	0.48	0.47	NS	10.98	10.68	NS
Sub plot						
NCC	6.88	6.04	4.76	118.5	99.6	74.5
H	7.99	6.70	4.93	150.5	116.1	78.6
T	7.43	6.52	4.86	137.6	111.6	77.8
LSD(0.05)	0.32	0.27	NS	9.32	4.44	NS
Interaction	NS	NS	NS	NS	NS	NS

CT: Conventional tillage, MT: Minimum tillage, M: Maize, C: Cowpea,
NCC: No cover crop, H: Horsegram, T: Mustard, SOC: Soil organic carbon, MBC: Microbial biomass carbon

The favorable effects of the organic binding that result from the accumulation of organic matter in the surface soil layers (0–5 cm) significantly increased the quantity of water-stable macro-aggregates (>0.25 mm) in plots receiving different MT treatments. MT resulted in a significant increase in the proportion of macro-aggregates present in the 0–5 cm and 5–10 cm soil layers (Table 2). This amounted to a gain of 14.9% and 11.9%, respectively, over the initial values (58.2% and52.88%, respectively). Soils managed with CT exhibited a reduction in macro-aggregates of 5.3% and 5.7%in the top and middle soil layers, respectively. Macro-aggregates in the 10–20 cm soil layer did not vary among treatments. At the end of the 3rd cropping year, treatments including the addition of cover crops elevated the proportion of macro-aggregates by 7.2% and 4.9% in the two top layers, respectively, over the proportions seen in plots receiving NCC treatment. On the other hand, the proportion of micro-aggregates (0.053–0.025 mm) in soils receiving MT treatments was reduced by 14.8% and 14.5% in 0–5 cm and 5–10 cm layers over the initial values of 17.7% and 19.2%, respectively. The concomitant increase of micro-aggregates in soils under CT treatment was 5.9% and 4.3% for the top and middle layers, respectively. The bottom layers (10–20 cm) of both MT and CT treated plots exhibited no noticeable variation in the proportion of micro-aggregates. Treatments involving the use of cover crops did not significantly influence the status of micro-aggregates.

Table 2: Effect of tillage and cropping systems on water stable aggregates and maize equivalent yield

Treatments	WSA % (>0.25mm)			WSA % (0.053-0.25mm)			MEY(q ha^{-1})
	0-5cm	5-10cm	10-20cm	0-5cm	5-10cm	10-20cm	
Initial	58.2	52.8	44.7	16.2	18.1	19.6	
Main plot							
CT-M	54.5	48.8	42.2	19.5	21.3	23.5	62.65
CT-M+C	55.7	50.8	43.6	17.3	18.7	21.7	103.7
MT-M	64.9	58.5	46.2	15.8	17.1	20.7	60.73
MT-M+C	68.9	59.8	46.0	13.9	15.7	20.1	105.84
LSD (0.05)	4.39	5.32	NS	1.75	1.75	NS	5.42
Sub plot							
NCC	58.2	52.5	42.9	17.4	19.2	21.9	71.29
H	63.3	55.8	45.7	16.0	17.3	21.0	85.76
T	61.6	55.1	44.9	16.5	18.1	21.6	92.64
LSD(0.05)	3.17	NS	NS	NS	NS	NS	2.83
Interaction	NS	NS	NS	NS	NS	NS	NS

CT: Conventional tillage, MT: Minimum tillage, M: Maize, C: Cowpea,
NCC: No cover crop, H: Horsegram, T: Mustard,
WSA: Water stable aggregates, MEY: Maize equivalent yield

The higher proportion of macro-aggregates (>0.25mm) concomitant with the lower proportion of micro-aggregates (0.053–0.25mm) observed in the soil sample surface from 0–5 cm and subsoil (5–10 cm) under MT is associated with reduced physical impact, leading to lower aggregate turnover rates. The intensive disturbance of the top layer of soils under CT, in contrast, disrupts macro-aggregates (Six *et al.*, 2000a). The higher input of fresh organic material under MT with cover crops increases microbial activity and the production of microbial binding agents, leading to the formation of higher proportions of macro-aggregates in the top two soil layers (Mikha and Rice, 2004). The proportion of soil aggregates in the 10–20 cm layers under both MT and CT is related to the organic matter content, rather than the particular tillage method used. This corroborates the findings of Zhou Hu *et al.* (2007).

Microbial biomass carbon (MBC) increases much more readily with changes in tillage system or residue supply, with the increase being more conspicuous in the top few centimeters of the soil, as C input is concentrated near the soil surface. The use of MT systems significantly enhanced the MBC of the treated soils (Table 1), by 115.1% and 66.5% over the initial quantities of 82.3 and 78.6 μg C g^{-1}, in the layers 0–5 cm and 5–10 cm below the soil surface, respectively. This also represented an increase over the SOC in plots receiving CT treatment, by 88.1% and 49.9%, respectively. The accumulation of crop residues through the use of cover crops elevated the MBC of soils by 21.5% and 14.3% as compared to NCC treatments (118.5 and 99.6 μg C g^{-1}) in the top two layers (0–5 and 5–10cm), respectively. The MBC of soils receiving different treatments decreased from the surface downwards. At the 10–20 cm layer, no significant variations could be observed. Microbial Biomass Carbon is considered to be a sensitive indicator of soil quality, and is closely related to soil fertility. In contrast to inversion tillage (CT), crop residues in soils managed with minimum tillage (MT) are left on or near the soil surface, where they undergo decomposition. The accumulation of concentrated C input in the top few centimeters results an increase in MBC near the soil surface.MBC increases much more readily with changes in tillage system or residue supply than SOC (Stock fiset *et al.*, 1999). From the contrasting profiles of MBC observed in the MT and CT systems, it may be deduced that a high proportion of SOC near the soil surface is maintained, or even that enrichment will be continued favoring higher microbial population and there by more MBC under MT. In the deeper soil layers of plots treated with MT (10–20cm), low MBC values is corroborated with a decline in SOM content in that layer of soil.

Significant variations in maize equivalent yield (MEY) among tillage and cropping systems were observed at the end of the 3rd cropping year (Table 2). Even though the maximum MEY was obtained from plots managed with MT and maize and cowpea intercropping (105.84 q ha^{-1}), there was no significant difference between MT and CT either maize-only or maize and cowpea intercropping. The inclusion of horsegram (HG) and mustard (T) as cover crops during the dry season increased the MEY significantly over that of plots allowed to lie fallow (NCC), by20.3%and29.9%, respectively. The maximum MEY, 105.84 qha^{-1},wasobtained from the plot treated with MT-M+C, followed by that treated with CT-M+C-T (103.70q ha^{-1}).The MEY in plots managed via MT systems, though statistically similar to that of plots managed with CT systems, was marginally increased. This may be related to the considerable buildup of SOM that moderates the aggregation, moisture retention, microbial profile, and the dynamics of nutrient mobility by the end of 3rd cropping cycle. The intercropping of maize and cowpea under both MT and CT systems increased MEY in comparison with plots planted with maize only, because of the additional gain in MEY provided by the added nutritional input provided by the planting of cowpea. The higher MEY due to inclusion of mustard is caused by its higher selling price.

The conservation agriculture production system (CAPS), comprising minimum tillage, maize–cowpea intercropping, and the use of a follow up cover crop of horsegram, enhanced the soil organic matter status considerably, via the accumulation of residue inputs at the soil surface and the decreased turnover of macro-aggregates caused by the reduced physical impact on the soils (Roul et al., 2015). The increased SOM, in turn, contributed significantly to the elevation of SOC, MBC, and macro-aggregates in the top soil layers. The redistribution of carbon in soils under conventional tillage, on the other hand, depleted the formerly incorporated soil organic matter, imparting negative effects on overall soil health. Minimum tillage in combination with maize and cowpea intercropping followed by the use of a horsegram cover crop appeared to be the CAPS most suited to this hilly, rain-fed agro-ecosystem, with respect to enhancing productivity and soil health.

Adoption of CA

Short-term solutions and immediate benefits always attract farmers and the full technical and economic advantages of conservation agriculture can be seen only in the medium and long-term run, when its principles (no-tillage, permanent cover crop and crop rotation) are well established within the farming system (Pradhan et al., 2015). In fact, if the two systems (conventional and conservation agriculture) are applied in two plots with the same agro-ecological and fertility conditions, no great differences in productivity during the first years are to be expected. However, after cultivating the same crops in the same areas for several years, the differences between the two systems become more evident. CA requires a new way of thinking from all concerned. Along with this "new way of thinking agriculture", there is already enough technical and agronomic evidence that could positively influence farmers contemplating the adoption of CA principles. It is, however, important to demonstrate to farmers that the technical and agronomic aspects are directly related to the management and economic ones and, therefore, any technical and agronomic improvements obtained by applying CA principles need to be quantified. Before analyzing the farm management and economic aspects of CA, it is illuminating to divide the adoption/adaptation process into four theoretical phases. These theoretical divisions, represented in Fig. 1, facilitates the reasoning when analyzing the farm activities and the impacts of new technologies in the production process.

- First Phase - Improvement of tillage techniques: During this first phase, no increase in farm output is foreseen. But decreases in: labour; time; draught animal or motorized power (reduction of production costs) would occur. An increase in agro-chemical use, especially to control weeds might be required. Furthermore, there may be an increase in family expenses to compensate a probable (but not certain!) reduction of production in comparison with the conventional agriculture;
- Second Phase - Improvement of soil conditions and fertility. Decreases in labour, time and draught animal and motorized power (reduction of production costs). Increases in yields and consequently increase in net farm income;
- Third Phase - Diversification of cropping pattern. Increased and more stable yields. Increased net farm income and soil fertility.

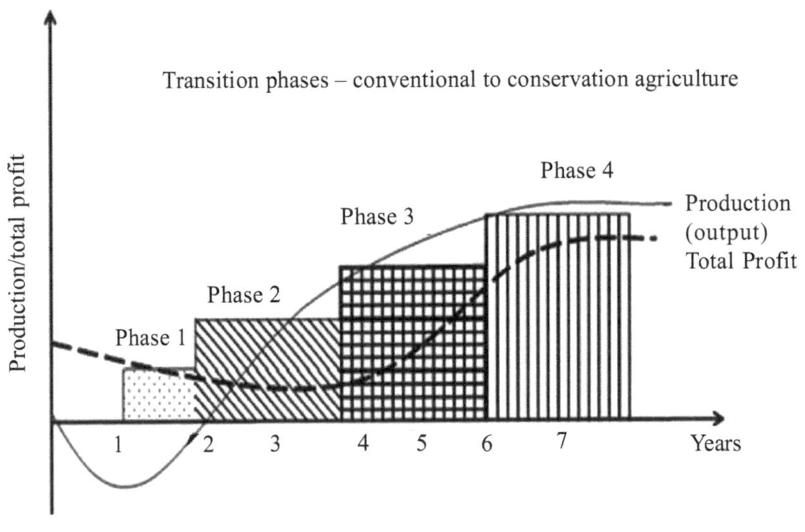

Fig. 1: The transition phases of CA adoption (Extracted from FAO. 2004)

- Fourth Phase - The integrated farming system is functioning smoothly. Stability in production and productivity. The full technical and economic advantages of conservation agriculture can be appreciated by the farmer.

Constraints in adoption

As much as conservation agriculture can benefit the world, there are some problems that come with CA. There are many reasons why conservation agriculture can't always be a win-win situation.

- There are not enough people who can financially turn from a conventional farmer to a conservationist. When a farmer first starts to process as a conservationist the results may be a financial loss to him.
- CA is based upon establishing an organic layer and producing its own fertilizer, which may take time. It can be many years before a farmer will start to see better yields than he/she has had previously before.
- A farmer may have to buy new implements and machineries *viz.*planters or drills in order to produce effectively.
- CA has not spread as quickly as most conservationists would like. The reasons for this is because there is not enough pressure for farmers in places such as North America to change their way of living to a more conservative outlook. But in the tropical countries like India, there is more of pressure to change to conservation areas because of the limited resources that are available. European countries have also started to catch onto the ideas and principles of CA, but still nothing much is being done to change due to their being a minimal amount of pressure for people to change their ways of living (FAO 2006).
- With CA, comes the idea of producing enough food, using less fertilizer, not tilling the field and among others comes the responsibility to feed the world. According to the Population Reference Bureau, by 2050 there will be an estimated 9.1 billion people. With this increase comes the responsibility for farmers to increase food supply with the same or even less amounts of land to do it on.

Each context brings different problems regarding the implementation of Conservation Agriculture, and many technologies can be adapted, including traditional ones. Technical solutions should be found through close partnerships between farmers, private sector industries, extension services and researchers. This approach has already permitted to develop and adapt Conservation Agriculture particularly for small and big farms, manual, animal traction and mechanized agriculture, in various agro-ecological zones and socio-economic contexts, and for many farming systems.

Requisites for the adoption of CA

1. It must bring the farmer a visible and immediate benefit, economic or otherwise.
2. The benefit must be substantial enough to convince the farmers to change their ongoing practices.
3. For the technology to be disseminated, the costs incurred must be affordable by the farmer.
4. The introduction of is time consuming and needs parsistant efforts.

Steps in adoption of CA

- Establish a detailed knowledge of the nature, properties, distribution, and potential uses of soils of the farm.
- Avoid mechanical soil disturbance to the extent possible.
- Avoid soil compaction beyond the elasticity of the soil.
- Maintain or improve soil organic matter during rotations until reaching an equilibrium level.
- Maintain organic cover through crop residues and cover crops to minimize erosion loss by wind and/or water.
- Maintain balanced nutrient levels in soils.
- Avoid contamination with agrochemicals, organic and inorganic fertilizers and other contaminants by adhering to appropriate quantities, application methods and timing to the agronomic and environmental requirements.
- Maintain a record of the annual use and inputs and outputs of each individual land-management unit.

Conclusion

Crop production in the next decade will have to produce more food from less land by making more efficient use of natural resources and with minimal impact on the environment. Only by doing conservation agriculture food production can be able to keep pace with demand and the productivity of land will be preserved for future generations. This will be a tough task for agricultural scientists, extension personnels and farmers. Use of productive but more sustainable management practices can help resolve this problem. Crop and soil management systems that help improve soil health parameters (physical, biological and chemical) and reduce farmer costs are essential. Development of appropriate equipment to allow these systems to be successfully adopted by farmers is a pre-requisite for success. Overcoming traditional mindsets about tillage by promoting farmer experimentation in a participatory way will help accelerate adoption. Encouraging donors to support this long-term applied research with sustainable funding is also an urgent requirement. In India, farmers are becoming more familiar with the potential benefits of CAPS, as they continue to work with research and extension professionals and witness first hand benefits of CA. In the near future CA will be acceptable to the smallholder farming communities for sustainable farming systems.

Acknowledgement

The authors thank USAID-SANREM Feed the Future Innovation Lab for funding this research through the LTRA 11 project: Sustainable Management of Agro-ecological Resources for Tribal Societies (SMARTS).

References

Ansari, I.Z., Akhtar, S. 2012. Incidence of poverty in India: issues and challenges. *International Journal of Multidisciplinary Research* 2(3): 443-449.

Food and Agriculture Organization (FAO). 2006. Agriculture and Consumer Protection Department. Rome, Italy Available from http://www.fao.org/ag/magazine/0110sp.htm (Accessed November 2007).

Food and Agriculture Organization (FAO). 2007. Agriculture and Consumer Protection Department. Rome, Italy Available from http://www.fao.org/ag/ca/ (Accessed November 2007).

Harris, D. 1996. The effects of manure, genotype, seed priming, depth and date of sowing on the emergence and early growth of sorghum (Sorghum bicolor (L.) Moench) in semi-arid Botswana. *Soil and Tillage Research* 40:73-88.

Hel, J., Kuhn, N.J., Zhang, X.M., Zhang, X.R. and Li, H.W. 2009.Effects of 10 years of conservation tillage on soil properties and productivity in the farming-pastoral eco-tone of Inner Mongolia, China.*Soil Use and Management*25: 201-209.

Herren, H.R. 2010. Moving in the right direction, In: Food Security: Creating an affordable, Sustainable, Economically Sound and Socially Acceptable Food Future for Everyone, London. Available at: www.feedingthefuture.eu.

Hobbs, P.R., Gupta, R., Sayre, Ken. 2007. The role of conservation agriculture in sustainable agriculture. The Royal Society. pp 1-13.

Ito, M., Matsumoto, T. and Quinones, M.A. 2007. Conservation tillage practice in sub-saharan Africa:the experience of Sasakawa Global 2000. *Crop Protection* 26:417-423.

Jaganthan, N.T., Morachan, Y.B. and Ramiah, S. 1974. Studies on the effect of maize and soybean association in different proportions and spacing on yield. *Journal of Potassium Research* 10(2):386-391.

Kalra, G.S. and Gangwar, B. 1980. Economics of intercropping of different legumes with maize at different levels of nitrogen under rainfed conditions. *Indian Journal of Agronomy* 25:181-185.

Kemper, W. and R. Rosenau. 1986. Aggregate stability and size distribution. In: Klute, A. (ed). *Methods of Soil Analysis, Part I*: Physical and mineralogical methods, 2nd edition., Agronomy Monographs, 9. Soil Science Society of America, Madison, Wisconsin, 425-442.

Koshta, A.K. and Karanjkar, S.V. 1986. Economics of intercropping system under rainfed farming. *Financing Agriculture.* 18(22):33-36.

Kushwaha, C.P., Tripathi, S.K. and Singh, K.P. 2001.Soil organic matter and water-stable aggregates under different tillage and residue conditions in a tropical drylandagroecosystem.*Applied Soil Ecology*, 16:229-241.

Lai, C., Chan-Halbrendt, C., Halbrendt, J., Shariq, L., Roul, P., Idol, T., Ray, C. and Evensen, C. 2012c. Comparative economic and gender, labor analysis of conservation agricultural practices in tribal villages in India. *International Food and Agribusiness Management Review* 15(1): 73 – 86.

Mikha, M.M. and Rice, C.W.2004. Tillage and manure effect son soil and aggregate associated carbon and nitrogen. *Soil Science Society of America Journal* 68:809-816.

Mloza-Banda, H.R. 2005. Integrating new trends in farming systems approaches in Malawi. African *Crop Science Society* 7:961-966.

New Standard Encyclopedia. 1992. Standard Educational Operation. Chicago, Illinois. pp A-141, C-546.

Ngwira, A.R., Aune, J.B. and Mkwinda, S. 2012. On-farm evaluation of yield and economic benefit of short term maize legume intercropping systems under conservation agriculture in Malawi. *Field Crops Research* 132:149-157.

Ofri, F. and Stern, W.R. 1987. Cereal-legume intercropping systems. *Advances in Agronomy* 41:41-90.

Page,A.L.;Millar,R.H.andKeeney,D.R.1982. *Methods of Soil Analysis, Part- II.* American Society of Agronomy, Inc. Publisher, Madison, Wisconsin, USA.

Pawley, W.H. 1963. Possibilities of Increasing World Food Production. Food and Agriculture Organization of the United Nations. Rome, Italy. pp 98.

Pradhan, A., Idol, T., Roul, P.K., Mishra, K.N., Chan, C., Halbrendt, J. and Ray, C. 2015. Effect of tillage, intercropping and residue cover on crop productivity, profitatbilits and soil fertility under tribal farming situations of India. In: C. Chan and J. Fenlle-lepczyk (eds) Frontiers in conservation Agriculture in South Asia and Beyond (F-ASA), CABI, Wallingford, UK. pp. 77-95.

Rao, M.R. and Willey. R.W. 1980. Evaluation of yield stability in intercropping studies on sorghum + pigeon pea. *Experimental Agriculture* 16:105-106.

Roul, P.K. Pradhan, A., Ray, P., Mishra, K.N., Dash, S.N., Chan, C., Idol, T. and Ray, C. 2015. Influence of maize-based conservation Agricultural Production System (CAPS) on crop yield, Profit and Soil fertility in rainfed uplands of Odish, Indian. In: C Chan and J. Fansle-lepczyk (eds) Frontiers in Conservation Agriculture in South Asia and Beyond (F-CASA), CABI, Wallingford, UK. Pp. 95-109.

Six, J., Elliott, E.T. and Paustian, K. 2000a. Soil macro aggregate turn over and micro aggregate eformation: a mechanism for Csequestrationunderno- tillage agriculture. *Soil Biology and Biochemistry* 32:2099-2103.

SMARTs. 2009 Annual report of SANREM funded research project on resistanable management of Agroecological Resources tribal societies.

Stockfisch, N., Forstreuter, T. and Ehlers, W. 1999.Ploughing effects on soil organic matter after twenty years of conservation tillage in Lower Saxony, Germany. *Soil and Tillage Research,* 52: 91-101.

Sustainable Management of Agroecological Resources for Tribal Societies (SMARTS). 2009. SMARTS Proposal: Response to EEP comments. University of Hawaii at Manoa, Honolulu, Hawaii.

Vance,E.D., Brookes, P.C. and Jenkinson,D.S.1987. An extraction method for measuring soil microbial biomass carbon. *Soil Biologyand Biochemistry*19:703-707.

Wang, E., Cresswell, H., Xu, J. and Jiang, Q. 2009. Capacity of soils to buffer impact of climate variability and value of seasonal forecasts. *Agricultural and Forest Meteorol*ogy 149: 38-50.

World Bank. 2008. World Bank Development Report 2008: Agriculture for Development. Washington, DC.

Zhou Hu, Yi-zhong Lu, Zhi-chen Yang and Bao-guoLi. 2007. Influence of conservation tillage on soil aggregates features in North China Plain, *Agricultural Sciences in China*6: 1099-1106.

Zotarelli, L., Alves, B.J.R., Urquiaga, S., Boddey, R.M. andSix,J.2007.Impactof tillage and crop rotation on light fraction and intra-aggregate soil organic matter in two Oxisols. *Soiland Tillage Research* 95:196-206.

Integrated Farming System Practices: Challenges & Opportunities, pp 471-495

New India Publishing Agency, New Delhi, India

19

Ethnic Practices on Integrated Farming Systems by Tribal

B.K. Mohapatra , P.K. Banerjee and C.R. Sarangi

Introduction

In India about 8.6 per cent of population enlisted in the category as 'Scheduled Tribes' under Article 342 of the Constitution of India, but they constitute almost 16 per cent of the poor. Lakshadweep is the most populous state with respect to proportion of tribal to total population (94.8 per cent) whereas, in Uttar Pradesh tribal population is lowest with only 0.6 per cent of the total. In Odisha tribal constitutes 22.8 per cent of total population. In India literacy rate of tribal is 59 per cent as against that of general category is 69 per cent. Tribal people are observed to be strongly associated with the forests, hills and remote areas, practising a unique life style, having a unique set of cultural and religious beliefs. Over the generations these communities have evolved time tested farming systems, which are purely subsistence in approach and are keeping pace in perfect harmony with nature. Over the recent few decades, however, population growth and lack of harmony between different agriculture and developmental policies have threatened these traditional agricultural systems for extinction. The encroachment of local Governments and private interests on forest and mineral resources has pushed the tribal groups on to increasingly degraded land, thus their survival became a constant struggle.

Directorate of Extension Education, Orissa University of Agriculture & Technology, Bhubaneswar-751003, Odisha, India

The primitive societies have passed through several stages of economic development everywhere in the world. Thus one can find the stages of food gathering, hunting and fishing, farming, etc. among the Indian tribes. The tribes living in the forests and hills usually earn their livelihood by means of food gathering, hunting and fishing. Such is the life of Kadar of Kerala, Birhor and Kharia of Bihar and other tribes. The tribes living in dense forests, full of wild beasts, live on hunting, such as the tribes of Naga, Kuki, Bhil, Santhal and Gond. The hunters leave the females to carry out household activities in the morning and return in the evening after hunting. In some tribes there is a usual custom of hunting collectively. The Nagas use spears, arrows and bows. The Bhils are very much specialized in shooting by arrow. The tribes living near rivers and seas usually earn their livelihood by catching fish. The hilly tribes rear the cattle, an example of which are Goojars and the tribes of Chamba. The Todas of Nilgiri rear buffaloes. Some tribes also carry out cultivation, but they are generally shifting from one place to another. So shifting cultivation is an intrinsic part of tribal system and tradition. Among the cultivating tribes are the Santhals and Gonds. Tribal groups mostly do not have land ownership and have a limited option to adopt agricultural operations in forest areas. Considering their sustenance Government of Odisha has passed an Act "Schedule Tribes and other forest dwellers Forest Right Act" in 2006. It permits ownership of tribals on forest land up to 4 ha. for forest use and subsistence agricultural activity without any penalty. Cottage industries, such as weaving cloths, preparing ropes and skins and utensils of different metals are prevalent in many tribes. The Kharia people are very much specialized to cottage industries. Their living standard is very poor in spite of their hard work.

Over the decades, the tribal economy and the livelihood strategies have undergone substantial changes. Since the tribals were traditionally dependent on natural resources, the change was all the more visible due to the depletion of resources. The shift in tribal economy and diversification of occupations has been corroborated in the people of India report by Anthropological Survey of India. The report maintains that " the number of communities practising hunting and gathering has declined by 24.08 per cent, as forests have disappeared and wild life has diminished. Ecological degradation has severely curtailed the related traditional occupations. For instance, trapping of birds and animals has declined by 36.84 per cent, pastoral activities by 12.5 per cent, and shifting cultivation by 18.14 per cent. However, there is a rise in horticulture (34.4 per cent), terrace cultivation (36.84 per cent), settled cultivation (29.58 per cent), animal husbandry (22.5 per cent), sericulture (82.6 per cent) and bee keeping (60 per cent). Many of the traditional crafts have disappeared and spinning, in particular, has suffered (25.38 per cent). Related activities such as weaving (3.32 per cent) dyeing (33.34 per cent) and printing (100 per cent) have similarly suffered. Skin and hide work as stone carving has declined "(Singh, 1997).

Biodiversity and traditional farming system of tribals

Traditional farming systems practised by tribals are associated with their degree of plant diversity in the form of polyculture and/or agro -forestry systems. It is a strategy of minimizing risk by planting several species and varieties of crops which stabilizes yields in long run, promotes diet diversity, and maximizes returns even with low levels of technology and limited resources. Such bio diverse farming are endowed with nutrient enriching plants, insect predators, pollinators, nitrogen fixing and decomposing bacteria, and a variety of other beneficial organisms performing various functions to maintain ecological balance. Traditional multiple cropping systems provide as much as 15-20 per cent of the world food supply. Polyculture are very common in Asia where upland rice, sorghum, millet, maize, and irrigated wheat are the staple crops. Lowland (flooded) rice is generally grown as a monoculture, but in some areas of Southeast Asia farmers build raised beds to produce dry land crops amid strips of rice. Tropical agro ecosystems composed of agricultural and fallow fields, complex home gardens, and agro-forestry plots, commonly contain plant species which are used as construction materials, firewood, tools, medicines, livestock feed, and human food.

Many traditional agro ecosystems are located in centres of crop diversity, thus containing populations of variable and adapted land races as well as wild and weedy relatives of crops (Harlan, 1976). Clawson (1985), describes several systems in which tropical farmers plant multiple varieties of each crop, providing both intra-specific and inter-specific diversity, thus enhancing harvest security. For example, farmers maintain a diversity of rice varieties adapted to a wide range of environmental conditions, and they regularly exchange seeds with each other (Grigg, 1974). The resulting genetic diversity heightens resistance to disease that attack particular strains of the crop, and enables farmers to exploit different microclimates and derive multiple nutritional and other uses from genetic variation within species.

Depending on the level of biodiversity in close vicinity, tribal farmers accrue a variety of ecological services from surrounding natural vegetation. Clearly, traditional agricultural production commonly reflects a total multiple-use system of both natural and artificial ecosystems, where crop production units and adjacent habitats are often integrated into a single agro ecosystem. It deals with biodiversity, food security and livelihoods, knowledge systems, landscapes (aesthetic values) and culture practised by tribal.

Farming systems practised by tribals of India

Some important farming systems practised by tribals of India are given below:

1. Benwar farming system of the baigas (Chhatisgarh)

The Baigas tribe of Chhattisgarh practise mainly shifting cultivation and locally this farming system is called *Benwar*. It is a shifting cultivation mostly practised on hilly slopes, where contour bunding cannot prevent soil erosion. The practice involves cutting bushes and branches of trees and laying them out on slopes for drying. The branches are then burnt, leaving behind a layer of ash on which seeds of crops are broadcasted a week before the rains are expected. The Baigas do not plough the earth. They believe that the lands are like mother's breast and they are not supposed to scratch their mother's breast again and again by ploughing the earth. Therefore, they practise broadcasting of seeds leave the land fallow for several years to ensure recycling of the land productivity. Therefore, the practice of *Benwar* has a strong relation with Baiga tradition and faith.

In 1893, when the colonial rulers banned shifting cultivation, the Baigas were allowed to practise their farming system in 7 villages. The forests surrounding these villages were still dense forests have disappeared and shifting agriculture was banned and the tribal people were forced to adopt modern farming practices.

The Baigas have mastered the art of mixed cropping. They grow coarse cereals, pulses, and vegetables. Mixed cropping provides multiple harvests, protection against pests and diseases, and replenishment of the soil with organic matter. The Baigas have a system of harvesting and storing the crops. *Neem (Azadirachta indica)* and *Kanchan (Bauhinia sp.)* leaves are used as natural insecticide to preserve the grains for a longer time.

The Baigas depend on the forests for survival but never exploit them. They have a unique knowledge of use of different plants and their use in cure of ailments. They have excellent knowledge of medicinal plants, especially their collection, harvest, and appropriate use. These medicines are used for curing in various diseases like fever, ulcers, wound, insect bite etc. The repository of traditional Baiga knowledge is enormous. There is, however, no system to recognize record and preserve it. The Baiga knowledge of agricultural practices, medicinal plants, diagnosis, and healing is oral and a non-literate tradition. Therefore, the challenge is to preserve this invaluable knowledge and ensure that the range and depth is not confined to a few households or people.

Baigas in third generation have started settled farming consistently. They classify the surrounding land on the basis of their fertility and land situation such as;

i) *Kali*-such heavy soils occurs on flat topography and the depth is up to 1.25 metres where suitable cropping systems are paddy-gram/paddy-lentil/ paddy-pea/paddy-local wheat/paddy-fallow.

ii) *Bahra* – Black, heavy soil, moderately flat tropography, depth is about 1.8m., where suitable cropping systems are paddy-pea/paddy-gram/paddy-lathyrus/paddy-wheat/paddy-some vegetables.

iii) *Bharra* – this type of soil occurs on undulated land, soil is light, shallow and redish with pebbles, the depth of soil is between 0.8 m to 0.6 m. Such lands gets eroded during the rainy season. Suitable cropping system are ; maize-fallow, maize +amta-fallow/ Kodo+amta-fallow/Kodo+redgram-fallow/Kulthi +redgram-fallow/paddy-fallow/paddy-pea.

2. Animal husbandry system of Raikas (Rajasthan and Gujurat)

The 'Raika' is a traditional nomadic pastoral community found mainly in Rajasthan and Gujarat. Traditionally, the Raikas are of 2 types – the *maru raika*, the camel breeders and the *godwar raika*, the sheep raisers. With time, this distinction is no longer maintained. Both groups breed and raise camels and sheep. However, their groups are different, organization, decision-making etc. is different, and they do not inter marry.

The Raika's innovativeness, flexibility, and specialized knowledge have enabled them to:

- Thrive in harsh, semi-desert environments, difficult migratory routes, and other problems like grazing in the light of diminishing grazing grounds / forests, diseases, and healthcare.
- Develop hardy livestock breed that are drought resistant, are capable of walking long distances, etc.
- Create a social web around the animals.

The camel herders are totally nomadic, however, they do not go off very far, and mostly they move within their district. They have voice commands for their herd, and trade the animals annually at the '*Pushkar*' fair, which is one of the largest animal fairs in Asia. The sheep raisers, on the other hand, could be either nomadic or engage in one-day grazing. Depending on the size of the herd, they could go off as far as Uttar Pradesh and Madhya Pradesh. They have a close voice communication system. They also engage in one-day grazing, which means their village remains their base and they take the animals for the day for grazing. Another method is setting up camps in the farmlands after harvesting. Sheep are sold in local markets.

The Raikas sell only male animals; they never slaughter nor sell for slaughtering; and they are strictly vegetarians. They also grow crops for their own consumption – corn (*Zea mays*), *bajra* (*Pennisetium glaucum*), *tiru, munga* (*Vigna radiata*), *urada* (*Vigna mungo*), and soora in the rainy season (July to October) and mustard (*Brassica campestris*), dill, and jowar (*Sorghum bicolor*) in winters (November to February).

The Raikas diversify within species to cope with ecological conditions and market demands. They also diversify between species in order to spread risks and maximize profits.

Some innovative farmers of raika community like Dharmram Raika had community led initiatives for conservation of native Bhagli(Sonadi) sheep in Pali district of Rajstan. He reared a flock of 30 Bhagli sheep (a local name for sonadi breed of sheep) comprising three rams and 27 ewes. In addition to him there are 35-40 households in his village also rear the Bhagli sheep. The raikas value this breed for its high milk and meat production. They migrate with their sheep flock for eight months each year to Néemuch in Madhya Pradesh. National Bureau of Animal Genetic Resources has awarded Dharmraman with Breed Saviour Award,2012 for his effort to protect and conserve the Bhagli sheep breed.

3. Paddy fish cultivation of the Apatanis (Arunachal Pradesh)

The Apatani are a tribal group located in Ziro valley in the Lower Subansiri district of Arunachal Pradesh.

The Apatani tribe is the only agrarian tribe practising settled agriculture in this part of the State. The neighboring tribes are involved in shifting agriculture (Nimachow *et.al.*, 2010)

The characteristics of the Apatani agriculture are:

- Permanent agriculture
- Wet rice cultivation
- Intricate system of canals and channels for irrigation
- Rice is the principal crop. Additional crops are millet and other grain crops
- In the 1980s they introduced the rice-fish culture.

By the month of April, the Apatanis start the preparation of the rice fields for fish culture. After 10 days of transplantation of the rice seedling, the fields are stocked with fish (15 to 20 mm). The fields are connected directly to the community irrigation channel. Water level in the field is maintained at 20 to 30 cm during culture period and the depth of the canal is maintained at 40 to 45 cm. No supplement feed is given during the culture period. No chemical fertilizers are used, sometimes cow dung (organic manure) is used at the time of field preparation. *Azolla and Lemna* are allowed to grow in the fields as nitrogen fixers. Three rounds of weeding is done round the rice-growing season. Fish is harvested after 3-4 months.

The indigenous technique:

- In order to provide movement and shelter to the fish (heat, etc.), refuge trenches are prepared in the rice fields.
- The refuge trenches are 25-35 cm deep and are constructed either perpendicular to each other or irregularly
- Two outlets are fitted on the trenches – one on top to release the excess water and one at the bottom to dry up the field water to harvest the fish
- Bamboo screens made of bamboo slits are used to prevent the fish from escaping.

In such poor areas, rice-fish cultivation is an excellent means of ensuring food security of the people. Fish provides the essential protein and rice, carbohydrates. The Apatani rice-fish cultivation is an eco-friendly system with high ratio of forest and agriculture land, multi-rice varieties, multi-bio-resources flow, and traditional biological pest control. The system involves conservation of ground water by conserving the forests. While certain trees like *Kiira* (*Castapnopsis* sp.), which have elongated deep root, cannot be cut at all, farmers also grow species such as *Terminalis marginalia*, *Ailanthus excels*, *Michelia sp.*, *Magnolia sp.*, pines and bamboo to maintain the ground water. The economy base of the Apatanis therefore comprises sustainable integration of land, water and farming systems. Apart from that the Apatani version of the rice cum fish cultivation offers agricultural sustainability and diversification unlike the jhum cultivation. *Mithun* (a breed of semi- domestic cattle) rearing and piggery culture also offer substantial support to this system.

4. Bamboo drip irrigation system of Jaintia and Khasi hills, Meghalaya

The Jaintias and Khasis, like most tribal groups, have a close association with nature, owing to which they have developed an indigenous knowledge of environmental protection, biodiversity conservation, and a pattern of agriculture.

The Khasis and Jaintias have evolved certain unique agriculture and land use systems. Among other systems, they have an innovative system of drip irrigation by using splited bamboo:

- Drip Irrigation: The Jaintias and Khasis have a very indigenous method of tapping stream and spring water and taking them to the plant on the field by using bamboo pipes. It is a perfect system, 200 year old. About 18-20 litres of water enter the pipes per minute, is transported over several hundreds of metres and finally gets reduced to 20-80 drops per minute at the site of the plant.

Other important land use and farming systems of the Khasis and Jaintias are as follows

- Sacred groves: All across Meghalaya, the tribal groups maintain sacred groves, which are virgin tracts of forests left untouched and undisturbed. The local belief is that their deities reside in these groves and would be offended if the trees were cut or twigs, flowers or seeds were plucked. Also, if left undisturbed, the deities would bless the people, their cattle and their land. The sacred groves are biodiversity rich communities that provide refuge to a large number of endemic, endangered and rare species of plants as well as animals. These are treasure houses of threatened species, dispensary of medicinal plants and gene bank of economic species. The concept of sacred grove is an indigenous knowledge, conceived, developed and perpetuated by the tribal communities of Meghalaya.
- Farming Systems are of mainly 2 types – slash and burn or shifting agriculture, known as *jhum* and terrace cultivation, locally known as *Bun*.
- The village council *Dorbar Shnong* owns and allots the forest land for *jhum* cultivation to the members of the tribe.
- In *Bun* cultivation, bench terraces are constructed on hill slopes with vertical interval of 1 metre, which helps retain rainwater within the benches disposes the excess runoff from the slopes to the lower bench and eventually to the foothill. This method improves production of crops, conserves soil moisture, and prevents land degradation and soil erosion.

5. Three tier agriculture system of Kondha Savara tribe of Seethampeta (Andhra Pradesh)

Seethampeta is a small village in the Eastern Ghats region of Andhra Pradesh inhabited by Kondha Savaras, tribal groups who have evolved an indigenous farming system for subsistence livelihood, which maintains ecological balance, and ensures food security, and economic returns. This system is a three-tier agriculture system: where the hills are divided into different land use classes based on the elevation, slope and ecological considerations. In this system the upland retains as forests,, the mid elevation lands are used for slash and burn agriculture, and in the plains, the tribal farmers grow the more economically viable varieties of crops. The tribal people cultivate their local varieties for their own consumption and also agriculture and horticulture varieties for the market and the agriculture almost entirely organic. Crop varieties and, landraces, of over 19 species of vegetables and around 150 varieties of paddy, sorghum, small millets and pigeon pea are grown. More than 95 wild relatives of crop plants are recorded in this area. Over 75 species of plants are endemic to this region out of the 359 endemic species recorded in the Eastern Ghats.

This region is a unique representative of traditional agriculture and land use, prasticed in harmony with nature since many generations Mixed Cropping: Crops are mixed to maximize production and hedge risks. The crops are rain-fed and moisture stress tolerant. Crops grown are cashew (*Anacardium occidcntale*), mango (*Mangifera indica*), turmeric (*Curcuma longa*), pineapple (*Anonas comosus*), banana (*Musa paradisica*), coconut (*Cocos nucifera*), custard apple (*Annona squamosa*), paddy (*Oryza sativa*), pulses – red gram (*Cajanus cajan*), green gram (*Vigna radiata*) and black gram (*Vigna mungo*), oilseeds – sunflower (*Helianthus annuus*) and ground nut (*Arachis hypogaea L.),* millets – jowar (*Sorghum vulgare*), bajra or pearl millet (*Pennisetum glaucum*), and ragi or finger millet (*Eleusine coracana*), spices, medicinal plants, and vegetables.

6. Zabo indigenous farming system, Nagaland

The Chakhachang tribe of Phek district of Nagaland practise a unique system of cultivation, called Zabo, which is a "combination of forestry, agriculture and animal husbandry with a well-founded conservation base for soil erosion control, water resources development and management as well as protection of environment."

- The agricultural system: On the hill top, the forests are protected and maintained. The forests act as catchments. At the middle, ponds are dug where water is harvested through inlet channels from the catchment. Below this are the terraced fields.
- The uniqueness of the system is the ponds. Water from the surrounding catchment of protected hill top areas are drained in to the pond through 2 channels of over a kilometer long each. The channel is made compact by hammering its base, thereby reducing water percolation. The ponds are constructed in such a way that the surplus water from one pond flows down to the pond below. The excess water is released through an opening at its lower end, which is otherwise blocked by a piece of wood.

- The terraced fields are of 3 different types, each designed to regulate water flow: (1) *Dzutse*, the water supply to the fields is regular throughout the year and water which is not needed in one terrace is conveyed to the next through channels, (2) *Khuso*, the water supply depends on streams or from overflows of *Dzutse,* and (3) *Vakhra*, water is regulated using labour.
- Another unique feature of the system is the cattle enclosures, which are built on the lower side of the ponds. Animals like buffaloes, cows and pigs are stocked on a rotation basis of 10-15 days. Water from the main pond passes through the cattle enclosure, collecting the dung and urine of the animals, and then goes to the fields providing water along with organic manure.
- The farming is purely organic. Leaves and succulent branches of *Alnus nepalensis* and *Albizia lebbeck* are added to the fields – they decompose and improve soil fertility.
- Farmers also do fish culture in the wet rice terraces. A small pit is dug in the middle of the rice field and fish fingerlings are put in the fields. When the water is drained out from the fields before intercultural operations and harvesting of paddy, the fish remain in the pit. Farmers get 50 – 60 Kg of fish per hectare.
- The system improves the environment, increases crop productivity on a sustainable basis, protects the forests, ensures water and soil conservation, increases the income of the tribal people, increases their quality of life, and reduces the dependence on outside resources.

7. Alder-based jhum system, Nagaland

The tribal people of Khonoma area of Nagaland practise *jhum* cultivation (slash and burn). Traditionally, this was a productive system but with the growth in population and the subsequent shortening of the *jhum* cycle, the practice is becoming more and more economically unproductive leading to ecological degradation. However, the people of Khonoma have incorporated one component in to the *jhum* system – the Alder tree (*Alnus nepalensis*) – a component which is native to the area – and the entire *jhum* system is revitalized.

- The Alder is a non-leguminous, large deciduous tree. It is a pioneer species of degraded lands and does not require fertile soil. It is a rapid colonizer of gravel-lands and old cultivated lands. It has root nodules, which improves soil fertility by fixing atmospheric nitrogen in to the soil. It sheds its leaves, which retain moisture and mulches and adds humus to the soil.

- The Alder tree is a traditional tree in the Angami system. Some of the trees present are more than 200 years old.
- In a *jhum* plot, normally 600 kg of nitrogen per hectare is lost in one year of cropping and it would take 10 years to recover it through the natural process of forest succession. The Alder tree can recover all of the 600 kg nitrogen during the five year period.
- The Angamis plant Alder saplings with a space of 3 – 4 metres between plants and 5 – 6 metres between rows. The trees are allowed to grow for 10 years or until they attain rough fissures on the bark. In the first year of *jhum*, the Alder trees are pollarded (cut off from the main trunk) at a height of 2 metres from the ground. Primary food grain crops and secondary crops such as vegetables are grown as mixed crops in the burned fields. The system provides at least 57 food crops to supplement the rice grown.

Taking advantage of nature of alder trees' ability to fix atmospheric nitrogen, the tribal farmers of Khonoma and other villages across Nagaland have incorporated this tree species to their jhum system. In this Khonoma farmers had evolved a system, where they could cultivate for two years followed by a fallow period for only two years and the same field was subjected to cultivation again. They were able to do this by pollarding the alder tree branches during the cropping period and carefully dressing the stump and nursing the sprouts so that each old stump was supporting only 3 to 5 coppices at the end of the cropping period. In this system area subjected to burning is minimal. By adopting this model in 970 KM^2 it shall be able to support a population of 1, 24, 444 persons and produce sufficient rice to meet their yearly requirements.

8. Hunting and gathering system of Horsley hill, Andhra Pradesh

The Chenchus tribe live in the deep forests. of horsley hill, Andhra pradesh They are not engaged in agriculture but survive by gathering food or hunting animals. The chenchus gather roots, fruits, tubers, beedi leaf, mohua flower, honey, gum, tamarind and green leaves. They make some money by selling these non-timber forest produce to traders and government co-operatives. They also hunt wild animals like boar, deer, rabbits, and wild birds. They live in complete harmony with the nature. They are known to identify trees by names. In the village there is a 150 year old eucalyptus tree named Kalyani. The forests have fauna such as leopards, bears, wild boar, etc.

The Chenchus are famous for the indigenous breed of cows known as Pungannur cows. These are small in size but possess high milk yield capability and have low appetite.

While the Chenchus preserve the forests, in the foot hills and the valley, there are farmers who practise agriculture. The system of agriculture is completely in harmony with the nearby forests, hills, and its people and is beautiful to see.

- Plots are cut on which tomatoes, rice, jasmine, tamarind and groundnut are grown. The cultivation is purely organic.
- The Bahuda river flows nearby. Small channels are cut from the river, which feed the crops.
- The river is used to recharge the ground water and in this region the wells are full of water even in the summer months.

9. Sikkim Himalayan traditional agriculture

In the Khanchendzonga landscape in the Eastern Himalayas of Sikkim agriculture is practised in landscapes called Demazong (the valley of rice) or the Shangrila (the hidden paradise on earth). This cultural-landscape is endowed with rich agro-biodiversity adapted and managed through traditional ecological knowledge of the culturally diverse ethnic communities. It comprises of trans-Himalayan agro-pastoral system of the *Dokpas* in the alpine plateaus, traditional agroforestry such as alder-cardamom and farm-based systems in the temperate zones, and terraced/valley rice systems in the lower sub-tropical zones. The traditional agriculture systems in Sikkim have evolved over 800 years through adaptation and innovation. The traditional shifting agriculture system gradually converted in to a sedentary system. The latter consists of , forestry, livestock, and agriculture. The agriculture is typically organic. The indigenous knowledge of steep slope conversion to and terrace farming constitutes its uniqueness. The tools, agricultural implements, cultivation patterns, irrigation systems, local crops and animals are traditional. The *Bhutias* and *Sherpas* have several ethnic sub-communities differentiated in dialects, social and cultural origins. The cultural diversity relates and contributes to deep agro-biodiversity and plural knowledge system.

The characteristics and functioning of the agricultural ecosystems can be categorized into three broad systems:

a) Agro-pastoralism of the Nomadic Tibetan herders

Agro-pastoralism in the alpine cold deserts of Lhonak valley, Chho Lhamo and Lashar valley above 4000 m sl in North Sikkim has been a part of human life support systems over several centuries. The trans-Himalayan nomadic Tibetans the Dokpas (graziers) herd yaks, dzos (cow-yak hybrid), sheep and goats (pasmina type) in the high altitude Tibetan plateaus and meadows adapted to sustain the harsh climatic conditions. The dry grasslands of these unique valleys of the world have been traditionally managed and inhabited by the Tibetan nomads, a unique example of how people survived in such drought and cold - stricken landscapes in the history of mankind through mobile livestock production systems tolerating all environmental fragility, marginality and poverty.

b) Traditional agriculture systems

The agro-ecosystems are innovated, adapted, managed and evolved over 800 years by traditional communities initially by aboriginal Lepchas and Limboos (now one of the Nepali ethnic group), followed by Bhutias after 1275 AD and later on by the other Nepalese (Rai, Yakha, Gurung, Mangar, Tamang, Sunuwar, Thakuri, Bahun, Chettri, Kami, Damai, Sarki, Majhi, Newar, Sherpa, Thami, Bhujel, Jogi) after 1774 onwards through the generations to the present state. The traditional shifting agriculture system over centuries is gradually conversed into sedentary system. This conversion is still continuing. The sedentary system is a combination of compartments such as agroforestry, forestry, livestock, and parcel of agriculture land forming altogether a unit mountain garden-based farming system. In the recent times, remains of shifting cultivation are rarely observed in the form of khoriya in Dzongu (traditional Lepcha area) and elsewhere. The microcosms of traditional agriculture are typically organic in Sikkim, the people and their knowledge of steep slope conversion to agroforestry and terraced productive zones (traditionally called pakho-khet, pakho-bari, birauto, sim-kholyang), ridges of operational fields, and management of very steep slopes as bhasmey and khoriya. The agricultural tools, farm implements and techniques, cultivation methodology, farm raised animals; local crops, irrigation system etc. are traditional. The Bhutias and Nepalese are composed of a number of ethnic communities with different languages/dialects, and social and cultural background. The cultural diversity attributes to a rich agro-biodiversity and traditional ecological knowledge.

c) Valley rice cultivation systems

The ancient Demazong vis-a-vis Sikkim is characterized by Ghyya-dhan (dryland paddy), which is followed by the valley rice cultivation along the river banks on flat lands traditionally called Thang (sometimes examples are Tarey-thang-byansi, Pi-thang-byansi) and along the typically terraced slopes in the lower hills

A large number of landraces of rice are cultivated. Some of the dryland paddy varieties were Ghyya-dhan, Takmari, Bhuindhan, Marshi etc. while the irrigated rice varieties now are Attey, Timmurey, Krishnabhog, Bachhi, Nuniya, Mansaro, Baghey-tulashi, Kataka, Champasari, Sikrey, Taprey, etc. adapted to agro-ecological zones between 300–1800 m. Krishnabhog, Nuniya and Kataka are famous for their aroma, medicinal importance and fine quality grain as good as Indian basmati rice. Almost all the dryland paddy landraces have now disappeared from the state. While many traditional irrigated land races have been disappeared already and others are on the verge of disappearance. The household mothers are generally the custodians of gene-banks of crop diversity; they keep the record of varieties/yields and preserve germplasm by growing them at different parcels of home garden allotted for growing heterogeneity of landraces. Farmers allow a variety of pulses and beans to grow along the raised bunds with terrace rice and follow maize, wheat, buckwheat etc. after rice is harvested. Legumes are the source of protein, household earning and enrich fertility to the soil through the roots. Between the terraced open rice fields, along the slopes, are the traditional agro-forestry systems, mostly cardamom-based and forest-based agro-forests. Such traditional landscape designing of multifunctional agro-forestry and protection to the terraced open rice cultivation system is characteristic of unique agro-ecosystem management in the mountains. The system in addition supports water conservation and flood control, provides nutrient and biomass to the rice field and farm.

10. Saffron cultivation, Kashmir

Saffron has several names – Zafran, Kesar, Kang, and Kang Posh. Saffron cultivation started three to four centuries back in Arabia and Spain; from where it spread as far as Iran, Sweden, and India. In India, Saffron cultivation is mainly found in Pampore region of Kashmir

Saffron cultivation is a traditional art. In India, Saffron is grown on 5,707 hectares of land, out of which 4496 hectares is in Jammu and Kashmir. In Pampore, region of Kashmir high grade variety of Saffron, which is famous world wide is grown.

Saffron flowers symbolize freshness and purity. The saffron field, lush with the flowers is a beautiful sight. In Kashmir, the people describe the sight as “a newly wedded bride, draped in a saffron shawl, taking a nap!”

Saffron in Jammu and Kashmir is sown in the months of August-September and the flowers are plucked in October-November. The flowers are plucked when it is cool so the ideal time is early morning. The plucked flowers are dried for 5 days and then kept in an airy container to retain its freshness and quality. One kg saffron means 160,000 to 170,000 tiny flowers. It is a time consuming, tiring, and delicate art.

Changes in the economic and climatic conditions is affecting this heritage –

- The investment that goes in to cultivation of saffron is much more than the returns for the poor farmers, which is forcing many farmers to shift to fruit cultivation

- With the average temperatures rising, saffron cultivation is affected
- Also the incidence of diseases is on the rise – common diseases being corm rot, dry tot, root rot, bacteria rot, ring rot, charcoal rot, mosaic, etc. Corm rot is the deadliest of the diseases. There is need for research on these diseases and assistance to the farmers to deal with them.

11. Paderu, farming system of Andhra Pradesh

The tribal groups in Visakhapatnam District raise mixed crops, ensuring minimum food security. This is the main feature of the traditional system of agriculture – maintenance of bio-diversity. Different local varieties are grown suiting the climate conditions and depending on their needs. The area is pest and disease free.

- The local varieties are not high yielding but are rich in nutrient content, are grown organically without any chemical fertilizer or pesticide, organic food increases the resistance power of the tribal people. The traditional cash crops taken like turmeric and ginger give them a secured income and these products possess great medicinal value

12. Traditional sugarcane cultivation, West Bengal

The process of sugarcane cultivation is unique and an age old tradition of tribals of Purulia area, West Bengal. The process is entirely rain fed, with no irrigation, minimum or no modernization, using available natural resources, and dependent exclusively on the indigenous knowledge. The system involves the optimum utilization of the soil-water-air relation and trapping of the autumn and winter dew drops. As the monsoon recedes, the tribal farmer ploughs his field daily in the afternoon, exposing the sub soil surface to dew drops throughout the night. This ensures better percolation of moisture into the soil. Then again the soil is leveled / laddered in the morning. This reduces the evaporation from the sub soil surface. In this way the plant gets the necessary moisture from the soil itself during the germination phase. This system is unique as it gives maximum results. In 1998-99, the yield rate in Purulia was 84,455 kg/hectare and the all India average was 69,647 kg/hectare

13. Traditional farming of Muthuvans, Kerala

Slash and burn method of agriculture is mostly practised by Muthuvans tribes of Kerla. In this system the community identifies open areas without trees but with lot of herbs and shrubs. The selection of land is based on ecological indicators – 2 grasses *Urukkala, Carex myosurus Nees* and *Scleria terrestris* (L.) Fasset. Once the land is selected, all the plants are cut and burnt. The land is then prepared and the seeds are sown. They usually cultivate 3 varieties of finger millet – of 3 months, 4 months and 6 months maturity. The seeds are sown with the onset of pre-monsoon showers i.e. May. The seeds are broadcast with cheera (*Amarantus caudatus* L.) followed by light hoeing. Vegetables and mustard are sown along the boundaries immediately after the broadcasting of finger millets. Cheera is harvested from the 30th day and used as vegetable. The first harvest of millets is in August, second in October, and the third in December. The seeds for the next season are selected from the fully matured and healthy plants. The heads along with the straw is stalked and hung down over the fireplace. The smoke and heat provides protection and preservation of the seeds. The Muthuvans claim, the seeds stored in this manner are good for 5 years. The land is then left fallow for 7 years for rejuvenation

The Muthuvan settlement is called *kudi* – which is a cluster of houses. The Muthuvans are the only tribe with a culture of dormitory system – for boys, girls and guests. It is in the dormitories that the young ones are taught and trained in the Muthuvan way of life. The traditional knowledge is passed on to the youngsters. Food is never a problem as the residents of the dormitory work together going from house to house and the house where they work take the responsibility of feeding them.

The Muthuvans are a storehouse of knowledge passed down generations – their traditional knowledge ranges from selection of site for cultivation; natural fertility enhancers; plants, animals and insects properties; sowing of seeds; knowledge of the lunar cycle; varieties of crops grown; natural manures; natural pest control; storage method; to herbal medicines.

14. Traditional rice-fish farming, Assam

Traditional Rice-Fish Farming system is practised by twenty five scheduled tribes in Assam The rice fields can be divided into 3 rice ecosystems – upland, lowland and deep water. Depending on the water source there are 2 types of fields – rain fed and irrigated fields. Traditionally, rice-fish cultivation is done on rain fed fields only. Depending on the area and topography, the practice of rice-fish farming differs – where there are unmanageable vast waterlogged rice fields or perennial waterlogged wet rice lands or oxbow type rice fields or flooded river basin rice fields, the practice is that of rice field capture fishery system. In this system the naturally occurring fish and prawns enter the field during the monsoon and grow together with the rice crop. During the monsoons, the field water overflows and connects with neighbouring watercourses – the fish enter the fields through these channels and form a vast sheet below the rice field. Fishing activities start with the floods i.e. June and continue till water recedes in November-December.

In rain fed lowland or closed deep water rice fields that are embanked all around and linked with canal systems, the tribes practice wild aqua cropping system wherein the farmers trap and rear the fish that enter the fields from the wild and intentionally utilize the rice environment for rearing fish.

In the rice fields on mountain valleys the system is called Mountain valley rice-fish system wherein dwarf varieties of rice are grown along with common carp *Cyprinus carpio*. The best example of the system is the Apatani rice fish cultivation of Arunachal Pradesh (explained above).

15. Terrace cultivation of Lanjia Saura, Odisha

Terrace cultivation of Lanjia Saura in Odisha represents an agricultural heritage system. The tribes' lives are based on forests, animals and crops which are interdependent. The terrace is also divided into 3 slopes – (1) the upper slope - 60p , where they cultivate forest plants and cashew nuts, Bageda/Swidden (mixed cropping of pulses and cereals), (2) 40p to 60p are the upper terraces with millets and orchards; in the slope between 20 to 40p , there are terrace fields of paddy, and (3) on the foothills at 20p , in the plain area, there are kitchen gardens, backyard plantation and paddy cultivation.

The bio-diversity in the area consists of forest plants that include *Azardirachta indica, Pongamia pianta, Madhuka indica, Annona squmosa, Amarinda indica, Shorea robusta, Terminalia glate, Caryota urens, Annona reticulata, Phaseolus humilis, Argemona mexicana, Coculus hisutus, Cassia samia* and cashew nuts etc.

They also grow small millets, pulses, and beans, horticultural crops like mango, jackfruit, coconuts, papaya, bananas, pineapples, cashew nuts, and peat palms.

The system is indigenous for the features including rich agri-biodiversity, stone bunding in rice terraces, complete check of soil and water erosion through sound soil and water management practices, resilient livelihoods system and labour co-operatives based on "lead and lag" approach.

The heritage value comes from its 200 year old culture, during which no drought has occurred, and rich bio-diversity reflects the traditions and culture. The rituals and festivities deeply linked with agriculture.

Indigenous knowledge is preserved and used in maintaining the system which continued to deliver environmental services – barren lands converted into cultivable ones, local resources are conserved, soil erosion checked through stone bunding, and water management through pits and fences, bio-diversity in agriculture maintains social integrity.

The threats facing the system are diminishing land sizes, climate change and onslaught of so-called high yielding mono-cropping agriculture, and disrupting traditional sustainable agriculture.

Some models of farming system adopted by tribals of odisha

Kalahandi: Pond based farming system for livelihood sustainability

Some tribal farmers of Kalahandi district adopted fingerling production and design and layout of fish pond, liming of the pond, aquatic insect control in nursery pond, Probiotics application procedure in fish pond, Stunted fingerlings production etc,. In those farming systems they have improvement of fish production through periphyton based composite carp culture practice, use of stunted fingerlings(yearlings) as stocking material in composite carp culture, application of soap oil emulsion to control aquatic insect in carp nursery etc. They also grow some seasonal vegetable and pulses on the pond dyke. Hence a particular piece of land can be optimally utilized. By adoption of this technology Jailal Kaibarta, a tribal farmer, earned his lively hood in a sustainable manner. Some of the photographs of his system are given below and cost- benefit analysis of his system and given in the Table1 and Fig. 1a,b,c & d.

Fig. 1a

Fig. 1b

Fig. 1c

Fig. 1d

Fig.1. Fingerling production systems in pond based farming system

Table 1: Cost-Benefit Analysis of Pond based Farming System in Kakhandi district

Sl No.	Enterprise	Area (ha.)	Yield	Cost of Cultivation	Gross Return	Net Return	B:C Ratio
1.	Fingerling production	3.2 ha.8 no of pond	10,00,000 no of fingerling per 0.4 ha. of pond	5,20,000	10,00000	4,80,000	1.92
2.	Tomato (VNR)	$400m^2$	12q	3500	9000	5500	2.57
3.	Brinjal (VNR)	$400m^2$	10q	3250	8500	5250	2.61
4.	Pigeon pea (ICPL-87-119)	$800m^2$	1q	2000	4400	2400	2.2
	Total			5,28,750	10,21,900	4,93,150	1.93

Mayurbhanj

Tribal farmers of periphery of Similipal bio sphere of Mayurbhanj district are practising Paddy-mustard-duckery farming system (Fig. 2 a & b) and paddy-maize-poultry farming system (Fig. 3 a, b & c)) for their livelihood sustainability with a good return being persued by the scientists of Krishi Vigyan Kendra, Mayurbhanj-II. Monocropping of paddy is a conventional practice there from which they get an average net return of Rs10000/ha. Average net return of Rs31000/ha is received from paddy- mustard-duckery farming system and an average net return of Rs33800/ha is received from paddy- -maize-poultry farming system

Fig. 2a

Fig. 2b

Fig. 2: Paddy-mustard-duckery farming system

Fig. 3a

Fig. 3b

Fig. 3c

Fig. 3: Paddy-maize-poultry farming system for better income generation

(iii) Integrated farming and farm mechanization

Some educated young tribal farmers like Mr. Radhanath Singh inclined to agriculture with an entrepreneurship spirit with the intervention of latest technologies provided by the KVK and optimum utilization of resources with use of farm machineries have excelled in the integrated farming system. He used Power Tiller, Mini Tractor, Power Thresher-cum-Winnower, Power Reaper, Sprinkler set, Axial flow thresher (power tiller operated) in his farming system. The income scenario from his farming system with adoption of mechanisation is given in table No.2

Kandhamal

Vegetable based IFS model

The district Kandhamal is favourable for vegetable cultivation with a congenial agro-climatic condition. The tribal farmers usually grow turmeric, rice, maize and groundnut in *Kharif* rain-fed areas and during *Rabi* season they grow vegetables in traditional methods. They generally grow vegetables by using local degenerated seeds without adopting INM and IPM practices which gives low productivity and less return.

Realizing the low productivity of vegetables KVK scientists selected the village Katadaganda for implementing the farming system models by including garden pea, cabbage, onion, poultry, mushroom under Tribal Sub Plan during *Rabi* 2013-14. Sri Nalaraj Pradhan is one of the interested innovative farmers. Sri Pradhan cultivated garden pea var. Azad P-3 in an area of 0.8 ha, cabbage in an area of 0.2 ha., onion in an area of 0.2ha. with INM and IPM practices. He also reared 15 nos. of 21 days improved Banaraja chicks. Mushroom cultivation finds place in his farming system (Fig. 4a, b, c & d).

Outcome

With these interventions Mr. Pradhan received a total annual net income of Rs 2,07,355/-, which is much higher as compared to mono cropping of rice or maize grown in the region. This farming system not only gave higher income but provided stability of production round the year (Table 3).

Impact

The outcome of the demonstrations has motivated the other near by farmers for adopting the IFS model. Sri Pradhan became a successful entrepreneur in his locality and eye opener for other farmers.

Table 2: Income analysis from a mechanized farming system in Mayarbhnja

Crop/ Enterprises	Area/ Unit	Produ-ction	Gross income	Expenditure	Net Profit
Paddy	5.2 ha	280 q	2,52,000/-	52,000/-	2,00,000/-
Mung	2.4ha	14 q	70,000/-	15,000/-	55,000/-
Banana	0.2ha	400 bunches	80,000/-	10,000/-	70,000/-
Vegetables	0.1ha	35 q	45,000/-	10,000/-	35,000/-
Dairy	2 nos.	8 lit	120/-	60/-per day	60/- per day
Duckery	108 nos.				
Fishery	0.4 ha	15 q	1,05,000/-	25,000/-	80,000/-

Table 3: Cost: Benefit Analysis of Vegetable based farming system in Khandhamal district

Farming system		Before KVK intervention Agri crops+Hort. crops			After KVK intervention Agri.+Hort.+poultry+ mushroom		
Crop/ Component	Season	Yield (q./ha)	Net return (Rs.)	B:C ratio	Yield (q./ha)	Net return (Rs.)	B:C ratio
Paddy	Kharif	30.8	16380	1.7	39.3	26750	1.9
Maize	Kharif	34.8	11415	1.5	42.6	19250	1.8
Cabbage	Rabi	190.3	52250	2.2	330.2	105750	2.8
Garden pea	Rabi	75.2	70300	2.6	119.3	120850	3.2
Onion	Rabi	-	-	-	326.3	185940	3.5
Poultry		-	-	-	4.5kg/bird/ year	7865/15 bird	6.4
Mushroom		-	-	-	1.9 kg/bed	3285/225 bottle	6.8

Fig. 4a

Fig. 4b

Fig. 4c

Fig. 4d

Fig. 4: Vegetable based farming system in Kandhamal District

Conclusion

- The tribals need to understand the biodiversity available in the region to follow up their own farming system with minimal expanses and higher return.
- Priority may be given to the diverse farming system keeping harmony with the ecosystem of the regions
- These in urgent need to standardize the farming practices of tribals living at different ecosystem under handicapped ecology for higher delivery of output.
- The socioeconomic conditions of the tribals needs to be characterized to provide correct technological support for IFS using extension support.

References

Beets, W.C. 1982. Multiple Cropping and Tropical Farming Systems. Westview Press, Boulder.

Chang, J.H. 1977. Tropical Agriculture: Crop Diversity and Crop Yields. *Econ. Geography*. 53:241-254.

Clawson, D.L. 1985. Harvest Security and Intraspecific Diversity in Traditional Tropical Agriculture. *Econ. Bot*. 39:56-67.

Francis, C.A. 1985. Variety Development for Multiple Cropping Systems. *CRC Cit. Rev. Pl. Sci.* 3:133-168.

Grigg, D.B. 1974. The Agricultural Systems of the World: An Evolutionary Approach. Cambridge University Press, Cambridge.

Harlan, J.R. 1976. The Possible Role of Weed Races in the Evolution of Cultivated Plants. *Euphytica,* 14:173-176.

Harwood, R.R. 1979. Small Farm Development — Understanding and Improving Farming Systems in the Humid Tropics. Westview Press, Boulder

Nimachow, G.,Rawat, J.S.,Dai, O. and Loder T.2010,A Sustainable Mountain Paddy-Fish Farming of the Apatani tribes of Arunachal Pradesh, India., Aquaculture Asia Magazine,XV(2) April-June,2010:25-28.

Singh, K.S.,1997. The Scheduled Tribes. New Delhi. Anthropological Survey of India.

Integrated Farming System Practices: Challenges & Opportunities, pp 497-512

New India Publishing Agency, New Delhi, India

20

Resource Recycling in Integrated Farming Systems

M. Mohanty

Indian agriculture has the challenge of providing national as well as household food and nutritional security to its ever increasing population with the average size of land holdings shrinking due to increasing fragmentation. Modern agriculture is facing multiple challenges of plateauing genetic potential of major crops, declining productivity in vast tracts of rainfed/ dryland areas, ill-effects of green revolution technologies in all intensively cultivated areas, climatic aberrations etc., which is threatening the sustainability of the important agricultural production systems and national food security. The assets of a small and marginal farmer is of various kinds – a small piece of land, a homestead garden, few chickens and ducks, 1 or 2 cows/pig etc which has to be managed judiciously to sustain the production productivity and livelihood of the farmer. Farmers take decisions as to crop and livestock management given their access to knowledge and information, personal circumstances, and in the context of the broader socioeconomic, institutional and political environment. Under such conditions, some farmers are painstakingly trying to make a livelihood out of their small pieces of land. However, unable to produce sufficiently, some have sold off or leased out their land to big commercial farmers. Such farmers have become a daily labour or shared cropper in their own land, or even migrated to cities in search of livelihood.

Directorate of Planning, Monitoring & Evaluation, Orissa University of Agriculture and Technology, Bhubaneswar-751003, India

To overcome the problems of small resource poor farmers under diverse and risk prone environments a more holistic, resource based, client oriented and interacting approach, popularly known as Integrated Farming System (IFS) has been developed. This system defines output as total biomass outcome of the system. Integrated farming system is a reliable way of obtaining high productivity with substantial nutrient economy in combination with maximum compatibility and replenishment of organic matter by way of effective recycling of organic residues/wastes etc. obtained through integration of various land- based enterprises.

An integrated farming system consists of a range of resource-saving practices that aim to achieve acceptable profits along with high and sustained production levels, while minimizing the negative effects of intensive farming and preserving the environment. The inter-related, interdependent and interlinking nature of IFS, involves in utilization of primary and secondary produce of one system as basic input of the other system, thus, making them mutually integrated as one whole unit. This helps to reduce the dependence on external inputs, making thereby the IFS a self-supporting and sustainable system over time.

The principles of the Integrated Farming System represents a winning combination that increases crop yields, soil biological activity and nutrient recycling, reduces cost of production and improves profits with optimum productivity with maximum input use efficiency.

The studies involving enterprises such as crop, fishery, poultry, duckery, apiary and mushroom production revealed that by-products of dairy i.e. cow dung forms a major raw materials for biogas plants, digested slurry of bio-gas plant forms a major part of feed of pisciculture for increasing plankton growth as well as supplying valuable manure to raise the productivity of field crops/enrich the soil. The by-products of field crops like paddy straw forms a major input of mushroom cultivation. Straw after use in mushroom production is utilized as cattle-feed and compost preparation. Similarly, the poultry droppings are used as fish feed, help in plankton growth as well as manure. Even apiary improves in pollination, apart from giving a wholesome product like honey to farmers (Singh, 2013).

Philosophy of resource recycling

- Reduces production cost of components while increasing farm income through proper residue recycling amongst allied components.
- Sustainable soil fertility and productivity through organic waste recycling.
- Helps in environmental protection through effective recycling of animal waste.

- Inclusion of biogas and agro forestry in integrated farming system can solve the energy crisis.
- Cultivation of fodder crops as intercropping and as border crop results in the availability of adequate. nutritious fodder for animal components like milch cow, goat / sheep, pig and rabbit.
- Firewood and construction wood requirements could be met from the agroforestry system without affecting the natural forest.
- To ensure optional utilization and conservation of available resources and effective recycling of farm residues within system.
- To maintain sustainable production system without damaging resources/ environment.
- Various tree leaves such as *Tephrosia candida*, *Gliricidia maculata*, *Indigoferra* spc., *Leucaenea leucocephala* can be successfully used for supplying the leaf biomass to maintain the soil fertility and moisture status.
- Sugarcane trash, compost, vermicompost , bio- gas slurry, industrial wastes, municipal and sewage wastes crop residues can be effectively utilized for enhancing soil fertility without any adverse effect on soil and environment.

Case studies

- The integrated farming system comprising of crop + pigeon + goat + buffalo + agro forestry + farm pond generated the highest residues of 11.58 t/year with a nutrient value of 105.9, 46.2 and 76.9 kg N, P and K, respectively in western zone of Tamil Nadu (Jayanti *et al.*, 2008).
- Integration of crop + fish + poultry resulted in higher fish productivity under lowlands. The poultry, pigeon and goat droppings were utilized as feed initially and at the end of a year after the fish harvest, about 4500 kg of settled silt from each pond were collected. The pond silt was utilized as organic sources to supply sufficient quantity of nutrients to the crops cultivated in pond dykes. Total quantities of 118.4 kg N,79.8 kg P and 108.12 kg K/year generated from various livestock components of IFS i.e., bullocks, cow, goats and poultry birds in Raichur, Karnataka.
- Nutrients recycled from poultry and duckery resulted in higher plankton development (256/ liter) in the ponds than cow dung in fish + duck + poultry farming systems in sloping hilly uplands of Andaman and Nicobar islands (Ravishankar *et al.*, 2007).

- It has been estimated that one tonne of deep litter fertilizer is produced by 25-30 birds in a year's time. As such 500-600 birds are adequate to produce manuring for a hectare of water area under poly fish culture (Esthen *et al.*, 2005).
- The coconut husk, peduncle and fonds, not used as fuel can be recycled as organic source of nutrients to provide 52-7-162 kg N-P-K /ha/years in terms of fertilizers in Kerala. The peduncle can also be used to support creepers at early stage (Thomas *et al.*, 2005).
- Well decomposed coir wastes @ 12.5 t/ha with recommended fertilizer to groundnut and maize has also increased crop yield in Kerala.
- Rice-sunhemp system supplemented with mushroom and poultry further enhanced the recycling potential by 14.9% over rice-sunhemp system alone in Goa because of the use of rice straw as substrate for mushroom cultivation (Korikanthimath and Manjunath, 2009).
- The supplemented nutritive value of used-paddy straw after the harvest of mushroom can be well utilized for cattle feed, which results in enhancing milk yield in rice-dairy-mushroom systems (Singh, 2013).
- Residue recycling in crop + fish +cattle+ poultry model revealed that integration of crop with fish and poultry resulted in higher fish productivity, which resulted in higher net returns of Rs. 7, 090/year from 0.06 ha of pond in Bihar. Recycling of organic manures obtained from different components added 506.2 kg N, 348.5 kg P_2O_5 and 341.7 kg K_2O into the system (Kumar *et al.*, 2013).
- Resource and energy flow in integrated farming system for small and medium farmers of Siruguppa, Karnataka, involved cropping (rice, maize, sunflower, vegetables in 0.73 ha), fishery (0.06 ha), thirty poultry birds and 12 goats including fodder area in one hectare land. The poultry shed was constructed on the fish pond. This was compared with the conventional rice-rice system. To sustain the productivity the residues obtained in the system was recycled. The integration of crop with fish, poultry and goat resulted in higher productivity than adoption of conventional rice-rice alone. Poultry droppings was allowed to drop into the pond directly which served as the source of food for fish. Fishes were harvested after completing one year using drag net. The nutrient rich pond water was used as source of irrigation for crops. Pond silts are used as manure in crop fields.
- Data generated through a survey from rural households in Ethiopian, where mixed cropping along with livestock rearing is a common practice revealed that large farm family size and membership to farmers

associations contribute to higher level of resource use efficiency as compared to small farm families and individual efforts. The findings of Nagoli *et al.* (2013) suggested that resource use efficiency would significantly improve through a better crop livestock association/integration and expansion of off farm activities.

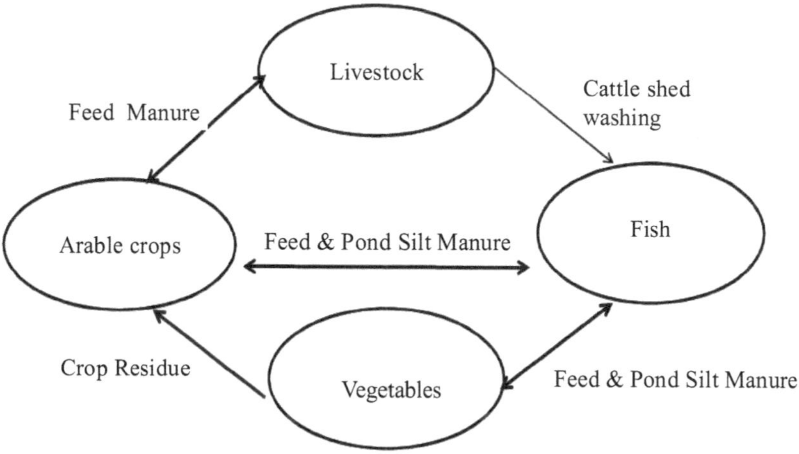

Fig. 1: Bio-resource flow model in crop- vegetable-livestock -fish farming system
Source: Nagoli *et al.*, 2013

Residue recycling in integrated farming system model for 1 ac land in Tamilnadu involving crop, livestock, goat and guinea fowl revealed that total quantity of bio-compost obtained from the integrated farming system was about 3.5 tonnes. Of this 2.6 tonnes was applied to annual crop which was raised in 0.60 acre and the remaining 0.9 tonnes was diverted to fodder crops raised in 0.2 acre of land area. About 1.5 tonnes of vermicompost was obtained by way of recycling the goat, guinea fowl manures and vegetable crop waste. Out of this, about 0.5 t of vermicompost was diverted to vegetable crop raised in 0.10 acre and the remaining one tonnes was sold to add the revenue of the farm. In the traditional cropping system, the residue generated was less as compared to integrated farming system. The nutrient content was 0.7 % N, 0.6% $P_2O_{5,}$ 0.7 % K_2O and 2.3 % N, 0.7% $P_2O_{5,}$ 1.2 % K_2O in bio-compost and vermicompost respectively. In vermicompost the nutrient content was slightly higher than the bio-compost because of higher nutrient content in goat and guinea fowl manures. The manure obtained were recycled as nutrient input to the crops after composting and served as organic manure for sustaining the productivity of crops. The system of crop + milch cows + goat + guinea fowl + bio-compost and vermicompost could provide better bio resource utilization and recycling

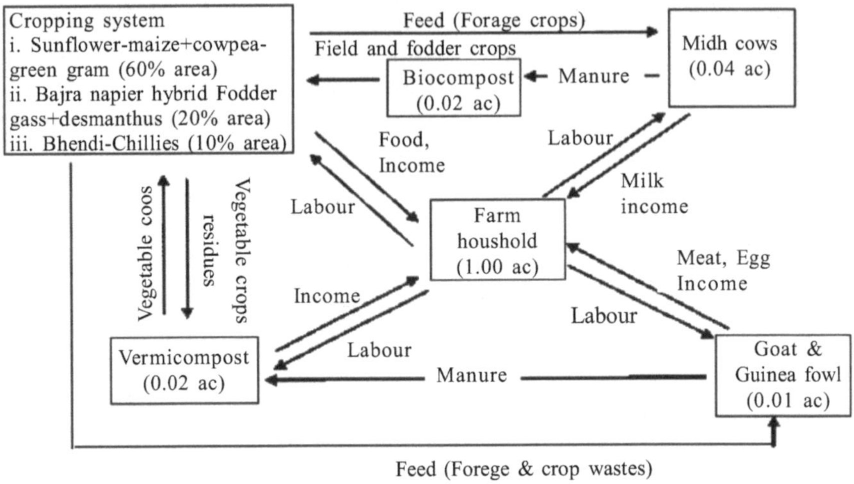

Fig. 2: Resource flow chart of integrated farming system for 1.0 acre land
Source: Jayanthi *et al.,* 2008

(Jyanti et al., 2008). Synergistic interaction of the farming system in terms of labour, resources and residue recycled is depicted in Fig. 2. The dependence on external inputs for all the systems decreased during the second year indicates that over long periods of time, the integrated farming systems will become more self supporting, self sufficient and sustainable.

Table 1: Nutrient content of different crop residues/organic inputs

S.No.	Crop residue	N%	P205%	K20%
1.	Rice-wheat	0.50	0.60	1.50
2.	Azolla	0.42	0.20	0.40
3.	Dhanicha	0.60	0.10	0.25
4.	Press mud	1.68	1.45	1.20
5.	Tea waste	0.35	0.40	1.50
6.	Coir pith	1.26	0.06	1.20
8.	Mushroom spent straw	1.90	0.40	2.4
9.	Litter fall of trees	150-300 KgN	10-20KgP	75-150Kgt, 100-300 kg ca

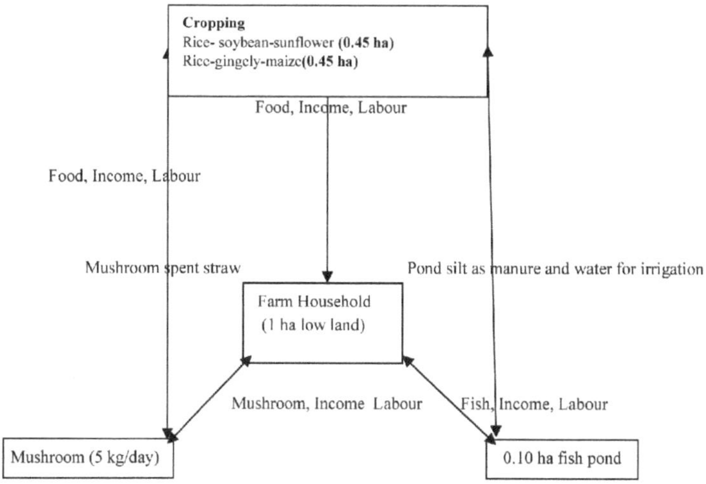

Fig. 3: Resource recycling in Crop + Fish + Mushroom farming system for 1.0 ha low land

Resource recycling in crop to fish to mushroom farming system for 1.0 ha low land generates food grains, mushrooms and fish which not only ensurces balanced food supply to the farm family but also generates additional income and employment opportunities round the year (Fig. 3). The pond silt and mushroom spent straw can be utilized as organic manure in the crop field (Rana, 2014).

Integrated farming system model with crop + goat + as grow forestry for 1.0ha land could generate food, feed and organic manures besides, employment and income generation for the farm family as depicted in Fig. 4. (Singh, 2011).

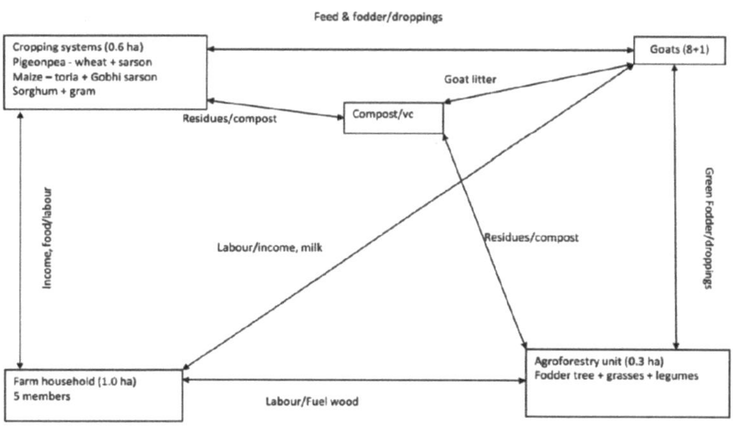

Fig. 4: Resource flow model in Crop + Goat + Agro-forestry system for 1.0 ha land
Source: Singh, S., 2011

Resource model in IFS designed for irrigated areas of Tarai region exhibits number of farm enterprises including field crops horticultural crops, animal and poultry components can effectively be incorporated along with on farm production of green and organic manures with available resource base of the farmer to make the systems sustainable and profitable (Singh, 2013) as debited in Fig. 5.

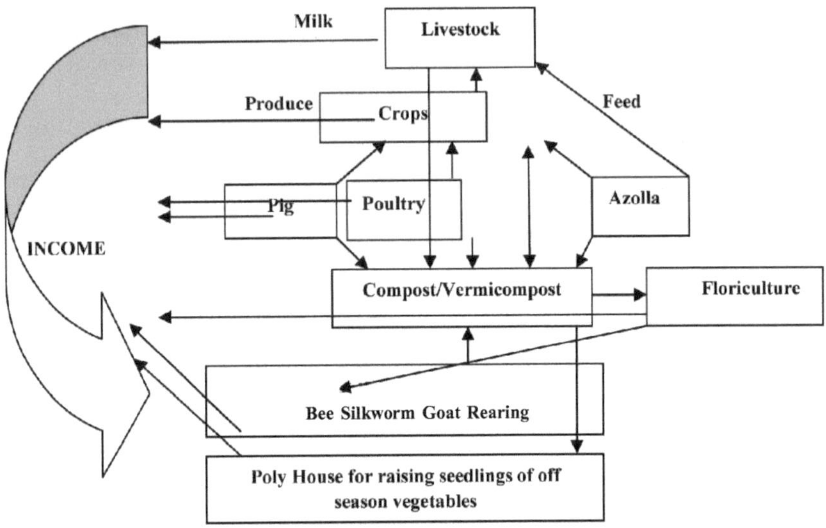

Fig. 5: Resource flow model in IFS for hilly areas

Success story of on-farm resource recycling

Integrated farming is a system which tries to imitate the nature's principle, where not only crops but, varied types of plants, animals, birds, fish and other aquatic flora and fauna are utilized for production. These are combined in such a way and proportion that each element helps the other; the waste of one is recycled as resource for the other.

Pravat Pal lives with his wife and son in Birbhum district of West Bengal, India. He owns an ancestral farm of size 0.81 acre. His son is physically challenged and Pal fell into debt for his son's treatment. The only way they could afford this was to take out loans and sell most of the produce from the farm, which was only rainfed paddy, along with his assets from time to time. This created a severe food and cash shortage for the family (Fig. 6.).

In the middle of this crisis, he switched over to IFS influenced by an orientation session. Slowly he developed nutrition garden in his fallow land to cultivate 8 to 10 vegetables in every season, mostly used for consumption. He also keeps seeds of all the vegetables. His 0.81 acre of land is cultivated 3 times now with paddy and black gram in rainy season, 0.37 acre for vegetables, pulses like lentils & field pea and oil seeds like mustard and 0.44 acre of land for winter paddy. He also grows vegetables, sesame, black gram and green gram in summer in his 0.3 acre of land. With his 3 cows, 1 bullock, 1 calf, 3 goats and 2 hens – he is now independent of external inputs. He also set up a biogas unit - slurry is used as manure and gas is used as fuel. The livestock feed is now managed with straw, mustard cake, pulses and other agri-waste, the hens are fed with the food waste. He does not have any pond, but has share in 10 ponds in the village from where he earns a portion of his annual income (Das, 2013).

Now, Pal's farm has 5 subsystems, which interact with each other positively, waste products are consumed entirely within the system. Around 70% of his farm inputs, 100% of fodder and 100% of the fuel needs are met from his farm, which amounts to Rs. 28000 per year. They have managed to pay off part of their debts. Their income has increased as surplus production, seeds and seedlings are sold in the market after meeting all their family needs. The Pals hardly have to buy food now – they can rely on a steady supply of rice, oil, pulses, fish, milk, egg and vegetables throughout the year.

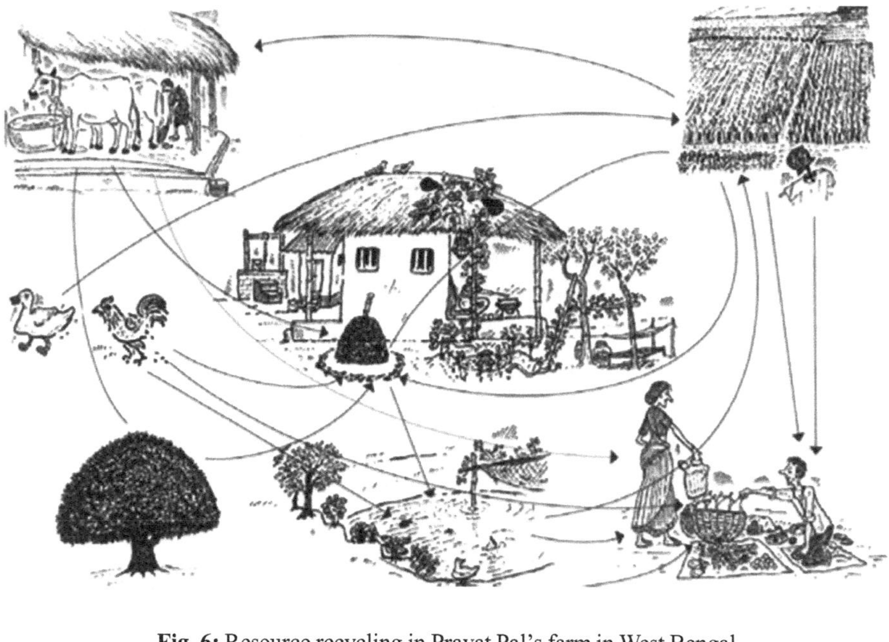

Fig. 6: Resource recycling in Pravat Pal's farm in West Bengal

Organic agriculture : A resource saving/ recycling way of farming

Organic farming is economically feasible in areas where resources are available within the farm and least dependent on external inputs. The areas like Eastern and North Eastern regions of India, where a lot of biomass is available from forest, weeds, crops, animal wastes etc. organic farming would be more economical and viable. By this practice, on-farm resource recycling can be effectively carried out with minimal dependence on off-farm inputs and management practices that restore maintain and sustain ecological harmony. In this method, production system relies on animal manures, organic wastes, crop rotations, legumes and aspects of biological pest control. It avoids the use of synthetically produced fertilizers, pesticides, growth regulators and livestock additives. Organic crop production is not only a holistic approach of production system that gives quality "Organic food", but helps to restore soil fertility on long term basis. It is relatively independent production system compared to conventional agriculture, which depends mostly on synthetically produced inputs i.e., fertilizers, fungicides, insecticides, herbicides, growth regulators, etc. Moreover, organic produce are expected to fetch premium price (Das, 2013) and therefore can provide more return to the farmers. It is the ecological production management system that recycles the farm wastes thereby promotes and enhances biodiversity, biological cycles and biological activity of the soil. Thus, it is an effective way of turning the farming system models self sustaining and least dependant on external inputs as well as eco-friendly.

Crops grown with high organic manure application could tolerate the pest and disease attack better. It is sound and sustainable way of growing more food. Organic recyclable waste include – crop residues, farm and industrial waste, municipal sewage. They are valuable sources of plant nutrient for tropical and subtropical soils found in India, where there is general deficiency of organic carbon and plant nutrients due to rapid loss of this component by bio- degradation. Kumar et al., 2013 observed the nutrient contents of crop residues and residues from other sources as follows.

1. **Sugarcane Trash:** Fresh sugarcane trash contains 0.36% N with a wide C: N ratio of 122:1.The composted trash contains – higher content of N (1.09) with reduced C:N ratio (20.1). Per hectare availability of trash is about 6-8 tonnes.

2. **Bio- gas Slurry:** Organic manures from animal wastes are very important nutrient sources in building up soil fertility. In India, estimated production of dung and urine is about 1002 and 658 million tonnes, respectively. They contribute about 5.7 million tonnes of N P K with proper utilization. Biogas

as fuel (gas) and fertilizer (slurry). The dry slurry contains about 1.8% N, 1.10 % P_2O_5 and 1.50% K_2O.

3. **Industrial Wastes:** Among the industrial by products, spent wash from distilleries and molasses and press mud from sugar factories have good manurial value. It is important to use only well decomposed press mud at 10 tones / ha. Addition of press mud improves the soil fertility it acts as a reclamation agent in saline and sodic soils. Well decomposed coir waste @12.5 t/ha with recommended fertilizer to groundnut and maize increases yield.

4. **Municipal and Sewage Wastes:** This is one of the important components of organic wastes. In India, the total municipal refuse is about 12 million tonnes / annum containing 0.5 %N.0.3%P and 0.3% K. Sewage (liquid portion) and sludge (solid portion) and available to an extent of 4 million tonnes /annum containing 3 % N, 2 % P and 0.3% K. Such organic waste can be used after proper treatment as it contains metals thus hazardous to plants, animals & human beings.

5. **Crop residues:** Residues left out after the harvest of the economic portion are called crop residues /straw. Paddy straw and residues contains about 0.5% N, 0.6% P and 1.5 % K. The crop residues can be recycled by way of soil incorporation, making compost or as mulch material.

6. **Rice Husk:** It is major by- product of the rice milling industries. It is poor in nutrient content (0.3 % N, 0.2 % P and 0.3 K) but can be incorporated into the wet soil and can be used in saline and alkaline soils to improve the physical condition. It can also be used as a bedding material for animals.

How to design IFS models for successful on-farm resource recycling

Farm designs are made for each farm based on the farm analysis, which is different from each other. Generally, to begin with, the focus is on improving crop diversity. The diversity of the farm land is increased as much as possible by introducing at least 5-6 types of cereals and pulses/oilseeds, 10-12 varieties of vegetables, 5-6 varieties of fruit, fuel wood and fodder, 5-6 types of spices or medicinal plants.

Fast growing trees and shrubs like Gliricidia, Sesbania, Bauhinia, Pigeon Pea are planted as they add high nutrient content to the soil. They can also be used as fodder for livestock as well as fuel. Once the crop diversity is enhanced and integration between existing components is ensured, 2-3 types of livestock, 3-4 types of birds and fish are added, depending on the carrying capacity of the farm judged during farm analysis.

Integration is designed based on the existing subsystems. The process is facilitated by using the resource flow diagram. Farmers are helped to design various components in a way that they integrate into each other – the output of one component becoming the input to the other. For example the agro/livestock waste gets recycled through a biodigestor for vermicompost or biogas. Some components are designed based on the need.

If there is livestock, then fodder crops are integrated on the farm. The decisions are taken by the individual farmers as the new design, at times calls for additional financial investment.

Another task is to create local resource persons to take IFS forward. The programme envisages to develop resource farmers who can train others in IFS, thus ensuring sustainability of the system.

Interactions and linkages of different enterprises in farming system

IFS deal with utilization of wastes and residues. It may be possible to reach the same level of yield with proportionately less input in the integrated farming and yield would be interestingly more sustainable because the waste of one enterprise becomes the input of another leaving almost nothing to pollute the environment or to degrade the resource base. To put this concept to practice efficiently it is necessary to study linkages and complementarities of different enterprises in various farming systems. The knowledge of linkages and complementarities will help to develop farming system (integrated farming) in which the waste of one enterprise is more efficiently used as inputs in another with the system (Fig.7).

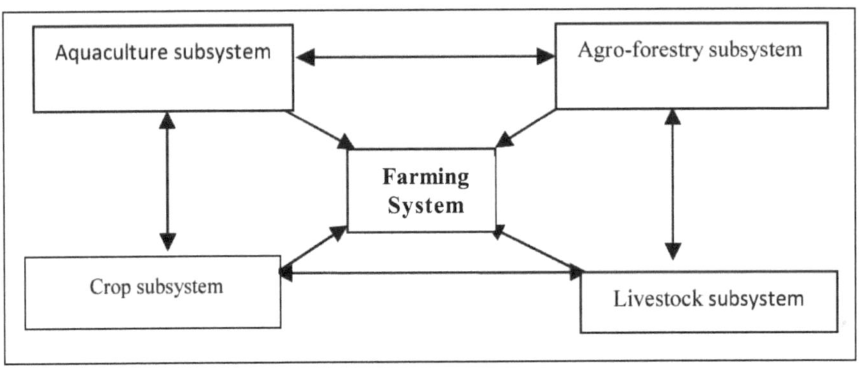

Fig. 7: Linkages of different enterprises in farming system

Benefits of resource recycling under IFS

a) Social benefits

a) This type of activity is labour intensive.

b) Creates round the year employment both on and off the farm.

c) It gives year round availability of nutritious and seasonal food.

d) Seasonal migration is reduced.

e) Fodder and fuel shortage is also minimized.

b) Ecological benefits

a) Soil organic carbon increased.

b) Fossil fuel dependency is practically zero as all the variable inputs are produced within the farm.

c) The crop diversity is huge as various types of crops, creepers, climbers, strategic crops etc. are cultivated within the farm.

d) Many numbers of local crops are introduced.

e) Soil micro/macro fauna increased.

f) Each and everything is recycled within the system resulting in zero waste farming system.

g) Diversification makes the system disaster resilient.

c) Economic benefits

a) The input cost is reduced, as dependance on extrenal input is lowered.

b) Net income increases.

c) Risk is reduced as scope for diversification in farm enterprises in creases.

d) The farmer gets income throughout the year from different sources.

Table 2: Average nutrient content (%) of some organic nutrients

Material	N	P	K
I. Bulky Organic manures :			
Farm yard manure	0.5-1.5	0.4-0.8	0.5-1.9
Compost (Urban)	1.2-2.0	1.0	1.0-2.0
Compost (Rural)	0.4-0.8	0.3-0.6	0.7-1.0
Green Manure fresh (Average)	0.5-0.7	0.1-0.2	0.5-0.6
Filter press cake	1.0-1.5	4.0-7.0	2.0-7.0
Water hyacinth compost	2.0-3.0	1.0-2.0	3.0-4.0
Pultry manure	3.72	2.68	1.23
Pond Manure	1.96	1.02	0.72
Goat manure	2.62	1.60	1.08
Vermi compost	2.50	2.12	2.24
II. Oil cake:			
a) Non-edible cakes-			
Castor cake	5.5-5.8	1.8-1.9	1.0-1.1
Cotton seed cake (Undecorticated)	3.9-4.0	1.8-1.9	1.0-1.1
Mahua cake	2.5-2.6	0.8-0.9	1.8-1.9
Karanja cake	3.9-4.0	0.9-1.0	1.3-1.4
Neem cake	3.5-5.3	1.0-1.1	1.4-1.5
Safflower cake (Undecorticated)	4.8-4.9	1.4-1.5	1.2-1.3
b) Edible Oil cake			
Coconut cake	3.0-3.2	1.8-1.9	1.7-1.8
Groundnut cake (Decorticated)	7.0-7.2	1.5-1.6	1.2-1.4
Groundnut cake(Undecorticated)	4.4-4.6	1.6-1.7	1.5-1.6
Rapeseed cake	5.1-5.2	1.4-1.9	1.1-1.3
Sesame oil cake	6.2-6.3	2.0-2.1	1.2-1.3
III. Manures of Animal origin			
Fish manures	4.0-10.0	3.9	0.3-1.5
Bird Guano	7-8	11-14	2-3
Raw bone meal	3-4	20	-
Steamed bone meal	1-2	22	-
Cattle Urine	0.9-1.2	-	0.5-1.0
Cattle dung and urine mix	0.6	0.1-1.5	0.4-0.5
Sheep dung and urine mix	0.95	0.25-0.35	0.4-1.0
IV. Plant residues			
Rice husk	0.3-0.5	0.2-0.3	0.3-0.4
Groundnut husk (Shell and Stalk)	1.6-1.8	0.3-0.5	1.1-1.7
Wheat straw	0.53	0.10	1.10
Sugarcane trash	0.35	0.10	0.60
Maize	0.42	1.57	1.65
Paddy	0.36	0.08	1.71
Cotton	0.44	0.10	0.66
V. Dry Tree leaves			
Careya arborea(kumbhi/kundhei)	1.67	0.40	2.20
Cassia occidentalis (Chakunda)	0.98	0.20	0.67
Dillenia pentagyna (Saharbaha)	1.34	0.50	3.20

Contd.

Material	N	P	K
Madhuca indica (Mahua)	1.66	0.50	2.00
Pongamia pinnata (Karanja)	3.69	2.41	2.42
Pterocarpus marsupium (Piasal)	1.97	0.40	2.90
Terminalia tomentosa (Sahaj)	1.39	0.40	1.80
VI. Weeds			
E.colonum (suan)	1.75	0.24	0.67
C.dactylon (duba)	1.00	0.19	0.37
C.rotundus (mutha)	2.25	0.27	0.37
D.aegypticum (kaugodia)	2.00	0.23	0.54
A.aspera (apamaranga)	2.21	0.71	1.09
A.spinosus (khada saga)	1.92	0.68	2.75
A.mexicana(agara)	1.01	0.60	1.10
C.album (bathua)	3.99	0.66	8.29
C.viscosa (ana sorisia)	1.96	0.67	2.55
C.benghalensis (kanasiri)	2.02	0.64	1.54
T.portulacastrum (puruni)	2.64	0.43	1.30
P.hysterophorus(gajar ghasa)	2.68	0.68	1.45
I.cornea (amari)	2.01	0.33	0.40
C. gigantea (Arakh)	2.06	0.54	0.31
C. fistula (bana chakunda)	1.60	0.24	1.20
L.camara (nagaairi)	2.50	0.25	1.40

References

Agronica,2005. Directorate of Agriculture and Food Production, Government of Odisha.

Das, A. 2013. Integrated Farming: An approach to boost up family farming. LEISA India 15 (4)

Esther Shekinah D, Jayanthi, C. and Sankaran ,N. 2005. Physical indicators of sustainability- A farming systems approach for the small farmer in rainfed vertisols of the Western zone of Tamil Nadu. *J. Sustainable Agriculture.* 25 (3): 43-65.

Gill, M. S., Singh, J. P. and Gangwar, K. S. 2010. Integrated farming system and agriculture sustainability. *Indian Journal of Agronomy* 54(2): 128–39.

Jamu, D. 2003. Identification of sediment processes regulating nitrogen use efficiency and retention in integrated aquaculture/agriculture systems. Final Report submitted to the African Career Awards, The Rockefeller Foundation, 43pp, Nairobi,Kenya.

Jayanthi, C., Rangasamy, A. and Chinnusamy, C. 1997. Integrated nutrient management in rice based cropping systems linked with lowland integrated farming system. *Fertilizer News.* 42 (3): 25-30.

Jayanthi, C. 2002. Sustainable Farming System and low land farming of Tamil Nadu. IFS Adhoc scheme Completion Report.

Jayanthi, C., Devasenapathy, P. and Vennila, C. 2008. Farming Systems Principles and Practices.

Jayanthi, C., Vennila, C. and Nalini, K. 2008. Farmer's participatory research on integrated farming system. "Global Issues. Paddock Action." Edited by M. Unkovich. Proceedings of 14th Agronomy Conference 2008, 21-25 September 2008, Adelaide, South Australia

J.,Nagoli, J., Valeta and F. Kapute.2013. Analysis of bio-resource utilization in integrated agriculture-aquaculture farming systems in zomba district, southern malawi. *Malawi J.aquac.fish.*, 2 (1): 15-19

Kumar, S., Singh, S. S., Meena, Shivani, M. K. and Dey, A. 2012. Resource recycling and their management under integrated farming system for lowlands of Bihar. *Indian Journal of Agricultural Sciences* 82 (6): 504–10.

Munda, G. C., Das Anup, and Patel, D. P. 2013.Organic Farming in Hill Ecosystems – Prospects and Practices. ICAR Research Complex for NEH Region, Umiam -793 103, Meghalaya.

Korikanthimath, V.S. and Manjunath, B.L. 2009. Integrated farming systems for sustainability in Agricultural Production. *Indian Journal of Agronomy*. 54(2):140-148.

Rana, S.S. 2014. Lecture note on resource recycling and flow of energy in different farming systems.

Rangasamy, A., Venkitasamy, R., Jayanthi, C., Purshothaman, S. and Palaniappan, S.P. 1995. Rice based farming system: A viable approach. *Indian Farming*. 44(11): 27-29.

Ravishankar, N., Pramanik, S.C., Rai, R.B. Shakila, N., Biswas, T.K. and Bibi, N. 2007. Study on integrated farming systems in hilly areas of bay island. *Indian Journal of Agronomy* 53(1):7-10.

Singh, R.P. 2013. Farming Systems: Concept, Scope and Modules. Power Point Presentation at GBPUAT, Uttrakhand.

Singh, S. 2011. Farming Systems Approach for Food Security and Sustained Rural Economy. Power Point Presentation at HPAU, Himachal Pradesh.

Thomas, G.V. Krishna Kumar, V., Maheswarppa, H.P. and Palaniswami, C. 2010. Coconut Based cropping farming systems. Central Plantation Crops Research Institute, Kasaragod, Kerala.

Integrated Farming System Practices: Challenges & Opportunities, pp 513-531

New India Publishing Agency, New Delhi, India

21

Participatory Extension Approaches for Technology Transfer

B.P.Mishra

Farming System Development (FSD) is better perceived as a means of facilitating interactive links between the actors (Table-1) who play significant roles in the process. A system perspective is needed in designing and evaluating suitable technologies as well as in designing policy / support system that improve both productivity and sustainability. Considerable challenges do exist in modifying the conventional top-down orientation characteristics of extension services. This requires the fundamental principle of FSD i.e. interactive mode of operation.

Table 1: Role of functions at 'actors' in Agricultural Development

Role	Functions	Actors
Implementing	-	Farmers
Supporting	Transmitting Input provision and market for products	Extension staff Development agencies, non-government agencies, commercial farmers
Providing	Technologies/policies	Research
Potential means	Support system	Planning

Directorate of Extension Education, Orissa University of Agriculture & Technology, Bhubaneswar-751003, Odisha, India

Principle of FSD in IFS

The FSD approach has been developed by Farm Management and Production Economics Services of FAO, more popularly known as Farming System Research and Extension (FSRE). This is based on the principles of development-improving productivity, increasing profitability, ensuring sustainability and guaranteeing an equitable distribution of the results of production (Fig. 1).

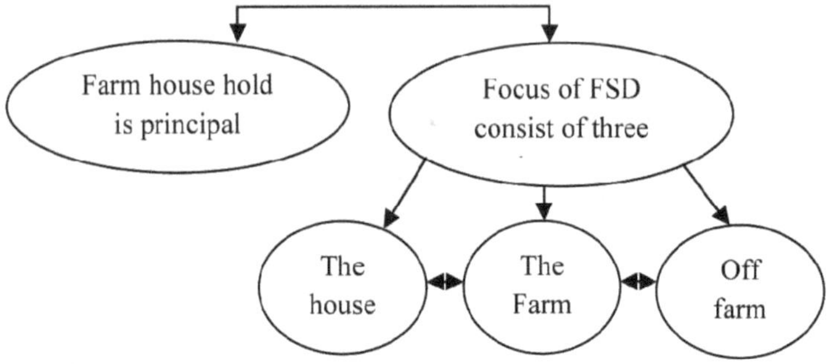

Fig. 1: FSD and Farm house hold interactive systems

The evolution of participatory approaches

This system of extension assumed that its primary task was to convey superior technologies into local practices seeing the farmers as the recipients or as the adopters or rejecters of innovations overlooking the local knowledge or indigenous practices. Participatory approaches, on the other hand are accepted not only with improving farming practices but also giving a "voice" to the farmers. Participatory approaches recognize the importance of all stakeholders in the generation and dissemination of knowledge and in removing systems constraints (Table 2).

Table 2: Conventional ToT model versus Participatory Extension Approaches

Feature	Transfer of Technology (ToT)	Participatory Extension (PEA) Farmers First (FF) approaches
Main objective	Transfer technology	Empowerment of farmers
Analysis of needs and priorities set by	Extension agents and researchers	Farmers facilitated by outsiders
Primary R & D location	Research station	Farmers' fields and conditions
Menu of technologies	Fixed or blanket recommendations	Baskets of options (flexible recommendations)
Dissemination process	Linear transfer of precepts, messages and technological packages	Dynamic process based on joint analysis and farmers' choices

Contd.

Feature	Transfer of Technology (ToT)	Participatory Extension (PEA) Farmers First (FF) approaches
Farmers behavior	Taught about the message: adopt or reject	Apply principals, use methods, choose from basket and experiment
Outsiders desired outcome	Widespread adoption	Wider choice for farmers and enhanced adaptability
Main mode of diffusion	Extension workers to farmers	Farmer to farmer
Role of extension agent	Provider of information, technical supervisor, teacher	Facilitator, catalyst, advisor

Participatory technology development

The technology innovation & transfer process require to be tuned better, to fit the requirements of the farmers in a given situation. It is technology integration process, which refers to the process or a set of activities, if new research information can be put to use, in the farm production system. This intermediary function has been poorly recognised in institution building effort. It is the function that can best relate the research and technology development efforts to extension. The aim of technology integration should promote the adoption of balanced production system, in which the requirements for productivity, stability, sustainability and equitability are met fully, with low use of external inputs. It has been reported in several studies at national and international levels that most of the available technologies are yet to be adopted by the farmers. The fact behind is; they either are not matching with the farmer's need or not compatible with the farmer's overall farming systems their by resulting in low acceptance of technologies. Generally Scenario of Research System hardly has scope to provide data base for realistic planning and emphasis on the need for change in research and development approach. These back ground studies led to the evolution of a more realistic and systematic approach called as Farming System Approach. This approach is farmers need based, farmers participatory, problem solving, gender sensitive, inter disciplinary, interactive, iterative and location specific.

The process of participatory research and extension approach for development of IFS

The process of Participatory Research and Extension Approach is required for IFS that scopes interaction among stakeholders which is different from conveying standard messages from the top down. The processes is to be carefully managed, facilitated, and stimulated with contributing partners, particularly where they have a unique role to play (Table 3).

Table 3: Changes from conventional to participatory approaches to development

Conventional approach (T&V)	Participatory approach (PREA)
Telling farmers what to do	Discussing with farmers
Making farmers change	Working with farmers to effect change
Extension workers and researchers know best	Learning from farmers
Modern methods are better than traditional ones	Building on local traditional knowledge
	Providing feedback to communities
Teaching	Promoting farmer- to-farmer extension
Use of contact farmers and demonstrations	

Gender dimensions of participatory research and extension approach

Participatory research and extension approach allows the mainstreaming of gender, providing an opportunity for encouraging gender equity through the inclusion of women as participants and promoting equal access to resources with benefits for both men and women. This requires commitment from implementing partners to ensure that gender issues are given appropriate consideration for farming system development.

Gender mainstreaming

- Discussion among partners on gender issues and development constraints.
- Capacity building for partners on gender interventions.
- Assessment of progress on lowering gender barriers.
- Commitment for gender mainstreaming from policymakers and decision takers. Policy changes towards the adoption of gender sensitive approaches.

The introduction of any intervention needs to be negotiated and integrated by both male and female farmers for it to be sustainable The Gender Equality Wheel is a tool that would be used in gender mainstreaming by the project which identifies four stages of the progress toward gender equality (Fig.2)

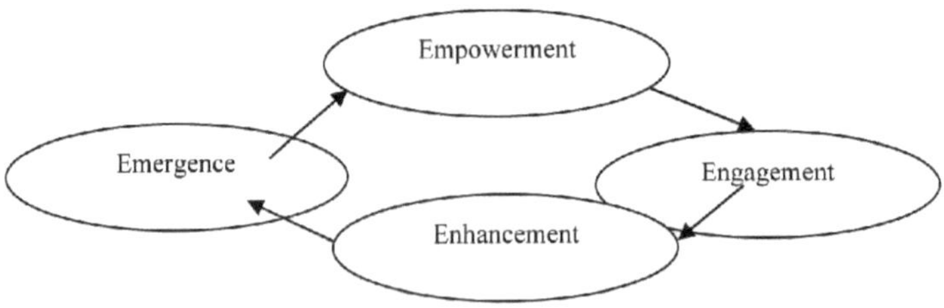

Fig. 2: Gender equality wheel

- **Empowerment:** Resources such as ideas, knowledge, and skills that are available to the community. These resources are the cornerstone of social capital, building self-confidence in women and men in exploring new ways of seeing and acting.
- **Engagement:** Attention to be paid to identify and develop lead farmers as well as to farmers' groups to ensure gender balance during the selection of participants; women come out of isolation, discover new possibilities for their lives, and begin to build mutual support.
- **Enhancement:** Process when women and men start applying the new ideas, knowledge, and skills to enhance the socio economic standard of family and community members. Trainings women farmers and women collaborators are necessary to process the technology for enhancing income and improve household nutrition and health.
- **Emergence:** It is the process when women and men move onto the public stage, to take social and political action that transforms their social, cultural, and political environment. This is in addition to conducting field days and mid-season evaluations on women farmers' fields to voice their experience to fellow women farmers.

Phases of participatory extension approach

Four phases of participatory extension approaches are important for the participatory extension linkage

Phase I Social mobilization : Facilitating the communities' own analysis of their situation.

Phase II Community-level : Action Planning.

Phase III Implementation and farmer experimentation.

Phase IV Monitoring the process through sharing experiences, ideas and self evaluation.

In each of the phases several steps are involved in implementation of participatory process.

Phase I: Social mobilization

The development activities should be owned by the community, for which two key conditions are needed to be in place:

- Real motivation and enthusiasm within the community; and
- Effective community organizations that would support the process for further replication.

Motivating people for learning and action the key concerns are to be identified and addressed by participants. Extension Scientists facilitate this analysis and prioritization process. It is important to understand that the community is not homogeneous and that it consists of several institutions with different roles and responsibilities. Hence key to the process is to identify institutions which can take a lead in catalyzing the development process within the community, and to build the capacity of these institutions and developing action plans.

Entering and building trust in the community

Every community is different. The way the community functions is to be understood before trying to introduce a process of transformation. So after an informal meeting the extension agents should spend some days in the village and learn about the perceptions of the local institutions and problems and needs of the people. Thus a relationship can be developed within the society to build trust. This facilitates the technology dissemination for improving integrated farming systems for better livelihood management.

Assessment of livelihoods

Assessing livelihoods provides an understanding of the different ways in which people derive a living;

- who in the community is involved;
- the relative importance of each in providing either food or cash;
- the extent of per centage involvement of the community;
- the trends over the years and
- the reasons for these trends.

Identification and support to effective organizations

Community members take the lead to

- identify the types of formal, informal modern or traditional institutions operating within the community;
- discuss and understand the role of these organizations in the community, their functions, strength and weakness;
- understand relationship between the institutions, the conflict and alliances and networks that determines how things work and
- identify the resources (human and material) for use in the developmental process.

The leaders, members and non-members of the various groups of the institutions are informally interviewed to understand feelings of people about the government and non-governmental institutions and agricultural technology development and management system in the area concerning the development and management of integrated system of farming with appropriate institutional support.

Feedback to the community

The findings of instituitional surveys are important to the extension workers as well as the community members. They get opportunity to be aware of and reflect about community organizations, perceptions of institutional functions, roles, strengths and weaknesses. At the end of the meeting an initial selection of possible partner institutions is made, and discussions occur regarding how final selection shall be made. The feedback mechanism facilitates the technology diffusion process for meaningful integration of various components of integrated farming system development.

Raising awareness in the community

The objective of the awareness in the community involves

- to motivate people for involvement in an action learning process to improve their livelihoods;
- to stimulate actions on issues such as how people see development, how they solve their problems and organize themselves to achieve their goals;
- to create space for the less powerful and poor groups to express their needs.

Training for transformation

Workshops are conducted to first strengthen peoples analytical and planning skills and ability to cooperate with each other. 'Training for Transformation' (TFT) (Hope and Timmel, 1984) is a key methodology for such workshop. TFT is a practical training for community development based on the philosophy for empowerment by strengthening people's awareness. Five important steps may help problem solving through analysis and self organization for proper action and reflection. The steps comprise (i) exploring views on development (vision), (ii) root cause analysis of problems in the community, (iii) self organization and leadership, (iv) improving leadership, (v) Openness, criticism and sharing.

Understanding needs

The next step is to hold intensive discussion with individual families of different wealth ranks to understand their needs that give scope to involvement of members of different categories with special emphasis to resource poor farm families. A representative sample of ten per cent of total number of households may be chosen from each category for survey to assess farming needs properly that facilitates in preparation of a practicable action plan on integrated farming system development.

Identification of coping mechanisms

To explore practical solutions to problems, the following questions may be asked with regard to coping mechanisms:

- What methods/system do people use for resolving the problem?
- What are the advantages and disadvantages of each method /system?
- What are the trends and reasons in the use of such practices?

Phase II: Community-level action planning

A community- level meeting is held after needs of individual households are explored.

The objectives

- To collect feed back of the issues and needs identified in the survey to the rest of the community
- To enable the community to prioritize the needs
- To analyse causes of the problems identified and to suggest possible solutions through participation of the whole community
- To identify possible local institutions to spread the solutions
- To frame a schedule of work to address the identified needs.

Prioritization of problems and needs

Problems identified here often need to be analyzed more deeply. Hence the root cause analysis is applied again, and problem trees help to visualize the cause and effect and to clarify in detail real underlying problems.

Searching for solution

Knowing the root causes of the priority problems help to identify possible solutions. People are asked to discuss possible constraints with these solutions. The solutions based on the available resources and skills can be developed to avoid the dependence on external assistance. Fresh solutions to problems need to be generated by blending solutions from local people with ideas from outside.

Appropriate extension material will assist in creating a better understanding. Outside help may be needed. For example, access to input and output markets may form part of a solution coupled with the adoption of an improved technology which gives increased productivity with same investment. Farmers would want to invest in the new technology if a market is guaranteed. The search for solutions should first focus on people's own knowledge.

Exposures (look and learn)

Exposure tours to innovative farms, neighboring communities or research stations are planned to find new and more ideas. The representative chosen by the farmers make visit and report back to the community

Mandating local institutions

After selection of possible solutions, local institutions are to be mandated to coordinate activities and to take responsibilities for progress initiating actions forward. Preferably farmers clubs may be chosen as local institution.

Local groups and lead farmers

Local groups and lead farmers are selected by the community-based organizations that they represent, for piloting and testing of new technologies. Regular feedback from the lead farmers to the groups will ensure the groups' involvement in planning and implementation, and encourage a process of further farmer-to- farmer testing and adaptation. The lead farmer is not a title but a role to be played by the actor with specific reference to the current activities. The lead farmer is also called a master/ volunteer/ demo/ farmer.

Potential role of local groups

- To formally adopt the program into their activities.
- To appoint a person or persons responsible for reporting on progress and identifying issues/problems that affect the program. This is likely to be done by the lead farmer.
- To encourage participation by other farmers in trying out the new techniques.
- To invite the extension worker to attend meetings on a regular basis.
- To arrange field days that cover all farmers.
- To evaluate the new methods at the end of the season and plan for the new season.

Potential responsibilities of lead farmers

- To motivate other farmers to try out new technologies.
- To assist with the project planning process using participatory methods.
- To assist the extension workers in training the group and other farmers.
- Hosing mid- and end-of-season evaluations of the test fields and demonstrations.
- To ensure that information is disseminated to the community at large.
- To hold regular meetings with other farmers and present concerns to the group and the extension workers.
- To facilitate coordination between the group and the extension agents.

Action planning (Hagman *et al*,1999)

After clarifying possible solutions and institutional responsibilities, concrete actions need to be agreed and planned. This takes place after getting back the report of **'look and learn'** tours. A good method of feedback is 'field day'. Here all the ideas gained from look and learn tours are explained and demonstrated. The most promising options are chosen. Then how and who should try and implement them is decided. Accordingly a time plan of action is developed by the community.. The Small steps and phases are needed in implementation so that the community is motivated by success and encourage further action.

Introducing competition for best ideas

The inter village and intra village competitions are arranged to find out the best innovator village (between the neighboring villages and individuals within the village).

Phase III: Implementation of farmers' experimentation

New ideas have to be tried, adapted and improved to suit local conditions. Such as natural pesticides, low cost methods of animal health care, alternative to labour-intensive conservation measures, or a social innovation like testing of new by-laws for resource utilization in the context of farmers' right etc. pertaining to integrated farming system development.

Learning through experimentation of new ideas

- The experimentation phase is also called as farmer experimentation to emphasize the learning process involved. It may be referred as the 'school of trying, or 'try and share' in participatory extension approach in project implementation.

- The farmers are encouraged to experiment with ideas and techniques emanating from their own knowledge or from outside sources.
- The process strengthens farmer's confidence in their capacity and knowledge and. increases their ability to choose the best options and to develop and adapt solutions appropriate to their specific ecological, economic and socio-cultural circumstances.
- Simple paired designs help farmers to observe, compare and analyse by themselves. The comparisons between conventional practice and new technique shall be a powerful tool for learning. The researchers and extension workers join the farmers and put 'check plots' in pairs to measure yield and growth parameters in detail.

Phase IV: Monitoring and evaluation

This phase consist of Joint learning by sharing ideas and experiences and by reflecting on the successes and failures of the action and the experiments carried out (self-evaluation) in the process of enhancing rural peoples problem solving capacity. The informal process of information sharing may not be adequate to make available the practices developed to everyone in the community and formal steps must be built into the process.

Two formal steps of monitoring and evaluation

- A 'mid-season evaluation' of agricultural innovations experimented
- An evaluation of the process, leading to the planning for the coming season.

The objectives

- Share knowledge among farmers
- Build confidence through presentations; and
- Encourage more farmer to farmer extension.

After seeing the experiments the farmers decide which techniques merit further research and/or promotion. The researchable techniques are put to more formal on farm trials, or fed back to research stations. The technologies ready for promotion are promoted to neighboring areas which is done jointly by farmers, researchers and extension workers?

Process review, self-evaluation and planning

One or two month s before the start of the following season, a feedback/review and planning workshop is organized by the community to review the process, evaluating planned activities and the indicators for success suggested during the planning phase. This includes criteria like leadership, strengthening of self-organisational capacities, as well as community participation. Successes and failure are assessed and analyzed to know strengths and weaknesses for further action. This analysis leads to the next cycle which starts again with issues of social mobilization.

Operational framework for participatory extension approaches

For concrete implementation of the process, a more structured operational frame work needs to be developed in collaboration with the implementing staff, in view of increasing in the conceptual understanding of participatory extension approaches. The guiding principles (Box-1) of participatory approaches must be followed to prepare an appropriate operational frame work.

Box-1: Guiding Principles of participatory approaches

Organising process	While participatory approaches make use of a flexible "basket" of different methodologies, those methodologies are guided by objectives and responsive facilitation.
Systematic learning	The methods used in participatory approaches emphasize learning about systems and the relations between different elements in those systems. Participatory approaches need to be "holistic".
Multiple perspectives	Participatory approaches take into account the fact that different people have different perspectives and try to accommodate these different perspectives.
Group learning	Participatory approaches emphasize the value of learning in groups as a means of coming to consensus decisions regarding action to address commonly identified objectives.
Context specific	Each community and its context are different. Participatory approaches accommodate their specific characteristics.
Facilitating	The use of participatory approaches by development practitioners involves the adoption of a facilitating or catalytic role, rather than a protagonists.
Leading of change	participatory approaches accommodate local knowledge and skills. They are focused on facilitating changes that people regard as appropriate.

Capacity building for implementing participatory extension

- It requires new skills, knowledge, and attitudes, for extension workers and the farmers who generally use conventional approaches for success.
- It is a process learning cycle having flexibility in its implementation.
- Extension agents are process leaders and facilitators than technical advisors.
- Dissemination begins within the project through a series of training activities.

- Specialists would organize workshops of cooperators, particularly the representatives of large farmers' associations and NGOs with equal proportion of men and women.
- Grassroots level of master farmers are created with equal proportion of women.
- These master farmers will, in turn backstop the participation of farm households.

Documentation

Record keeping and documentation are very important. The researchers, extension agents as well as the lead farmers should endeavor to document every major and minor activity.

- The attendance of participants during meetings, field-days, etc., should be recorded with a note of the number of males and females present.
- During visits to the field, anything abnormal or outstanding noticed should be recorded; these details are very important in this approach as they will ultimately help in interpreting results.
- Records is kept for all meetings, stating the decisions and recommendations made as well as the follow up activities.
- During field days and other meetings or when interacting with farmers, open-ended and probing questions are asked to fully understand their opinions and experience.
- Publication of success cases.
- The reports are made available to partners, farmers, neighboring communities, projects, etc., to appreciate and share experiences.

Criteria for successful implementation of participatory extension approaches

Expected output of participatory extension may be (Hagman et al, 1999):

- **Farmer participation/involvement in extension activities**: Farmers participate fully in decision makings, meetings and training sessions and implementation as a whole.
- **Empowerment**–increased farmer-articulation, confidence and decision making.
- **Active farmer experimentation with ideas and innovations**: the number of farmers' own experiments in the area increases and new farmer innovations are developed;

- **Process documentation**: the learning experience and farmers' ideas / knowledge are well documented. Indigenous knowledge is also documented and made available.

Participatory technology development in tropical tuber crops (Anantharaman *et al*, 2006)

The study was conducted in Chenkal village Tiruvananthapuram district as a part of Institution Village Linkage Programme (IVLP) implemented by Central Tuber Crop Research Institute (CTCRI), Trivandrum. IVLP was implemented with principles of farmers' participation in technology development. The various steps followed in IVLP were selection of operational area, constitution of multi disciplinary team of scientist, agro eco system analysis, problem diagnosis, action plan, formulation and technology assessment refinement. Four local sweet potato varieties were subjected to farmers evaluation under rice fellows. All the verities performed well. The farmers opined that the technology of using rice fallows for sweet potato cultivation is profitable in the cropping pattern of rice production system in upland conditions. Similarly farmers assessed groundnut and cow pea as ideal inter crops in cassava and arrowroot and white yam, greater yam and elephant fruit yam as suitable inter crops in coconut plantations in terms of profit and market demand.

Farming system research and extension (FSRE) approach (Mohapatra and Behera, 2004)

Number of constitution studies on integration of different enterprises : Low land rice cum pisciculture farming system, rice-poultry-fish-mushroom integrated farming systems for low land, alternate system of land use through diversification of farming system etc. have been conducted in different parts of the country just by simulating the small and marginal farms situations.

Participatory generation, assessment, refinement and dissemination of rural technology (1996-99) by Central Soil and Water Conservation Research and Training Institute, Dehradun , India (samra *et al*, 2000).

The participatory research and technology transfer was carried out under the IVLP in three Doon valley micro water sheds of the Himalayan of North West India. Biophysical and Socio-economic constraints for low productivity of wheat, rice, maize, pulses, horticulture and fodder were analysed to agro-eco system analysis based on ERA. The participatory technology development and dissemination resulted in :

- Construction of a special tank for harvesting inter- flow with community contribution of 35%. The water from the tank was used to irrigate wheat crop in a community managed system thereby increasing the productivity by 20% as compared to that of rainfed conditions.

- Average yield of rice increased by 96% with application of herbicide, 45% with fertilizers, 30% with improved varieties and 12% with zinc application.
- The productivity of maize improved by 31% due to inter cropping and 7% with application of atrazine.
- Inclusion of *toria* (Brassica compestris var., *toria*) in maize wheat rotation improved total yield in terms of wheat by 39%.
- Pusa-256 gram produced 53% higher grain yield over local variety.
- Diversification to mango plantation was highly ranked priority of the farmers followed by Anla and Papaya.

IARI model village (IMV) approach (Vijayaraghavan, 2013)

The IARI Model Village (IMV) approach was developed with an objective of transforming of few villages in and around Delhi through agricultural technologies to act as model for other surrounding villages. This model has helped to enhance agricultural productivity and income through adoption of modern IARI technologies. For instance crop diversification was increased through introduction of vegetables and horticultural crops. The introduction of summer mung has helped to improve soil fertility and introduction of Integrated Pest Management (IPM) practices have resulted in reduction of pesticide use.

Rural bio-resource complex (RBRC) implemented by University of Agricultural Sciences, Bangalore (Gowda, 2013)

Rural Bio-Resource Complex (RBRC) was implemented by University of Agricultural Sciences, Bangalore in a pilot project funded by DBT-GOI in Tubagere Hobley of Bangalore rural district in Karnataka state from April 2005. It was attempted to address problems of farmers in holistic manner with active involvement of farmers. The project was formulated to provide sustainable and profitable technologies on field crops, horticulture crops, animal based enterprises and natural resource conservation and management as well as seed production activities in selected crops, sericulture and value added products in ragi, red gram and jack fruit as per the available site specific rural bio-resources. The farming community in the project area have realized overall net income from increased productivity of the existing crops, shifting from low income generating crops/enterprises to high income generating crops, introduction of new crops like rose, sweet corn, baby corn, increasing the are under mulberry (138 acres to 425 acres), introduction of tissue culture banana, seed production and nursery multiplication and subsidiary rural enterprises like sheep, fish farming, vermin-compost and value addition to agricultural based products (VAPs). Further, introduction of integrated farming system demonstrations, improving water use efficiency measures, cultivation of bio-fuels in non-arable land, promotion of custom hire services of tractors and improved agricultural implements and start of 10 different commodity based associations had cumulative impact on enhancing the income of stakeholders.

NAIP on technology adaption through participatory technology development in Odisha (Dash, 2014)

The NAIP on value chain on ginger and ginger products implemented at Orissa University of Agriculture and Technology (OUAT) successfully developed and validated cultivation and value added products of ginger with involvement of farmers. The project was operated in two blocks, one from each district, viz. Pottangi from Koraput and Daringbadi from Kandhamal of Odisha state during 2009- to 2013. The significant outcome of the projects are :

- Adoption of improved cultivation technology of ginger with high yielding variety Suprava.
- Two hundred thirty six farmers have adopted of organic farming of ginger and have been registered under organic certification.
- Raising of ginger seedlings in nursery is a new technology for the ginger farmers in the operational area (the crop is grown in rainfed conditions).
- Farmers' were trained on preparation of value added products like dehydrated ginger flex and powder, ginger candy, ready to serve beverage and paste in scientific and safe manner developed by the project.
- The project had continuous and regular interface with the marketers and potential traders for sale of the products.
- Around 1.38 lakh man days/year employment could be generated in project locations.

Thus project has brought socio-economic upliftment of the tribal of operational area.

Action research at DRWA, Bhubaneswar (Sarangi *et al*, 2011)

The Directorate of Research on Women in Agriculture (DRWA) has developed gender sensitive extension model with a focus on involvement of a group of Village level Para Extension Workers (VPEWs). The model aimed at involving public institutions like extension agencies of the State Government, research

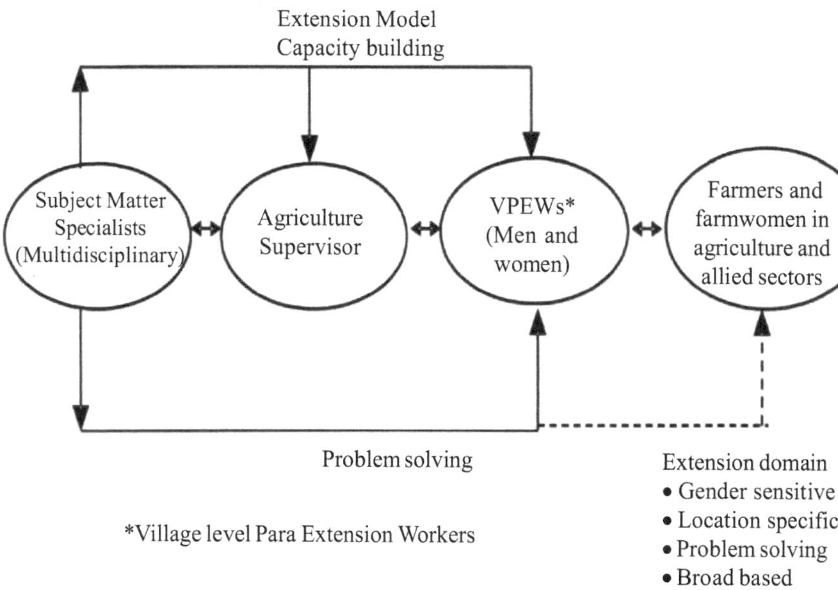

institutions and private actors namely Village level Para Extension Workers to manage agriculture extension services at the village level. There is positive change in following areas through implementation of the model

- Gender awareness
- Sensitization on gender mainstreaming
- Contact with change agent for farm advisory
- Learning farming skillsenhancing farming activity
- Capacity to identify problems
- Adoption of new farming skills
- Participation in extension programmes
- Cooperation among stake holders
- Conflict management
- Demand for farm information
- Extent of success
- Meeting current needs in farming.

On farm trial / testing by Krishi Vigyan Kendras (Nanda *et al*, 2014)

On Farm trial is an adaptive trial which is conducted on farmers' fields by the farmers with the support of KVK scientists. It is conducted within a farming system perspective. The main objective of OFT is to identify existing inputs or practices that might help to solve major problems of the farmers in an operational area of the KVK. Farming system perspective that implies seeing things from the farmers view point are used. Assessment and refinement of various available technologies at farmers' field in their own micro situation by farmers are very adaptive in farming community as main principle of KVK is based on 'learning by doing' and 'seeing is believing'. Farmers can learn by themselves and able to adopt that particular technology which has been found based at their own field. Several technologies in the field of cereals, oil seeds, pulses, millets, commercial crops, fruits and vegetables, livestock, poultry and other agri-allied enterprises have been successfully tested through OFTs by KVKs of Odisha and other states of the country which has brought increase in yield and profit. Many of such technologies have been established in the respective farming systems through farmers' participatory research and field level demonstrations.

Conclusion

Participatory extension approaches are key components in promoting improved agricultural technologies to farmers. The Participatory extension approaches focuses on encouraging the active participation of local leaders, farmers, researchers, extension agents and the private sector in identifying the problems of local communities and their possible solutions. It involves communities in analyzing their situation, planning, trying, monitoring, and evaluating new technologies. This will result in the widespread adoption of identified/relevant new practices and new varieties as a result of farmer-to-farmer extension and several different partners are implementing extension activities in each country. Many Countries have been successful disseminating encouraging continuing using extension methods that have worked well in the past and incorporating some of the Participatory extension concepts and approaches. The diversity of dissemination approaches should make it possible to compare the efficiency and impact of the different methods.

References

Ashby, J. A. and Sperling L. 1995. Instititionalising participatory, client- driven research and technology development in agriculture. *Development and Change* 26:753-770

Chambers, R. (1993) Challenging the profession, frontiers for rural development. London :IT publications.

FAO, 1994. Farming Systems Development: A Participatory Approach to helping small scale farmers. Rome, FAO of the United Nations.

Hagman J., Chuma E., Conolly,M, and Murwira. K., 1998 Client- driven change and institutional reform in agricultural extension: an learning experience from Zimbawe. ODI Agricultural Research and Extension network, AGREN. Network Paper No.78.

Hope, A and Timel, S. (1994) Training for Transformation, a hand book for community workers. Gweru: Mambo Press.

Mangan, J. and Mangan M. (1998). Acomparision of two IPM training strategies in China: The importance of the rice ecosystem for sustainable pest management. Agriculture and human Values,15,209-221. Netherlands, Kluwer Academic Publishers.

Ryan J.G. and Spencer D.C., Future challenges and opportunities for agricultural R & D in the semi-arid tropics. Patencheru 502 304, India: ICRISAT.

Van der Fliert, E. and A.R. Braun. 2001. Conceptualising integrative farmer participatory research for sustainable agriculture: from opportunities to impact. Working document No. 16. CGIAR system-wide.

Dash, S.K.(2014). Final Report, NAIP (National Agricultural Innovation Project) Compmnent II (Indian Council of Agricultural Research), Subproject on A Value Chaina on Ginger & Ginger Products, April, 2009-Dec, 2013, OUAT, Bhubaneswar-751003, Odisha,

Samra J.S. Mishra A.S., Mohan S.C., Dhyani B.L. and Tomar D.S., (2000). Participatory Generation, Assessment, Refinement and Dissemination of Rural Technology (1996 to 1999). Central Soil & Water Conservation Research & Training Institute, Dehradun-248195 (India).

Mohapatra I.C. and Behera U.K., (2004). National Symposium on Recent Advances in Rice-based farming Systems, November 17-19, 2004 organised on Association of Rice Research Workers, Cuttack (Odisha)-753006, p.79-113.

Sarangi B.N. Mishra S. Dash S.K., Sahoo P.K., Srivastav S.K., Sahoo L.P., Singh A., Behera B.C., (2011). DRWA, Gender Sensitive Extension Model, Directorate of Research on Women in Agriculture, (Indian Council of Agricultural Research), Bhubaneswar-751003, Odisha.

Anantharaman M. Ramanathan S. and Santoshkumar K.P. (2006). Root and Tuber Crops, In Nutrition, Food Security and Sustainable Environment. Published by Regional Centre of central Tuber Crops Research Institute, Indian Council of Agricultural Research, Bhubaneswar-751019, Odisha, India, p.315-321.

Choudhury S. and Ray P. (2011). Farm Sector development Emerging Issues. Published by Agrobios (India), Jodhpur. P.365-376.

Nanda S.S. Raj R.K. Mohapatra B.K., Mohapatra S.C., Mishra B.P., Banarjee P.K., Behera M., Sarangi C.R., Sahoo S.C., (2014). Technologies assessed by KVKs of OUAT (2008-09 to 2012-13). Directorate of Extension Education, OUAT, Bhubaneswar-3

Gowda, K.N. (2013). Led paper presented in international conference on Extension Educational strategies for sustainable agricultural development, held during 5-8 December, 2013, UAS, Bangalore, India p.6-11.

Vijayaraghavan. K. (2013). Led paper presented in international conference on Extension Educational strategies for sustainable agricultural development, December, 5-8, 2013, UAS, Bangalore, India p.157.

Integrated Farming System Practices: Challenges & Opportunities, pp 533-544

New India Publishing Agency, New Delhi, India

22

Farmer-led Approaches in up Scaling Farming System Research and Extension

M.P. Nayak and *M.R.Mohapatra

Introduction

Farming system conceptually is an integrated set of activities that farmers perform in their farm with available resources and circumstances to maximise productivity and net farm income on a sustainable basis. The system approach suggest a judicious mix of enterprises and activities based on the cardinal principle of minimising the competition and maximising complimentarity taking into account the components of soil, water, crops, livestock, labour, capital, energy and other resources. With all the resources, enterprises and external forces interacting in the system, the farmer operate at the centre of the farm household to take decisions observing variability, profitability, stability and sustainability. It is not the technology alone, but a set of social, political, institutional and economical forces influence the decision making process in a farming system leading to a model or acceptable combination of enterprises for overall development of the farm household.

Though science and technology have played a vital role in finding out innovations in the farming system approach through addition of new enterprise leading to profitability; new technologies ensuring stability and sustainability, the basic

Krishi Vigyan Kendra, Sundargarh
*Krishi Vigyan Kendra, Dhenkanal, Orissa University of Agriculture & Technology, Bhubaneswar - 751003

philosophy of the approach based on set principles is being pursued since initiation of agriculture in the civilisation.

In general farming system approach is more of a thought, a movement, a philosophy than simply choosing enterprise or combination of so to improve upon the standard of living. The interaction of bio-physical and socio-economic values plays an important role in up-scaling farming system approach and the adoption behaviour largely depend on the human values rather than material gain.

With the changing scenario, several factors like increasing instability in production, market price, high cost of production inputs, labour and negative impact of climate change, degradation of natural resources directly or indirectly warrants the use of sustainable rather regenerating technologies, compatible and non-competitive farm enterprises to bring in sustainability and enhancement in production and income. For more than a century, programmes and activities have been designed to address the vulnerability of rural poor and are often targeted towards improving agricultural practices as means of increasing productivity, efficiency and ultimately income. In majority of the TOT cases technology has taken the centre stage even though the basic fundamental question like 'whether the farmer adopt the offered technology and, if they do, whether they do it in ideal combination and for a considerable length of time needed to produce designed result remains unanswered. More importantly, in general the technology in isolation can not be recommended to all categories of farmers operating under varying situations. On the other hand acceptance and continuity of the technology is a dynamic process and varies both with the quality of the technology being promoted and the adoption behaviour of the individual farmer.

Technology and technology dissemination

The concept of technology can be defined simply to mean the application of science in the development, production, utilisation or application of materials or things or methods of undertaking certain activity or work (Hashim, 1998). In this context agriculture technologies include both physical objects like seeds and fertiliser as well as new farming methods. The degree of acceptance of a technology depends on five important characteristics like; 1) relative advantage which explains how the technology is subjectively perceived to be superior to the previous one; 2) compatibility that reflects how the technology is perceived to be consistent with the existing values, experience and need of the adopter; 3) complexity reflecting the perceived difficulty to understand and use the technology; 4) triability i.e the degree to which the technology can be tried with on a limited scale; and 5) obseravability which reflects the visibility of the results of the technology to others in the social system. Experience has show that technology may not be appropriate in every context, but its suitability depends on how well it fits the particular farming context (CIMMYT, 1993). However, much of the focus of adoption dynamics has been on the individual farmer (i.e the attitude of the farmer or their socio economic characteristics, such as wealth, land holding or education) and the characteristics of the technology, rather than the context in which technology adoption and diffusion takes place.

Scaling-up

Scaling-up leads to "*more quality benefits to more people over a wider geographic area more quickly, more equitably and more lastingly*". Scaling-up can occur both vertically and horizontally. Vertical processes involve expansion from the level of grassroots organizations to national institutions and policies. Horizontal processes refer to geographical spread or replication on a larger human scale, from hundreds to thousands or millions of people.

Critical factors influencing the scaling-up process

Scaling up from thousands to millions of people is an enormous challenge and depends on many factors. The willingness and capacity of potential beneficiaries to take risks and test and adopt new practices, the availability of critical inputs such as seeds or livestock, and the presence of market demand and incentives can also determine whether scaling-up efforts will be successful. Whether the development agency responsible has a sound technical knowledge of the practice and sufficient funds and time to ensure its uptake can also influence scaling-up outcomes. Scaling-up efforts are more likely to succeed in contexts where governments, the private sector, development organizations and communities are already engaged in effective partnerships or networks, informal groups are present and active, or rural people are involved in local and national decision making processes.

Scaling-up is complex and there is no single process appropriate for all situations. Depending on the context, policy-makers may consider the approaches based on past experience and feasibility. In one hand supportive policy to ensure better institutional integration, broad-basing and closer linkage among the partners involved in technology development transfer as well as policy makers can ensure vertical scaling up. On the other hand adopting approaches to involve more people in decision making as well as processes of technology transfer will help horizontal scaling up in a sustainable way. This not only leads to developed ownership of the peasantry over the technology but also results in multi functional empowerment.

Extension strategy in scaling up FSR-E

During the Green Revolution only limited attention was paid to the complexity and diversity of the farmer's physical, economic and social environment while gradually the attention has been shifting on farming system. (Barker, R., Herdt, R.W. & Rose, B., 1985).

After decades of using Transfer of Technology models, extension theorists and practitioners began to notice that the ToT approach proved successful only under certain conditions such as homogenous production environment, for larger commercial units, and where innovation is driven by strong market forces. The approach has been inadequate in more complex situation, particularly heterogeneous environment. To deal with the inadequacy, the importance of developing technologies within the context in which it is to be used made its way into the research, development and extension agenda under the label of Farming System Research. This approach has been seen as more holistic shifting the decision centre both in research and extension agenda more towards farmers. The farmers have seen as a system manager and the extension agent as a diagnostic partner and promoter of new technologies and practices. A gradual realisation came to make extension more human development driven rather than only technology transfer to achieve higher productivity.

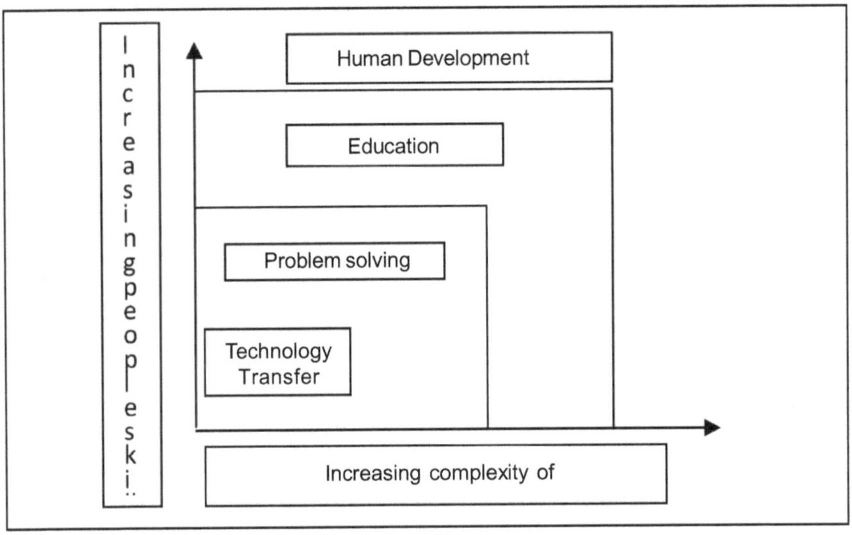

Fig. 1: The four paradigms of extension (*Source*: Coutts, 1994)

The model allows the domain of extension to include more of human development activities with the increasing complexity of situation. With the changing time and realisation of the importance of human values, indigenous knowledge, local experience the approaches in extension has more become farmers led. Farmers led extension is a multidimensional communication process between extension staff and farmers involving the sharing, sourcing and developing of knowledge and skills in order to meet farming needs and develop innovative capacity among all the actors in which farmers have a controlling interest in the "center stage" and play a key role in technology development and delivery; involving farmers in sharing, sourcing and transferring knowledge and skill. Extension workers take a role to facilitate the farmers in extension activities,; to serve as a key link between some sources of knowledge and skills and help develop capacity of the farmers within extension activities.

Farmer to farmer extension within research-extension paradigms

	Technology Generation (research)	Technology Transfer (extension)
Process is led by "outsiders"	Formal research	Formal Extension
process is led by "insiders"	Indigenous "expert"	Indigenous communication
Process is operationalised by "insiders" but facilitated by "outsiders"	Farmers Participatory Research, participatory technology development	Farmer – to – farmer extension

Principles of farmer led extension

The basic principles of farmer-led extension approach 1) suggest active participation of farmers and their communities in agricultural extension activities; 2) emphasise on empowerment and developing capacity of local people to lead their own development efforts; 3) provide a link between research and extension through the use of small scale experimentation; 4) combination of outside technical knowledge and practice based farmers experience; and 5) offers technological approaches which address broader natural resource management and household economic concerns – not just pursuing yield and income maximisation.

Methods and approaches

Several farmer-led approaches have been tested worldwide and have been found potential to contribute to scale up pro poor, sustainability oriented and at the same develop human capabilities to manage both technology and process.

Farmer field school

Farmer Field Schools (FFS) consist of groups of farmers who get together to study a particular topic. The topics covered vary from conservation agriculture, organic agriculture, animal husbandry, and soil husbandry, to income generating activities. It teaches basic agricultural and management skills that make farmers experts in their own farms. FFS is a forum where farmers and trainers debate observations, experiences and present new information from outside the community.

Farmer field school is described as platform and "school without walls" for improving the decision making capacity of farming communities. Starting farmer field school at field level serve as a "learning center" to the farmers' situation and for technology transfer through demonstrations which is based on the principle of discovery based learning. As the demonstrations/conducting session is at farmers' field it gives better understanding and applicability of the technologies to local situations (Nehve and Wani, 2006; Singh *et al.*, 2006). Further there is a scope to fine tune the technologies to suit the local conditions. Farmers field school concept ensures participation of farmers in identification of production constraint, formulation of technology module, demonstration of technology, observing and analyzing data and decision making. Farmers field school empowers participating farmers with scientific information/know how which is the driving force to produce quality product at minimum cost of production to compete in the international market.

The first wave of FFS was conducted in 1989 in the rice fields of Indonesia. By 1990, the Indonesian National IPM Programme scaled up and launched 1,800 FFSs for rice IPM in six provinces in Java, Sumatra and South Sulawesi. Around 1991, the pilot FFSs in IPM for rotation crops (mainly soybeans) was initiated while the FFS Programme spread out to different countries in Asia (CIP-UPWARD, 2003).

Central to the popularity of FFS programmes is an appropriate topic and methodological training of the people who organise and facilitate farmer field schools. To be a successful FFS trainer/facilitator, one must have skills in managing participatory, discovery-based learning as well as technical knowledge to guide the groups' learning and action process. Without an adequate training of trainers (ToT) programme, the subsequent FFS programme will fall far of its potential (Luther *et al.*, 2005).

In general, farmer field schools consist of groups of people with a common interest, who get together on a regular basis to study the "how and why" of a particular topic. The farmer field school is particularly suited and specifically developed for field studies, where hands-on management skills and conceptual understanding (based on non-formal adult education principles) is required.

Essential elements of FFS

1. The group

The group comprises of individuals (20-30 in number) who have a common interest, forming the core of a Farmer Field School. The FFS tends to strengthen existing groups or may lead to the formation of new groups

2. The Field

The field is the teacher. It provides most of the training materials like plants, pests and other facilities. In most cases, communities provide a study site with a shaded area for follow-up discussions.

3. The Facilitator

The facilitator is a technically competent person who leads group members through the hands-on exercises. The facilitator can be an extension agent or a Farmer Field School graduate.

4. The curriculum

The curriculum follows the natural cycle of the subject, be it crop, animal, soil, or handicrafts. This allows all aspects of the subject to be covered in parallel with what is happening in the FFS field.

5. Programme leader

The programme leader is essential to support the training of facilitators, get materials organized for the field, solve problems in participatory ways, and nurture facilitators. The programme leader should be a good leader who empowers others.

The basic approach of farmers field school encourage and allows the questions of how and why all through the sessions that results in observing the problems, identifying the cause, deciding and proposing solutions and again observing the effect. This cause effect relationship gives rise to discovery based learning and the process help to develop the management capacity of the participants. Further there is a scope for fine tuning the technologies to suit the local conditions. Farmers field school concept ensures participation of farmers in identification of production constraint, formulation of technology module, demonstration of technology, observing and analyzing data and decision making. Farmers field school empowers participating farmers with scientific information/know how which is the driving force to produce quality product at minimum cost of production A FFS on any problem, crop or enterprise run for a crop period to observe and learn.

A FFS consisting of 20 to 30 members having homogeneity in enterprise, problem/opportunity and thought undergo periodic participatory trainings in their own farm situation with a demonstration as the observation field. The facilitator help the FFS students to lead the demonstration and learn the principles and practices involved in problem solving.

A typical FFS session starts in the morning hour and run as per the schedule

8:00	Opening (often with prayer), attendance call, day's briefing of activities, stretching exercises.
8:30	Go to the field in small teams, make observations that are noted by the facilitator and one other person in the group records. Facilitator points out interesting new developments.
9:30	Return to shade. Begin making agro-ecosystem analysis drawing and discuss management decisions.
10:15	Each team presents results and the group arrives at a consensus on management needs for the coming week.
11:00	Short tea/coffee/water break.
11:15	Energiser or group building exercise.
11:30	Special study topic related to the FFS subject.
12:30	Closing (often with prayer).

The Agro-Ecosystem Analysis (AESA) is the FFS's core activity, and other activities are designed to support it. The agro- ecosystem analysis process sharpens farmers skills in the areas of observation and decision-making, and helps develop their powers of critical thinking. The process begins with small group observation of the treatment and control plots. During the observation process participants collect field data, observe variations and effect of the treatments. They critically analyse the observation and make a formal presentation of the analysis before other members which further enrich their knowledge. One member of each small group then presents these findings and decisions to the larger group. After this brief presentation of results there is time for open questions and discussion. Good large group discussions often involve posing alternative scenarios.

The topics covered in Farmer Field Schools can vary considerably – originally it was IPM in rice, now organic agriculture, animal husbandry, soil husbandry, groundwater management, cropping system to income-generating activities are also covered effectively. The most interesting fact about the outcome of this approach is that the first generation students of FFS can very effectively run

FFS as FFS graduate facilitator hence the spread appears to be in a arithmetic progression.

Farm school

Farm school has been taken as another farmer-led approach in technology upscaling. Many researches have indicated that the farmers learn the most from his neighbourhood. The proximity in thought, language, socio economic and bio-physical situations, closer realisation of the problems and moreover the appearance of similar set of external forces like input- output market, availability of services create greater realisation of the success and motivate for adoption. Over a period of time with adoption of technologies several exemplary farms have been developed. The farmers or farmwomen behind these bright spots may not necessarily be resource rich but hard working, innovative and open minded who not only have the attitude to develop their own enterprise but also have certain level of inclination for the community to develop. Such achiever or innovative farmer who has achieved enhanced productivity and profitability in their farm with scientific and sustainable technology management, are designated as farmer researcher / farmer scientist/farmer professor in their respective field of expertise and the farm school is established in the model farm of that farmer. These farm schools have been accepted as powerful methods in promoting farmer – to – farmer learning and at the same time create an opportunity for fine tuning of the scientific know how. Farm school in specific is more powerful a method to up scale farming system approach in similar agro-ecological situations.

The basic philosophy eternalised for development of farm school is to develop a model of technology management at the village level and provide scope for the participating farmers to learn at their own vicinity accepting the farm, the farmer and the technology as suitable for them.

Basic activities carried out in operationalising the farm school are,

- The achiever farmers are selected based on their present performance in achieving excellence in technology management, past experience with farming and attitude to share.
- The farmer is provided technology back stopping by both research and extension system for developing his skill not only in technology use but also facilitation and training skills.
- A set of technologies as suitable for the farm are demonstrated with technical knowhow and critical inputs.
- Minimum support is provided to create a learning atmosphere with physical facilities and teaching aids.

- The students of the farm school will visit the farm as per specific schedule, observe activities, problems, innovations and interact with the farmer professor to learn the principles of technology applications.
- The farmer will provide scope for field interaction, learning by doing in possible areas.
- The extension staff will work as facilitator at regular interval and clarify the learning and application of the technology.
- The students apart from exposure at the farm school will also be provided with training and exposure at district and state level.
- After completion, the students would have the responsibility to provide extension support to other farmers in their own village or area.

Farmer promoter approach

Despite the transformation, extension in India is still faced with many challenges which have been accelerated by structural adjustment reforms aimed at reduced public spending. Some of the challenges include low budgetary allocation and understaffing. Passivity at the community level and a tendency to treat all farmers, their contexts and needs as homogenous are additional invisible contributions to the underperformance of state extension programmes. It is against this background many a times extension professional and policy makers are advocating for participatory, demand-driven, client-oriented and farmer-led agricultural extension systems, with emphasis on targeting women and disadvantaged groups. These approaches focus on farmers as the principal agents of change in their communities and, therefore, enhance their learning and empowerment, thereby increasing their capacity to adapt/innovate and train other farmers. The role of extension officers is also changing from agents of technical messages to facilitators who train farmers on entrepreneurship, and link them to markets and credit institutions. One such farmer- led approach is the farmer promoter approach that is being used by many extension system as formal or informal mechanism of technology transfer.

The volunteer farmer promoter is a form of farmer-to-farmer extension where farmers take centre stage in information sharing. It is envisaged that the farmer-to-farmer extension is a more viable method of technology dissemination as it is based on the conviction that farmers can disseminate innovations better than extension agents because they have an in depth knowledge of local conditions, culture, practices and are known by other farmers. In addition, they live in the community, speak the same language, use expressions that suit their environment and instil confidence in their fellow farmers. It works on the basis that the model is able to achieve, economies of scale in technology dissemination by reaching more farmers more quickly. The farmer-to-farmer extension has its origins in Guatemala in the 1970s, spreading to Nicaragua in the 1980s, then Mexico and Honduras. It is currently practiced widely in many other countries in Latin America, Asia and Africa in different forms. Farmer-to-farmer extension emerged as a reaction to the top down transfer of technology model that left very little possibility for farmers' participation and initiative, did not address farmers' needs, was inefficient, was biased towards well-to-do farmers and extended inappropriate technologies, leaving behind disinterested farmers and demotivated extension officers. The most famous and well known farmer-to-farmer extension is the "*Campesino a Campesino*" movement in Nicaragua.

At the centre of this approach are farmer promoter who are known by many names like farmer trainer, farmer teacher, farmer friend in different countries and projects. A farmer promoter is defined as an individuals with little or no formal education who through a process of training, experimentation, learning and practice, increase their knowledge and become capable of sharing it with others, functioning as extension workers.

Although the farmer-trainer approach differs from country to country due to the conditions under which it takes place, the organisational set up and management, they all have one thing in common: farmer promoters are trained by external agents ; they in turn share their knowledge and skills with other farmers in the community. The role of farmer promoters/trainers varies from project to project depending on how they are selected to become trainers, the mode of operation and whether they are volunteers or are compensated for the time they spend promoting other farmers, whether they work with groups or individuals, whether they are trained as specialists in one subject or as generalists and whether they work in their own community or also conduct trainings outside their community, whether they only provide technology support or even input and market facilitation.

Essential characteristics of farmer promoter

The farmer promoter should necessarily be; 1) one from the grass-root; 2) having practical experience of farming; 3) interested in both learning new technology, information and sharing with the fellow farmers; 4) active individual dedicated to hard work; 5) socially acceptable.

The farmer promoters are given technical backstopping by the extension or research institutions, hand holding support to facilitate linkage with input, market and even financial institutions. The multiagency affiliation of the promoters help them to facilitate schematic linkage.

These promoters may operate voluntarily of their own interest with any financial gain or even it will not be unwise to help them gain service charges which will make the process sustainable.

Conclusion

It is increasingly acknowledged that public extension services in developing countries are no longer able to meet the changing needs of farmers. As a result, the sector has, over the last decade, been going through a transformative process from the linear model of technology transfer to the more pluralistic demand-driven extension. Moreover the emphasis of extension system has been more on isolated enterprises rather than promoting a system approach in reality. The visibility of impact of new enterprises, technologies when pay off, is more than resource conserving, regenerating and sustainable ones. On the other hand the present day challenges equally emphasise the use of system approach and appropriate technology. Both the need and challenge requires an in depth thinking to allow and operate with farmer-led approaches which will again be stable, sustainable and develop the capability of rural poor not for the present but to take the future day challenges.

References

Barker, R., Herdt, R.W. & Rose, B., 1985. The rice economy of Asia, Washington D.C and Manila: Resources for the future and The International Rice Research Institute.

Biggs SD. 1989. A multiple source of innovation model of agricultural research and technology promotion. Agricultural Research and Extension Network (AgREN) Paper 6. London: Agricultural Administration (Research and Extension) Network, ODI.

CIMMYT, 1993. Adoption of agricultural technology: A guide for survey design, Mexico D.F.:international maize and wheat improvement center.

CIP-UPWARD, 2003. Farmer Field Schools: From IPMN to Platforms for Learning and Empowerment. International Potato Center – Users' Perspectives with Agricultural Research and Development, Los Banos, Laguna, Phillippines. 87 pp.

Coutts, J. 1994. ' process papare policy and practice - a case study of the introduction of a formal extension policy in Queensland Australia.' Wageningen Agricultural University, The Netherlands

Hashim, M.Y. 1998. "The Development and Transfer of New or Improved Technology and Implications for Extension Training" in Radhakrishna and M M Singh (eds.). Technology for rural Development (COSTED), Institute of Science, Bangalore.

Luther, G.C., C. Harris, S. Sherwood, K. Gallagher, J. Mangan and K. Touré Gamby, 2005. Developments and Innovations in Farmer Field Schools and Training of Trainers. In: Norton, G.W., E.A. Heinrichs, G. C. Luther and M.E. Irwin. Globalizing Integrated Pest Management - A Participatory Research Process. Blackwell Publishing, Iowa, USA. pp. 159-190.

Nehve, F. A. and Wani, G. M., 2006, Transfer of technologies for maize production and their impact under actual farm situations. *Agric. Extn. Rev.,* 21(1): 42-46.

SUSTAINET EA 2010. Technical Manual for farmers and Field Extension Service Providers: *Farmer Field School Approach.* Sustainable Agriculture Information Initiative, Nairobi, Kenya

FSC
www.fsc.org
MIX
Papier | Fördert gute Waldnutzung
FSC® C083411

Zeitfracht Medien GmbH
Ferdinand-Jühlke-Straße 7
99095 Erfurt, Deutschland
produktsicherheit@kolibri360.de